ENVIRONMENTAL ASSESSMENT & IMPACT STATEMENT HANDBOOK

PAUL N. CHEREMISINOFF, P.E.
Associate Professor of Environmental Engineering
New Jersey Institute of Technology
Newark, New Jersey

ANGELO C. MORRESI, P.E.
Hoffman La Roche Inc.
Nutley, New Jersey

ANN ARBOR SCIENCE
PUBLISHERS INC / THE BUTTERWORTH GROUP

Fourth Printing, 1980
Third Printing, 1979
Second Printing, 1977

Copyright © 1977 by Ann Arbor Science Publishers, Inc.
230 Collingwood, P. O. Box 1425, Ann Arbor, Michigan 48106

Library of Congress Catalog Card No. 76-050989
ISBN   0-250-40158-4

"Concern for man himself and his fate must always form the chief interest of all technical endeavor. Never forget this in the midst of your diagrams and equations."

Albert Einstein

FOREWORD

The environmental impact statement (EIS) is a result of provisions in the National Environmental Policy Act of 1969 (NEPA), which formulated national policy to protect the environment. NEPA also established the Council on Environmental Quality and says that "major Federal action significantly affecting the quality of the human environment" must be accompanied by a "detailed statement" of potential impacts (favorable and unfavorable) of any irreversible commitment of resources. The law does not prohibit anything. All it says is *look before you leap*.

Environmental impact statement requirements have had far-reaching effects. Copious documentation running into several thousand pages is not extraordinary with some statements. The documentation on the environmental impact assessments on the Alaska pipeline, for example, was 10-feet thick and cost $9 million to compile. Federal agencies now produce over 1000 statements per year. "Federal actions" has been construed to include any undertaking in which the federal government has a financial or jurisdictional role.

The NEPA has thus far survived all attacks, and 14 states have adopted counterparts with a dozen others practicing the principle administratively. In California the requirements have been extended to local government, and that state now outdoes the federal government with over 4000 environmental assessments per year.

The EIS process has obtained generally positive results. A thorough environmental assessment of the Alaska pipeline, though it took several years, averted serious construction mistakes. Government departments such as Interior, Transportation and other agencies and bureaus have abandoned or modified hundreds of projects as a result of environmental impact studies. Such far-reaching legislation would of course be expected to draw criticism, particularly from industry because of delayed ventures and increased surveillance and regulation. The process generally has not been overly cumbersome, time-consuming or expensive. The quotation

v

from Albert Einstein at the beginning of this book aptly expresses the intent of the authors.

Most projects require an assessment of environmental changes caused by the project. Frequently, and foolishly, this desirable requirement is viewed as an exercise in "creative writing." Increasing activism by citizen and community groups makes it somewhat dangerous to become too creative. Precise, logical engineering and legal minds may be expected to dissect and examine the material presented. It is necessary to walk a rather fine line in the preparation of these documents. The best case must be presented and both the good and bad aspects of the proposal must be covered. The basic purpose of an environmental impact statement is to avoid possible mistakes.

The purpose of this book is to provide a guide of informational requirements for those in industry involved in evaluation and/or preparation of environmental assessments. It is intended as an introductory handbook for the evaluation process and supporting rationale required.

Paul. N. Cheremisinoff
Angelo C. Morresi

CONTENTS

CHAPTER 1

THE NATIONAL ENVIRONMENTAL
POLICY ACT OF 1969

INTRODUCTION

Up to the last decade man has continuously defiled his habitat without considering the consequences of his actions nor, it seems, caring about them. In many instances misguided values have justified these actions as necessary in that they provided more wealth for the population and a better standard of living than they detracted from the environment. This viewpoint is still very much apparent even with environmental legislation, strict regulations, enforcement policies and public awareness, and detrimental health effects such as cancer and respiratory ailments.

There are still many persons who fail to understand the concept of protecting the environment. They are found throughout levels of government and industry and, although few in number, make up too large a portion of the public. They tend to believe that when they flush the toilet, take out the garbage, dump waste into streams, or affect stack emissions, a strange metamorphic process mysteriously diffuses their wastes into the biosphere. To quote an unknown author, "Dilution is no solution to pollution."

We now know that enough water and atmospheric exchange with the pollutant does not necessarily render it harmless. Substances such as mercury, DDT and Kepone, to name a few, build up in the food chain and have cumulative detrimental effects on man. The regenerative capacity of the atmosphere and rivers is not infinite, and natural removal processes have been overworked for decades.

Table 1 gives a summary of the origin and nature of some atmospheric pollutants. It lists the annual emission rates, background concentrations, and the major sinks for these pollutants as generated naturally and by

1

Table 1   Summary of Sources, Annual Emission, Background Concentration and Major Sinks of Atmospheric Gaseous Pollutants

Pollutant	Major Source Anthropogenic	Major Source Natural	Estimated Emission (kg) Anthropogenic	Estimated Emission (kg) Natural	Background Concentration ($\mu g/m^3$)	Major Identified Sinks
SO_2	Combustion of coal and oil	Volcanoes	65×10^9	2×10^9	1-4	Scavenging; chemical reactions; soil and surface water absorption; dry deposition
H_2S	Chemical processes; sewage treatment	Volcanoes; biological decay	3×10^9	100×10^9	0.3	Oxidation to SO_2
N_2O	None	Biological decay	None	590×10^9	460-490	Photodissociation in stratosphere; surface water and soil absorption
NO	Combustion	Bacterial action in soil; photodissociation of N_2O and NO_2	53×10^9 combined with NO_2	768×10^9	0.25-2.5	Oxidation to NO_2
NO_2	Combustion	Bacterial action in soil; oxidation of NO			1.9-2.6	Photochemical reactions; oxidation to nitrate; scavenging
NH_3	Coal burning; fertilizer; waste treatment	Biological decay	4×10^9	170×10^9	4	Reaction with SO_2; oxidation to nitrate; scavenging
CO	Auto exhaust and other combustion processes	Oxidation of methane; photodissociation of CO_2; forest fires; oceans	360×10^9	3000×10^9	100	Soil absorption; chemical oxidation
O_3	None	Tropospheric reactions and transport from stratosphere		(?)	20-60	Photochemical reactions; absorption by land surfaces (soil and vegetation) and surface water
Nonreactive hydrocarbons	Auto exhaust; combustion of oil	Biological processes in swamps	70×10^9	300×10^9	$CH_4=1000$ non-$CH_4<1$	Biological action
Reactive hydrocarbons	Auto exhaust; combustion of oil	Biological processes in forests	27×10^9	175×10^9	<1	Photochemical oxidation

man.   In most cases the natural pollution is far greater than the anthro-pogenic emission rate, but throughout earth's history natural removal processes have been able to cleanse the environment.   These processes include foliar absorption, soil absorption, absorption by natural water bodies, absorption by natural rock, rainout and washout, and chemical reactions in the atmosphere.   It appears that these removal processes are balanced against the earth's emission rates, and the additional waste prod-ucts of man cannot be accommodated by nature's "treatment facilities."

This natural capacity to regenerate also applies to the aquatic environ-ment.   Rivers, lakes, streams, estuaries, marine and other waterways can effectively treat large quantities of many pollutants and naturally render them harmless.   Basically, man has destroyed these natural processes with the intrusion of vast amounts of diverse wastes that have singularly and collectively erased the ability of many of our water resources to sup-port life.

Historically, the environment has been abused at the expense of tech-nological and industrial advances which were considered the ideal and much more desirable than ecological factors.   This philosophy, along with the effects of overpopulation, has resulted in the inability of some areas to support life.   More and more air is being breathed.   More water, both domestic and industrial, is being consumed.   Land adjacent to landfills and landfills themselves are no longer considered undesirable by developers. Waters—previously misused, wasted and polluted—have become necessary sources for thirsty individuals.   In New Jersey, which has one of the worst air qualities in the nation, the region from its southern tip (Cape May) to the George Washington Bridge (New York City) is known as "Cancer Alley."[1]

Our natural resources have been misused to the extent that they are on the verge of depletion.   Industry and government up until recent times have shown no inclination for environmental protection.   Only a massive demonstration of public concern in the 1960s resulted in pressure for initiating a governmental movement to clean up the environment through legislation and appropriation of funds.

On January 1, 1970, the "National Environmental Policy Act (NEPA) of 1969" was enacted establishing a national environmental policy and the Council on Environmental Quality (CEQ) to assist and advise the President on environmental and ecological matters.   It wasn't the first piece of environmental legislation (see Table 2), but it may turn out to be the most significant in terms of the quality of life in recent history. The President of the United States stated upon signing the bill, " . . . the 1970s absolutely must be the years when America pays its debt to the past by reclaiming the purity of its air, its waters, and our living environment.   It is literally now or never."

Table 2   Major U.S. Environmental Legislation

---

*   The River and Harbor Act of 1899

*   Atomic Energy Act of 1954

*   The Federal Insecticide, Fungicide, and Rodenticide Act of 1964

*   Solid Waste Disposal Act of 1965

*   The National Environmental Policy Act of 1969

*   The Clean Air Act of 1970

*   The Federal Water Pollution Control Act of 1972

Not everyone in government and industry shares this view. For many, the establishment of pollution abatement criteria resulted in forecasts of doom for the American economic system. However, since the enactment of the various antipollution acts, not only has the quality of the nation's air and water shown improvement, but few detrimental effects on the economy have been witnessed, resulting in the withholding of less than 2% of capital expenditures according to Environmental Protection Agency (EPA) estimates.

Alarmists' concern for environmental protection-induced unemployment, plant closings, and inhibition of plant expansions has been shown to be invalid. As a result of their failure to meet pollution regulations, 76 plants have been closed since 1971 affecting 10,520 jobs. However, quite a few of these jobs were not eliminated due to pollution regulations alone; many of the closed plants were operating at their margin of profitability and probably would have soon closed in any case.

On the other hand, the environmental protection industry, according to the CEQ, has generated nearly 1 million jobs with a net reduction in national unemployment of 0.4%. Over 100,000 jobs have been created by the nationwide expansion of wastewater facilities alone![2]

The economic stimulus provided by the environmental protection industry is exemplified by the EPA's estimated national cost for pollution compliance between 1974 through 1985 to be nearly $220 billion. This figure includes all industrial and governmental capital, operating and maintenance, and administrative costs.[3]   Table 3 gives the capital spending of nonfarm industries in their effort to control air, water and solid waste pollution. Expenditures for 1974, 1975 and those planned throughout 1976 are shown to be increasing to about $7.3 billion in 1976.[3]

Table 3   New Plant and Equipment Expenditures by Nonfarm Business for Pollution Abatement[4]   (Millions of Dollars)

	1974 Total	1975 Total	Total	Planned 1976 Air	Planned 1976 Water	Planned 1976 Solid Waste
All Industries	5,617	6,549	7,346	3,860	3,042	444
Manufacturing	3,656	4,475	4,488	2,157	2,074	257
Durable goods	1,648	1,775	1,762	1,020	668	73
Primary metals	798	1,012	1,007	675	300	31
Blast furnace, steel works	245	396	540	340	197	3
Nonferrous metals	500	546	396	272	98	26
Electrical machinery	207	136	158	32	116	11
Machinery, except electrical	77	83	106	42	61	2
Transportation equipment	140	116	137	51	68	17
Motor vehicles	115	86	114	39	59	16
Aircraft	22	26	20	11	8	1
Stone, clay and glass	191	198	164	118	42	4
Other durables	235	229	191	102	81	8
Nondurable goods	2,008	2,700	2,726	1,137	1,405	184
Food, including beverage	150	175	203	90	93	20
Textiles	28	31	46	14	32	1
Paper	491	489	502	213	274	15
Chemicals	469	684	786	247	478	61
Petroleum	796	1,239	1,100	530	490	81
Rubber	47	41	54	28	22	4
Other nondurables	28	41	34	15	16	3
Nonmanufacturing	1,961	2,074	2,859	1,703	968	187
Mining	57	73	99	44	44	11
Railroad	29	35	35	8	24	3
Air transportation	7	11	14	7	5	2
Other transportation	46	41	58	14	28	15
Public utilities	1,622	1,700	2,431	1,557	747	128
Electric	1,576	1,650	2,386	1,547	715	124
Gas and other	44	50	45	10	32	3
Communication, commercial and other	201	214	221	74	119	28

We have seen, then, that the fight against pollution will not cause the downfall of the U.S., nor will it produce dire economic consequences. Strong legislation and enforcement policies are our only means of continued ecological progress as our national goals are shifted to mesh with the priorities of our ecosystem. The goals do not have to be a hindrance to economic expansion, but can result in a more orderly utilization and development of our natural resources.

THE NATIONAL ENVIRONMENTAL POLICY ACT OF 1969 (NEPA)

The orderly utilization and development of our natural resources has become national policy with the enactment of the NEPA. The act does not specifically provide for enforcement policies but, through the Environmental Impact Statement, establishes a reviewing process of all major federal actions which would have a significant effect on the environment.

This review process requires an analysis by all concerned federal, state and local agencies affected by the proposed action. It also secures input from concerned citizenry, thus giving each individual the right to comment on any major federal action. This right should be exercised frequently to give planners insight into the needs of the people and problems that could be encountered.

The NEPA's purposes include:

1. the declaration of a national policy that will encourage productive and enjoyable harmony between man and his environment;
2. the prevention or elimination of damage done to the environment while stimulating the health and welfare of man;
3. the creation of interest and understanding of our nation's natural resources and ecosystem;
4. the establishment of a Council on Environmental Quality.

The NEPA is the first full-scale recognition of the problem of man and his effects on the environment and biosphere. It acknowledges the fact that these effects can be detrimental to the survival of mankind.

All aspects of the environment are interrelated. An action at one end of the environmental spectrum, although thought to be harmless, can produce an adverse reaction at the other end. Thus, not only must a particular action be considered, but also the chain reaction it generates.

Specifically, large-scale actions acknowledged by Congress to affect the national environment include: (1) a high population growth rate, (2) high-density urbanization, (3) industrialization at all costs, (4) resource exploitation, and (5) unchecked technological advances.

National goals reflect appreciation of the need for restoring and maintaining the environment, and recognize this need as beneficial to the development of man. These national goals, although set at the federal level, can only be accomplished through the cooperation of federal, state and local governments, public and private organizations, and individuals. A united effort by these diverse groups, which often have conflicting objectives, using the best practicable means and measures will produce a harmonic coexistence of man and nature to fulfill the social, economic, and other requirements of present and future operations. The federal government's role will ensure this cooperation through technical and financial assistance.

Federal Activities

The federal government establishes national policy so that all plans, functions, programs and resources of the government are directed to:

1. fulfill the responsibilities of each generation as trustee of the environment for succeeding generations;
2. assure for all Americans safe, healthful, productive, and aesthetically and culturally pleasing surroundings;
3. attain the widest range of beneficial uses of the environment without degradation, risk to health or safety, or other undesirable and unintended consequences;
4. preserve important historic, cultural and natural aspects of our national heritage and maintain, wherever possible, an environment which supports diversity and variety of individual choice;
5. achieve a balance between population and resource use which will permit a high standard of living and a wide sharing of life's amenities; and
6. enhance the quality of renewable resources and approach the maximum attainable recycling of depletable resources.

This is probably the most significant statement issued by government with regard to the relationship between the environment and man. It recognizes that man has more than a responsibility to himself alone. He is responsible for his actions and the effects of his actions. Government has finally put an end to the Machiavellian doctrine that has guided us in past years.

Progress and technological advances in the context of increased energy consumption, unacceptable aesthetics, and environmental disruptions may have to be reevaluated to determine their ecological and social desirability.

NEPA Administration

In the writing of any law, Congress first defines its objectives and gives reasons for its necessity. Congress then writes into the legislation the administrative directives that enable a law to become effective. Therefore, in the next section of the NEPA, the administrative responsibilities are defined:

Congress directs all policies, regulations and public laws of the United States to be interpreted in accordance with the policy of this act, and further all federal agencies will consider the environmental impact in their planning and design processes.

In conjunction with the CEQ, procedures are identified and developed by which environmental effects can be assessed relative to economic and technical factors.

For every recommendation or report on proposals for legislation and other major federal actions significantly affecting the quality of the human environment, *a detailed statement (EIS) must be included by the responsible official.* The EIS will include a discussion of the following and will be covered in greater detail in later chapters.

1. the environmental impact of the proposed action;
2. any adverse environmental effects which cannot be avoided should the proposal be implemented;
3. alternatives to the proposed action;
4. the relationship between local short-term uses of man's environment and the maintenance and enhancement of long-term productivity;
5. any irreversible and irretrievable commitments of resources which would be involved in the proposed action should it be implemented.

After it is written, the EIS is first submitted in draft form by the responsible agency to other agencies with jurisdiction, to an agency with a special expertise in the proposed project, to other local levels of government, to the President of the U.S., the CEQ, and to the public. A public hearing must also be held. All comments, pro and con, must be considered before the issuance of the final statement. The final EIS is again made available to the above organizations and individuals. Further, the Congress orders the federal government to:

• Study, develop and describe alternatives to the recommended courses of action in any proposal which involves unresolved conflicts concerning alternative uses of available resources.

• Recognize and make a commitment to world-wide and long-term environmental problems. An example of the workings of this provision of the NEPA is the bilateral agreement between the U.S.S.R. and the U.S. to cooperate on environmental protection concerns. Table 4 outlines the agreement which was signed on May 23, 1972. Meetings of scientists and administrators have been taking place for the last several years between the two countries allowing for the exchange of specialists and information.

Table 4   The U.S. - U.S.S.R. Bilateral Agreement on Cooperation
in the Field of Environmental Protection (May 1972)[5]

Points of Cooperation:

* Working out measures to prevent pollution
* Studying pollution and its effects on the environment
* Developing the basis for controlling the impact of human activities on nature

11 Specific Areas of Cooperation Involving 39 Projects:

* Prevention of air pollution (6)
* Prevention of water pollution (4)
* Prevention of pollution related to agricultural production (4)
* Enhancement of the urban environment (5)
* Protection of nature and the organization of preserves (6)
* Protection of the marine environment from pollution (2)
* Biological and genetic effects of environmental pollution (2)
* Influence of environmental changes on climate (3)
* Earthquake prediction and tsunami warning (5)
* Arctic and subarctic ecological systems (none) (Arctic-related work is done under projects of other areas)
* Legal and administrative measures for environmental protection (2)

Meetings of the Joint Committee to Date:

First meeting in Moscow	September 18-21, 1972
Second meeting in Washington, D.C.	November 13-16, 1973
Third meeting in Moscow	December 9-12, 1974
Fourth meeting in Washington, D. C.	October 28-31, 1975
Fifth meeting scheduled for Moscow	September or October 1976

Some Highlights of 1976:

* Discussion of air pollution modeling, instrumentation, and measurement methodology
* Joint testing of electrostatic precipitators, wet scrubbers, raw materials, and catalysts for process improvement and modification
* Exchange of cement, iron and steel, and chemical industry pollution control experts
* Joint symposium on modification of biological and chemical wastewater treatment methods
* Possible new project on environmental education
* Development of proposals for motor vehicle engine emission control technology, and a uniform test cycle procedure for trucks
* Exchange of specialists on river basin water quality planning and decision-making processes

Table 4, continued

- Integrated Pest Management conference, exchanges of beneficial organisms, and joint IPM research trials
- First meeting on interreaction between forests, plants and pollutants
- First symposium on forms and mechanisms by which pesticides and chemicals are transported
- First meeting on effects of agricultural chemicals on fauna
- Negotiations on migratory bird convention
- First exchange on environmental impact of pipeline construction and operation in permafrost areas
- First exchange on mine-spoil restoration and reclamation
- First symposium on biosphere reserves
- First meeting on arid ecosystems
- Joint research cruises on walrus and ice seals
- Environmental workshops on mutagenesis and carcinogenesis
- Joint symposium on numerical modeling of atmospheric heat balance
- Joint balloon measurements of stratospheric aerosols
- Seismological investigations of San Andreas fault, Garm Region (U.S.S.R.), and Nurek Reservoir (U.S.S.R.)
- Second meeting of tsunami experts
- Discussion on harmonization of measurement and methodology standards

---

- Make available to the general public information useful in restoring, maintaining and enhancing the environment.
- Initiate and utilize ecological information in the planning and development of resource-oriented projects.
- Assist the Council on Environmental Quality.

THE COUNCIL ON ENVIRONMENTAL QUALITY

A report to Congress by the President on the state of the environment is to be made each year. This report shall be divided into five parts describing various aspects of environmental concern.

1. The status and condition of the major natural, man-made, or altered environmental classes of the nation, including air quality, aquatic quality (marine, estuarine, fresh water), and terrestrial environment (forest, drylands, wetlands, range, urban, suburban and rural environment).

2. The current and foreseeable trends in the quality, management and utilization of such environments and the effects of those trends on the social, economic, and other requirements of the nation.

3. The adequacy of the available natural resources for fulfilling human and economic requirements of the nation in light of the expected population growth.

4. A review of the activities of the federal government, state government, local government, private corporations and organizations, and individuals with respect to their effect on the conservation, development and utilization of the natural resources.

5. A program for remedying the deficiencies of existing programs and activities, together with recommendations for legislation.

To aid the President in an advisory capacity, the Council on Environmental Quality (CEQ) was created and made part of the executive branch. It consists of three qualified members whose functions and duties are to:

1. analyze and interpret environmental trends;
2. appraise federal programs and activities in relation to the NEPA;
3. be conscious and responsive to the scientific, economic, social, aesthetic and cultural needs of the nation;
4. formulate and recommend national policy to foster and promote the improvement of environmental quality;
5. conduct investigations, studies, surveys, research and analyses relating to ecological systems and environmental quality;
6. document and define changes in the natural environment;
7. report at least once a year to the President on the state of the national environment;
8. make available such studies, reports and recommendations with respect to matters of policy and legislation as the President requests,
9. assist and advise the President in the preparation of the aforementioned annual Environmental Quality Report.

As previously noted, the CEQ reviews all EIS's and is the motivating force behind them. The CEQ interprets the NEPA for each agency and considers agency actions in accordance with NEPA requisites. Further, it helps set up agency NEPA compliance procedures and solves problems as they arise.

The CEQ is not a regulatory commission, agency or officer. It is an organizing and advisory council aware of government's role in protecting the ecosystem. The CEQ does not necessarily comment on all EIS's it reviews, nor does it approve or disapprove a particular EIS. The EIS is used by the CEQ in its role as advisor to the President and federal agencies.

CEQ guidelines issued in the *Federal Register* (April 1, 1973) for the preparation of EIS's are included in the Appendix.

CHAPTER 2

THE ENVIRONMENTAL IMPACT STATEMENT

PURPOSE AND DEFINITION

The NEPA has established national environmental goals. At the foundation of this national commitment is the environmental impact statement (EIS). For matters under their jurisdiction, all federal agencies must prepare EIS's, detailing the potential environmental impact of a proposed action.

The purpose of the EIS is to determine before implementation the environmental effects of a proposed action. The EIS notifies the public, Congress, the President, and the various decision-making agencies of an action's environmental consequences. It defines and evaluates a proposed action's environmental effects, bringing together all facets of a project in a form that allows a decision to be made logically and rationally. Negative environmental impacts are exposed, but attempts are made to alleviate them through the defining of possible alternatives or the creation of favorable trade-offs.

The writing of an EIS is unlike writing a piece of legislation. It is not a law and by itself does not decide the undertaking of a project. It is used with other data to assist the decision-maker. Other considerations may include: the project's cost, its necessity, and the benefits derived. In certain instances these factors could far outweigh some detrimental environmental effects. This has certainly been true in Alaska and off the eastern coast, where our national need for oil overshadows certain ecological concerns.

On the other hand, a project that might seem timely, justified and beneficial at its initiation could result in environmental havoc. This has been evident in the case of a proposed coal gasification plant in New Mexico and a proposed southwest power plant. In light of environmental considerations these projects have been discontinued.

The EIS must also describe the effects of nonimplementation of a project. A description of this alternative gives evidence to the decision-maker of the actual overall benefits of a project.

The statement creates an environmental conscience at the highest level in federal agencies, thus ensuring that national environmental goals are adhered to in all agency policies, plans and actions.

CRITERIA FOR DETERMINING WHEN TO PREPARE AN EIS

Each federal agency in conjunction with the CEQ prepares its guidelines for the preparation of an EIS. In general the guidelines are similar and should not vary greatly from the mechanisms described herein.

An EIS is written when a "major federal action significantly affecting the quality of the human environment" is proposed. If an action involves no federal monies or involvement or licensing or permits, then it is not federal and the action is not covered by the NEPA. An action is federal when federal monies support it and federal responsibility controls it. Federal monies do not firmly establish federal actions, as with the case of revenue sharing to states and municipalities. The federal government does not exercise direct control over the spending of these funds; therefore, there is no federal responsibility.

A "major action" and "significant effect" are not two separate criteria established to be met before an EIS is prepared. They tend to complement one another. When an action significantly affects the environment, no matter how small, it is a major action. The reverse is true also: Any action that is major will have to affect the environment in some manner. During the initial stages of any project, the agency with jurisdiction begins its environmental assessment. Dealings with contractors or licensees usually require a filing of environmental impacts along with applications. Thus, the burden of protecting the environment is spread over the public and private sectors. In requiring an environmental assessment at the time of application, the agency is forcing the applicant to consider his environmental impacts early in the planning process.

Similar logic follows for a governmental agency such as the Army Corps of Engineers proposing a dam across a river. Before the damming is fully conceived the environmental aspects may alter its location, size or feasibility. With regard to this assessment, the controlling agency then establishes a project as a "major federal action" and/or "environmentally significant which affects humans or the quality of human life."

The agency then reviews this assessment of the proposed project with these two standards in mind and considers the following general criteria [6, 7, 8] before requiring an EIS.

The EIS should be prepared in light of the overall cumulative impact of the proposed action. The action in itself might be small, but combined with several other "minor" federal projects could constitute a major action.

Beneficial as well as detrimental effects should be considered environmentally significant. Even if the net effects are beneficial, an EIS may be required. However, preference should be given to preparing EIS's on those actions which have adverse effects.

Both the long-term and short-term effects of a project are considered. Beneficial effects may be generated in the short-term, but in the long-term adverse effects could develop. An action's primary and secondary effects are also examined. In observing these effects the agency should consider: changes in land use patterns; changes in energy supply and demand; increased development in floodplains; significant changes in ambient air and water quality, or noise levels; potential violations of air quality, water quality and noise level standards; significant changes in surface or ground water quality or quantity; and encroachments on wetlands, coastal zones, or fish and wildlife habitat, especially when threatened or endangered species may be affected.

A particular action by an agency may be rather insignificant as it stands by itself, but the involvement of several agencies could make the action major. Further, an action in an urban area or on the east coast could have relatively few environmental repercussions. However, the same action in a suburban or rural region or on the west coast could result in phenomenal altercations. The converse is also true.

An EIS should be prepared when the environmental impact of a proposed action will probably be highly controversial. A relatively minor action that could set a precedent for future larger actions would require consideration for an EIS. Uniqueness of site location—such as historic sites, parklands, wild and scenic rivers, archaeological sites—results in a project being classified as having significant environmental impacts regardless of its size.

Using the above criteria, an agency can determine when a federal action is major and environmentally significant. Examples of when an EIS would be required include:[6]

> agency recommendations on their own legislative proposals;
>
> agency reports on legislation initiated in another part of government but which deal with the agency's subject of primary responsibility;
>
> agency projects and activities that may be directly acted upon by the agency; or that are supported (even partially) with federal contracts, grants, subsidies, loans, or other funding assistance; or that are part of a federal lease, permit, license, certificate or other entitlement for use;
>
> agency decisions of policy, regulation and procedure-making.

Figure 1 gives a typical agency flowsheet outlining procedures for determining if an EIS is required for agency projects.

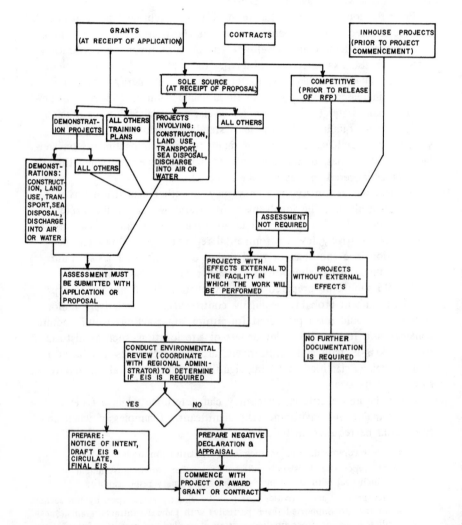

Figure 1   Procedures for determining if an EIS is required on solid waste management project operations.[7]

EIS DEFINITIONS

Environmental Assessment—is a written analysis describing the environmental impacts of proposed actions of an agency.

Environmental Review—is a formal evaluation undertaken to determine whether an action may have a significant impact on the environment.

Notice of Intent—is a memorandum prepared after the environmental review, announcing to federal, regional, state and local agencies, and to interested persons, that a draft EIS will be prepared.

Environmental Impact Statement—is. a report which identifies and analyzes in detail the environmental impact of a proposed agency action and feasible alternatives.

Negative Declaration—is a written announcement, prepared after the environmental review, which states that an agency has decided not to prepare an EIS and summarizes the environmental impact appraisal.

Environmental Impact Appraisal—is based on an environmental review and supports a negative declaration. It describes a proposed agency action, its expected environmental impact, and the basis for the conclusion that no significant impact is anticipated.

Responsible Official—is the agency official with the responsibility of making project decisions.

NEPA-Associated Documents—are any one or combination of notices of intent, negative declarations, exemption certifications, environmental impact appraisals, news releases, EIS's, and environmental assessments.

Interested Persons—are individuals, citizen groups, conservation organizations, corporations, or other nongovernmental units that may be interested in, affected by, or technically competent to comment on the environmental impacts of the proposed action.

HOW AN EIS IS PREPARED

Following the previously mentioned criteria an agency establishes which of its actions require an EIS. The agency then proceeds to follow formal procedures it has promulgated for EIS preparation and review. These procedures may vary from agency to agency, but all must include:

> a time period before an EIS decision is made that would enable other agencies to comment;
> a means of EIS availability to the public which would be described in detail;
> an outline of the information services and methods to be used in preparation of an EIS;

the designation of the responsible official or officials;

measures to be pursued for other agency consultation and EIS incorporation of these comments, with the EPA in particular; and

a system for public announcement of agency plans and projects with environmental impacts.

THE DRAFT EIS AND PUBLIC PARTICIPATION

Upon the decision to prepare an EIS, the responsible agency issues a notice of intent. This notice of intent informs the public of the decision to prepare an EIS in a preliminary form or "draft EIS." This statement would include all the requirements of a final EIS. The notice solicits comments from individuals and other agencies with jurisdiction or expertise related to the action. Table 5 is a CEQ list of governmental agencies and their fields of expertise which should be used in the review process.

The preliminary suggestions of the commenting agencies are incorporated into the draft EIS. The draft EIS is then made available to the CEQ, agencies with the desired expertise, and the public for review.

In the event that several agencies may hold jurisdiction over a single or multi-faceted project, the CEQ decides which agency is responsible for the draft EIS.

Circulation of and reaction to the draft EIS, project controversy, project magnitude, and project significance are factors determining the necessity of a public hearing. One or more public hearings may be necessary at several locations. The responsible official decides on the need for the hearing and the suitability of its site.

To allow for proper public absorption of information in the draft EIS, no public hearing may be held prior to 15 days after its issuance, and usually a longer time period is involved. Commenting agencies and individuals are allowed at least 45 days to review a draft EIS, and extensions of 15 or more days may be granted upon request.

One of the most enlightening facets of the NEPA and EIS procedures is the constant encouragement of public participation in the preparation of an EIS. Comments of citizens are always solicited and public hearings are held. The availability of an EIS or the announcement that a public hearing is being held is often published in local newspapers and publicized in other media.

At the public hearing, the proposed project is described and the draft EIS is presented. Individuals, organizations, and public officials are allowed to present their views on the project and the draft EIS.

Table 5 Areas of Environmental Impact and Federal Agencies and Federal State
Agencies[a] with Jurisdiction by Law or Special Expertise to Comment Thereon[b]

Air

Air Quality

Department of Agriculture - Forest Service (effects on vegetation)
Atomic Energy Commission (radioactive substances)
Department of Health, Education, and Welfare
Environmental Protection Agency
Department of the Interior -
  Bureau of Mines (fossil and gaseous fuel combustion)
  Bureau of Sport Fisheries and Wildlife (effect on wildlife)
  Bureau of Outdoor Reacreation (effects on recreation)
  Bureau of Land Management (public lands)
  Bureau of Indian Affairs (Indian lands)
National Aeronautics and Space Administration (remote sensing, aircraft emissions)
Department of Transportation -
  Assistant Secretary for Systems Development and Technology (auto emissions)
  Coast Guard (vessel emissions)
  Federal Aviation Administration (aircraft emissions)

Weather Modification

Department of Agriculture - Forest Service
Department of Commerce - National Oceanic and Atmospheric Administration
Department of Defense - Department of the Air Force
Department of the Interior -
  Bureau of Reclamation

Water Resources Council

Water Quality

Department of Agriculture
  Soil Conservation Service
  Forest Service
Atomic Energy Commission (radioactive substances)
Department of the Interior -
  Bureau of Reclamation
  Bureau of Land Management (public lands)
  Bureau of Indian Affairs (Indian lands)
  Bureau of Sports Fisheries and Wildlife
  Bureau of Outdoor Recreation
  Geological Survey
  Office of Saline Water
Environmental Protection Agency
Department of Health, Education, and Welfare

Table 5, continued

Department of Defense -
   Army Corps of Engineers
   Department of the Navy (ship pollution control)
National Aeronautics and Space Administration (remote sensing)
Department of Transportation -
   Coast Guard (oil spills, ship sanitation)
Department of Commerce -
   National Oceanic and Atmospheric Administration
Water Resources Council
River Basin Commissions (as geographically appropriate)

Marine Pollution, Commercial Fishery Conservation, and Shellfish Sanitation

Department of Commerce -
   National Oceanic and Atmospheric Administration
Department of Defense -
   Army Corps of Engineers
   Office of the Oceanographer of the Navy
Department of Health, Education, and Welfare
Department of the Interior -
   Bureau of Sport Fisheries and Wildlife
   Bureau of Outdoor Recreation
   Bureau of Land Management (outer continental shelf)
   Geological Survey (outer continental shelf)
Department of Transportation -
   Coast Guard
Environmental Protection Agency
National Aeronautics and Space Administration (remote sensing)
Water Resources Council
River Basin Commissions (as geographically appropriate)

Waterway Regulation and Stream Modification

Department of Agriculture -
   Soil Conservation Service
Department of Defense -
   Army Corps of Engineers
Department of the Interior -
   Bureau of Reclamation
   Bureau of Sport Fisheries and Wildlife
   Bureau of Outdoor Recreation
   Geological Survey
Department of Transportation -
   Coast Guard
Environmental Protection Agency
National Aeronautics and Space Administration (remote sensing)
Water Resources Council
River Basin Commissions (as geographically appropriate)

Table 5, continued

Fish and Wildlife

Department of Agriculture -
  Forest Service
  Soil Conservation Service
Department of Commerce -
  National Oceanic and Atmospheric Administration (marine species)
Department of the Interior -
  Bureau of Sport Fisheries and Wildlife
  Bureau of Land Management
  Bureau of Outdoor Recreation
Environmental Protection Agency

Solid Waste

Atomic Energy Commission (radioactive waste)
Department of Defense -
  Army Corps of Engineers
Department of Health, Education, and Welfare
Department of the Interior -
  Bureau of Mines (mineral waste, mine acid waste, municipal solid waste, recycling)
  Bureau of Land Management (public lands)
  Bureau of Indian Affairs (Indian lands)
  Geological Survey (geologic and hydrologic effects)
  Office of Saline Water (demineralization)
Department of Transportation -
  Coast Guard (ship sanitation)
Environmental Protection Agency
River Basin Commissions (as geographically appropriate)
Water Resources Council

Noise

Department of Commerce -
  National Bureau of Standards
Department of Health, Education, and Welfare
Department of Housing and Urban Development (land use and building materials aspects)
Department of Labor -
  Occupational Safety and Health Administration
Department of Transportation -
  Assistant Secretary for Systems Development and Technology
  Federal Aviation Administration, Office of Noise Abatement
Environmental Protection Agency
National Aeronautics and Space Administration

Table 5, continued

Radiation

Atomic Energy Commission
Department of Commerce -
  National Bureau of Standards
Department of Health, Education, and Welfare
Department of the Interior -
  Bureau of Mines (uranium mines)
  Mining Enforcement and Safety Administration (uranium mines)
Environmental Protection Agency

Hazardous Substances

Toxic Materials

Department of Agriculture -
  Consumer and Marketing Service (meat and poultry products)
Department of Health, Education, and Welfare
Environmental Protection Agency

Pesticides

Department of Agriculture -
  Agricultural Research Service (biological controls, food and fiber production)
  Consumer and Marketing Service
  Forest Service
Department of Commerce -
  National Oceanic and Atmospheric Administration
  Department of Health, Education, and Welfare
Department of the Interior -
  Bureau of Sport Fisheries and Wildlife (fish and wildlife effects)
  Bureau of Land Management (public lands)
  Bureau of Indian Affairs (Indian lands)
  Bureau of Reclamation (irrigated lands)
Environmental Protection Agency

Transportation and Handling of Hazardous Materials

Atomic Energy Commission (radioactive substances)
Department of Commerce -
  Maritime Administration
  National Oceanic and Atmospheric Administration
    (effects on marine life and the coastal zone)
Department of Defense -
  _ Armed Services Explosive Safety Board
  Army Corps of Engineers (navigable waterways)

Table 5, continued

Department of Transportation -
  Federal Highway Administration, Bureau of Motor Carrier Safety
  Coast Guard
  Federal Railroad Administration
  Federal Aviation Administration
  Assistant Secretary for Systems Development and Technology
  Office of Hazardous Materials
  Office of Pipeline Safety
Environmental Protection Agency

Energy Supply and Natural Resources Development

Electric Energy Development, Generation, and Transmission, and Use

Atomic Energy Commission (nuclear)
Department of Agriculture -
  Rural Electrification Administration (rural areas)
Department of Defense -
  Army Corps of Engineers (hydro)
Department of Health, Education, and Welfare (radiation effects)
Department of Housing and Urban Development (urban areas)
Department of the Interior -
  Bureau of Indian Affairs (Indian lands)
  Bureau of Land Management (public lands)
  Bureau of Reclamation
  Power Marketing Administrations
  Geological Survey
  Bureau of Sport Fisheries and Wildlife
  Bureau of Outdoor Recreation
  National Park Service
Environmental Protection Agency
Federal Power Commission (hydro, transmission, and supply)
River Basin Commissions (as geographically appropriate)
Tennessee Valley Authority
Water Resources Council

Petroleum Development, Extraction, Refining, Transport, and Use

Department of the Interior -
  Office of Oil and Gas
  Bureau of Mines
  Geological Survey
  Bureau of Land Management (public lands and outer continental shelf)
  Bureau of Indian Affairs (Indian lands)
  Bureau of Sport Fisheries and Wildlife (effects on fish and wildlife)
  Bureau of Outdoor Recreation
  National Park Service
Department of Transportation (Transport and Pipeline Safety)
Environmental Protection Agency
Interstate Commerce Commission

Table 5, continued

Natural Gas Development, Production, Transmission and Use

Department of Housing and Urban Development (urban areas)
Department of the Interior -
  Office of Oil and Gas
  Geological Survey
  Bureau of Mines
  Bureau of Land Management (public lands)
  Bureau of Indian Affairs (Indian lands)
  Bureau of Sport Fisheries and Wildlife
  Bureau of Outdoor Recreation
  National Park Service
Department of Transportation (transport and safety)
Environmental Protection Agency
Federal Power Commission (production, transmission, and supply)
Interstate Commerce Commission

Coal and Minerals Development, Mining, Conversion Processing, Transport, and Use

Appalachian Regional Commission
Department of Agriculture -
  Forest Service
Department of Commerce
Department of the Interior -
  Office of Coal Research
  Mining Enforcement and Safety Administration
  Bureau of Mines
  Geological Survey
  Bureau of Indian Affairs (Indian lands)
  Bureau of Land Management (public lands)
  Bureau of Sport Fisheries and Wildlife
  Bureau of Outdoor Recreation
  National Park Service
Department of Labor -
  Occupational Safety and Health Administration
Department of Transportation
Environmental Protection Agency
Interstate Commerce Commission
Tennessee Valley Authority

Renewable Resource Development, Production, Management, Harvest, Transport, and Use

Department of Agriculture -
  Forest Service
  Soil Conservation Service
Department of Commerce

Table 5, continued

Department of Housing and Urban Development (building materials)
Department of the Interior -
   Geological Survey
   Bureau of Land Management (public lands)
   Bureau of Indian Affairs (Indian lands)
   Bureau of Sport Fisheries and Wildlife
   Bureau of Outdoor Recreation
   National Park Service
Department of Transportation
Environmental Protection Agency
Interstate Commerce Commission (freight rates)

Energy and Natural Resources Conservation

Department of Agriculture -
   Forest Service
   Soil Conservation Service
Department of Commerce -
   National Bureau of Standards (energy efficiency)
Department of Housing and Urban Development -
   Federal Housing Administration (housing standards)
Department of the Interior -
   Office of Energy Conservation
   Bureau of Mines
   Bureau of Reclamation
   Geological Survey
   Power Marketing Administration
Department of Transportation
Environmental Protection Agency
Federal Power Commission
General Services Administration (design and operation of buildings)
Tennessee Valley Authority

Land Use and Management
Land Use Changes, Planning and Regulation of Land Development

Department of Agriculture -
   Forest Service (forest lands)
   Agricultural Research Service (agricultural lands)
Department of Housing and Urban Development
Department of the Interior -
   Office of Land Use and Water Planning
   Bureau of Land Management
   Bureau of Indian Affairs (Indian lands)
   Bureau of Sport Fisheries and Wildlife (wildlife refuges)
   Bureau of Outdoor Recreation (recreation lands)
   National Park Service (NPS units)

Table 5, continued

Department of Transportation
Environmental Protection Agency (pollution effects)
National Aeronautics and Space Administration (remote sensing)
River Basins Commissions (as geographically appropriate)

Public Land Management

Department of Agriculture -
   Forest Service (forests)
Department of Defense
Department of the Interior -
   Bureau of Land Management
   Bureau of Indian Affairs (Indian lands)
   Bureau of Sport Fisheries and Wildlife (wildlife refuges)
   Bureau of Outdoor Recreation (recreation lands)
   National Park Service (NPS units)
Federal Power Commission (project lands)
General Services Administration
National Aeronautics and Space Administration (remote sensing)
Tennessee Valley Authority (project lands)

Protection of Environmentally Critical Areas -- Floodplains, Wetlands, Beaches and Dunes, Unstable Soils, Steep Slopes, Aquifer Recharge Areas, etc.

Department of Agriculture -
   Agricultural Stabilization and Conservation Service
   Soil Conservation Service
   Forest Service
Department of Commerce -
   National Oceanic and Atmospheric Administration (coastal areas)
Department of Defense -
   Army Corps of Engineers
Department of Housing and Urban Development (urban and floodplain areas)
Department of the Interior -
   Office of Land Use and Water Planning
   Bureau of Outdoor Recreation
   Bureau of Reclamation
   Bureau of Sport Fisheries and Wildlife
   Bureau of Land Management
   Geological Survey
Environmental Protection Agency (pollution effects)
National Aeronautics and Space Administration (remote sensing)
River Basins Commissions (as geographically appropriate)
Water Resources Council

Table 5, continued

Land Use in Coastal Areas

Department of Agriculture -
  Forest Service
  Soil Conservation Service (soil stability, hydrology)
Department of Commerce -
  National Oceanic and Atmospheric Administration
    (impact on marine life and coastal zone management)
Department of Defense -
  Army Corps of Engineers (beaches, dredge and fill permits, Refuse Act permits)
Department of Housing and Urban Development (urban areas)
Department of the Interior -
  Office of Land Use and Water Planning
  Bureau of Sport Fisheries and Wildlife
  National Park Service
  Geological Survey
  Bureau of Outdoor Recreation
  Bureau of Land Management (public lands)
Department of Transportation -
  Coast Guard (bridges, navigation)
Environmental Protection Agency (pollution effects)
National Aeronautics and Space Administration (remote sensing)

Redevelopment and Construction in Built-Up Areas

Department of Commerce -
  Economic Development Administration (designated areas)
Department of Housing and Urban Development
Department of the Interior -
  Office of Land Use and Water Planning
Department of Transportation
Environmental Protection Agency
General Services Administration
Office of Economic Opportunity

Density and Congestion Mitigation

Department of Health, Education, and Welfare
Department of Housing and Urban Development
Department of the Interior -
  Office of Land Use and Water Planning
  Bureau of Outdoor Recreation
Department of Transportation
Environmental Protection Agency

Table 5, continued

Neighborhood Character and Continuity

Department of Health, Education, and Welfare
Department of Housing and Urban Development
National Endowment for the Arts
Office of Economic Opportunity

Impacts on Low-Income Populations

Department of Commerce -
  Economic Development Administration (designated areas)
Department of Health, Education, and Welfare
Department of Housing and Urban Development
Office of Economic Opportunity

Historic, Architectural, and Archeological Preservation

Advisory Council on Historic Preservation
Department of Housing and Urban Development
Department of the Interior -
  National Park Service
  Bureau of Land Management (public lands)
  Bureau of Indian Affairs (Indian lands)
General Services Administration
National Endowment for the Arts

Soil and Plant Conservation and Hydrology

Department of Agriculture -
  Soil Conservation Service
  Agricultural Service
  Forest Service
Department of Commerce -
  National Oceanic and Atmospheric Administration
Department of Defense -
  Army Corps of Engineers (dredging, aquatic plants)
Department of Health, Education, and Welfare
Department of the Interior -
  Bureau of Land Management
  Bureau of Sport Fisheries and Wildlife
  Geological Survey
  Bureau of Reclamation
Environmental Protection Agency
National Aeronautics and Space Administration (remote sensing)
River Basin Commissions (as geographically appropriate)
Water Resources Council

Table 5, continued

___

Outdoor Recreation

Department of Agriculture -
   Forest Service
   Soil Conservation Service
Department of Defense -
   Army Corps of Engineers
Department of Housing and Urban Development (urban areas)
Department of the Interior -
   Bureau of Land Management
   National Park Service
   Bureau of Outdoor Recreation
   Bureau of Sport Fisheries and Wildlife
   Bureau of Indian Affairs
Environmental Protection Agency
National Aeronautics and Space Administration (remote sensing)
River Basin Commissions (as geographically appropriate)
Water Resources Council

Advisory Council on Historic Preservation

Office of Architectural and Environmental Preservation, Advisory Council on Historic
   Preservation, Suite 430, 1522 K Street, N.W., Washington, D.C.   20005

___

Regional Administrator, I
   U.S. Environmental Protection Agency
   Room 2303, John F. Kennedy Federal Bldg.
   Boston, Mass.   02203

Regional Administrator, II
   U.S. Environmental Protection Agency
   Room 908, 26 Federal Plaza
   New York, N.Y.,   10007

Department of Agriculture

Office of the Secretary, Attn: Coordinator Environmental Quality Activities
   U.S. Department of Agriculture, Washington, D. C.   20250

Appalachian Regional Commission

Office of the Alternate Federal Co-Chairman, Appalachian Regional Commission,
   1666 Connecticut Avenue, N.W., Washington, D.C.   20235

Department of the Army (Corps of Engineers)

Executive Director of Civil Works, Office of the Chief of Engineers, U.S. Army
   Corps of Engineers, Washington, D.C.   20314

Table 5, continued

Atomic Energy Commission

For nonregulatory matters: Office of Assistant General Manager for Biomedical and Environmental Research and Safety Programs, Atomic Energy Commission, Washington, D.C. 20545

For regulatory matters: Office of the Assistant Director for Environmental Projects, Atomic Energy Commission, Washington, D. C. 20545

Department of Commerce

Office of the Deputy Assistant Secretary for Environmental Affairs, U.S. Department of Commerce, Washington, D.C. 20230

Department of Defense

Office of the Assistant Secretary for Defense (Health and Environment), U.S. Department of Defense, Room 3E172, The Pentagon, Washington, D.C. 20301

Delaware River Basin Commission

Office of the Secretary, Delaware River Basin Commission, Post Office Box 360, Trenton, N.J. 08603

Environmental Protection Agency

Director, Office of Federal Activities, Environmental Protection Agency, 401 M Street, S.W., Washington, D.C. 20460

---

[a]River Basin Commissions (Delaware, Great Lakes, Missouri, New England, Ohio, Pacific Northwest, Souris-Red-Rainy, Susquehanna, Upper Mississippi) and similar federal-state agencies should be consulted on actions affecting the environment of their specific geographic jurisdictions.

[b]In all cases where a proposed action will have significant international environmental effects, the Department of State should be consulted and should be sent a copy of any draft and final impact statement which covers such action.

Opposing and supporting views at the hearing, like all agency and individual draft statement comments, are incorporated in a separate section of the final EIS.

In many instances it is the public which forces consideration of project alternatives or overlooked environmental aspects. Projects have been held up indefinitely and stopped completely because of public disagreement with the conclusions of the draft EIS.

THE ENVIRONMENTAL PROTECTION AGENCY REVIEW

EIS's are reviewed by the EPA because of its role as protectorate of the environment. EIS's, whether state or federal in origin, are reviewed by the EPA with regard to possible air and water pollution, contamination of water supplies, solid waste disposal, pesticides, radioactive material and radiation, and noise.

The process of the EPA's EIS review is mandated in Section 30a of the Clean Air Act of 1970. The Act calls for the EPA to comment in writing on the environmental impact of any regulation, action or legislation effected by any federal agency as it affects the duties and responsibilities of the EPA.

Based on its review, the EPA publishes its comments pertaining to a particular EIS and action in the *Federal Register.* These comments may (1) approve of a project, (2) require additional information for adequate environmental assessment, (3) oppose the project based on ecological factors.

The authority of the EPA, in most cases, ends here without provisions to stop a proposed project. In reviewing environmental assessments, the EPA is not an enforcement agency but an advisor to the CEQ, other agencies, and the President.

THE FINAL EIS

The writing of the final EIS will consider all substantial comments made by reviewing agencies and individuals, and at public hearings. Additional investigation and analysis may be required before the statement is complete.

In its final form, the EIS includes the draft EIS, all project modifications, comments, testimony, and responses to inquiries. A copy then goes to the CEQ and the EIS is publicly distributed.

A negative declaration by the EPA means that no significant or adverse impacts will be made as the project now stands. A declaration of this sort must allow 15 days (and extensions) for individuals who disagree with this appraisal to respond. After these responses are reviewed, administrative action on the project may begin.

THE COURTS

The NEPA requires the preparation of an EIS for all major federal actions significantly affecting the environment. Each agency uses its own judgment in determining if it will prepare an EIS for its actions and the adequacy of that statement. As a result, much controversy has been created. This controversy has erupted in many court cases. Questions arise as to whether an EIS should be prepared when the responsible agency has decided against it (*i.e.*, is the action major and environmentally significant?), or whether a particular statement is adequate. The courts have ruled on several occasions on the inadequacies of EIS's, requiring that they be modified or rewritten.

In the U.S. District Court, Tampa Division, *Save Crystal Beach Association vs. Callaway* decision, the Corps of Engineers prepared a statement that considered the environmental impact of maintenance of the navigable waterways of the region in Florida including the St. Joseph Sound. This statement was ruled inadequate for the project (which involved dredging and filling the sound) because it did not specifically consider the project.[9]

The U.S. Court of Appeals in a decision handed down for the case *City of Davis vs. Coleman* ruled that state and local governments are covered by the NEPA. Based on this ruling, the California Division of Highway's (CDHW) negative declaration on the environmental impact of a proposed freeway exchange was set aside. The decision stated that the CDHW made the declaration without the knowledge of the project's economic, social and environmental effects.[10]

EIS IMPACT

The EIS and therefore the NEPA and CEQ are soundly criticized as being too bureaucratic, costly and unmanageable. Many in industry and government alike consider the EIS process too long and complicated and, in many cases, believe it can unnecessarily delay projects for up to two years. The preparation of an EIS can take from 9 to 12 months, increasing costs and placing a heavy burden on the resources of the EPA and other agencies.

On the other hand, a study done by the CEQ encompassing 70 federal agencies showed that with the EIS, agency management can make clearer and more logical decisions. Alternatives and problems are outlined by the EIS in the early stages of a project effecting a sleek and systematic planning process. In some instances questions pertaining to a project's feasibility and value are answered before large sums of money are wasted on a worthless or environmentally suspect project. The criticism presently leveled at the EIS program is due primarily to agency delay in preparing the initial environmental assessment, by court challenges and by poor agency planning.

However, recent history has shown a more efficient EIS program. Agencies now understand that environmental assessments must be prepared

early in a project's conception stages to avoid delays. Presently, guidelines for each agency's EIS are readily accepted, environmental analyses
are coordinated between agencies as is the issuance of draft EIS's and the
solicitation of comments. This has greatly reduced the time involved in
the EIS process. In the early years, court orders sometimes resulted in
the delay of EIS-related projects. However, of the 6000 draft EIS's
written since 1970, only about 5% (291) were challenged as inadequate
in court.[11]   A more efficient coordination mechanism between agencies
and the public will enable the goals of NEPA to be readily accomplished.

CHAPTER 3

ENVIRONMENTAL ASSESSMENT OF THE EIS

EIS requirements are distinctly outlined by the NEPA. These provide an assessment of the environment to ensure a planning process that benefits mankind through the reduction or elimination of the detrimental effects of a project.

The EIS identifies, develops and analyzes all relevant aspects and alternatives of a proposed project. These factors include a project's physical, economic and even spiritual effects. The physical and economic parameters are often easily determined. In the past the economic benefits of a project and its growth potential have usually far outweighed other considerations. In fact, our high standard of living has been rated on an annual income per capita basis.

This basis of life quality assessment has slowly been changing, and the EIS reflects these changes in government attitudes. A price cannot be placed on the beauty of a mountain range, the sound of birds, or the capability to do some fishing. These other indicators of life quality are considered by the EIS and may be the determining factors as to whether a project is implemented.

Five points are discussed below pertaining to the preparation of EIS's for major federal actions. Detail is incorporated into the EIS according to a particular project's dimensions and the scope of its impacts. Data should be presented in a clear manner and evaluated and appraised impartially.[7]

Comprehension of an EIS should be accessible to the general public. Expertise in a particular scientific or technical field should not be required to assess an EIS. For example, a PhD in biology should not be necessary to interpret the EIS description of the impact on a particular biosystem.

THE BACKGROUND AND DIRECT
ENVIRONMENTAL IMPACTS OF
THE PROPOSED ACTION

The proposed action should be described in detail. The statement should begin with the action's purpose and reasons for its implementation. Its location, size and the surrounding region are to be included along with other relevant information such as maps, diagrams, and a summary of technical data to facilitate a comprehensive assessment of the area's environment.

Figure 2 is a typical map that may accompany an EIS. It depicts the offshore oil-drilling tracts of New Jersey and Delaware. Table 6 summarizes each tract (not all are shown) by water depth and distance from shore. For the various significant resources that could be affected by oil drilling at a particular tract, a proximity value has been assigned to denote the distance of each tract from these resources, where 0.9 = drilling structure onsite and 0.0 = drilling structure totally outside the resource area.

Assorted detailed maps and much resource data, such as geological, geophysical and geochemical information, are available from the U.S. Geological Survey (USGS), the U.S. Department of the Interior, and other governmental agencies.

In this manner the project's scope and environment before implementation are catalogued. A brief historical account of the project could in many instances enlighten reviewers as to its necessity and desirability.

The initial stage of any project would involve site selection. Depending on the type of project, whether it be an offshore drilling platform, coal gasification plant, river dam, or highway construction, the resource potential and environmental conflicts must be evaluated at each site.

The following factors,[11] although recognizably varying with each project, should be considered in any site evaluation:

industry (present and future)

geological and geophysical data

resource availability and estimates

natural hazards (faults, earthquake potential, volcanic action, etc.)

current technology (both of pollution and process control)

general climatology and seasonal weather patterns (visibility, temperature, winds, and rainfall)

physical oceanography if applicable (sea temperatures, surface circulation, i.e., waves, currents, tides)

physical hazards that may disrupt project (military installations, air or shipping lanes, ocean or land dumping sites, proximity to urban areas, sites of unexploded ammunition)

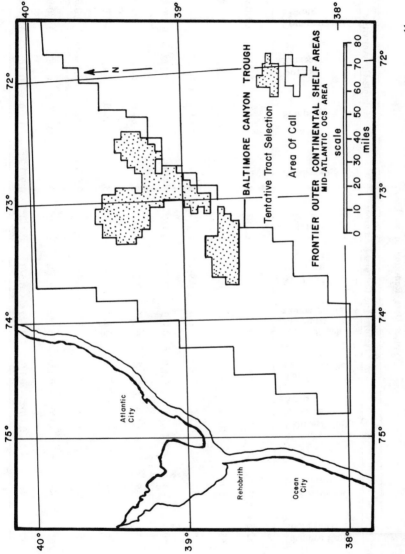

Figure 2. Map identifying location and size of offshore oil-drilling tracts, New Jersey and Delaware[11]

Table 6.   Summary of Effects on Significant Resources by Each Offshore Drilling Tract (Typical Data Required in an EIS)[12]

Block Data			Natural and Cultural Resources						Multiple-Use Activities			
Protraction Diagram and Block Number	Distance From Shore (mi)	Approximate Depth (m)	Refuges/Wildlife Management Areas	Marsh Communities	Sandy Beach Intertidal Communities	Sand and Gravel Deposits	Aesthetics	Archaeology	Shipping	Outdoor Recreation	Commercial Fishing	Sport Fishing
NJ 18-3 411	52.5	45	0.1/0	0.1/0	0.1/0	0.5/0	0.1/0	0.9/0.9	0.4/0	0.1/0	0.9/0.9	0.1/0
NJ 18-3 412	55	50	0.1/0	0.1/0	0.1/0	0.4/0	0.1/0	0.9/0.9	0.4/0	0.1/0	0.9/0.9	0.1/0
NJ 18-3 453	47.5	40	0.1/0	0.1/0	0.1/0	0.4/0	0.1/0	0.9/0.9	0.3/0	0.1/0	0.9/0.9	0.1/0
NJ 18-3 454	50	40	0.1/0	0.1/0	0.1/0	0.5/0	0.1/0	0.9/0.9	0.3/0	0.1/0	0.9/0.9	0.1/0
NJ 18-3 455	53	45	0.1/0	0.1/0	0.1/0	0.5/0	0.1/0	0.9/0.9	0.4/0	0.1/0	0.9/0.9	0.1/0
NJ 18-3 456	56	55	0.1/0	0.1/0	0.1/0	0.4/0	0.1/0	0.9/0.9	0.4/0	0.1/0	0.9/0.9	0.1/0
NJ 18-3 457	59	55	0.1/0	0.1/0	0.1/0	0.4/0	0.1/0	0.9/0.9	0.4/0	0.1/0	0.9/0.9	0.1/0
NJ 18-3 458	62	60	0.1/0	0.1/0	0.1/0	0.3/0	0.1/0	0.9/0.9	0.5/0	0.1/0	0.9/0.9	0.1/0
NJ 18-3 497	49	40	0.1/0	0.1/0	0.1/0	0.5/0	0.1/0	0.9/0.9	0.3/0	0.1/0	0.9/0.9	0.1/0

upset conditions and timetables of their effects
sport and commercial hunting and fishing grounds
recreational facilities
biological communities (marine, estuarine, stream, land, soils, vegetation)
rare and endangered species
archeological and historic sites
regional socioeconomic characteristics
transportation network
air and water quality
land use

Thus, this portion of the EIS must explain and describe in detail the proposed project. Included in this account are all data pertaining to air emissions, water and solid/liquid waste discharges, economic, social and other factors that can directly or indirectly affect the ecological status of a region.

The project's processes are to be summarized. Each unit operation should be developed and defined. Product and raw material handling, transportation, preparation, resource requirements, waste control measures, safety and process upset control measures, and resource recovery measures are some of the operations that should be considered. Related operations such as pipelines, docking, transmission lines, road and rail requirements must also be included in the summary.[13] The unit operations should be described from their initial points to their termination. This is important when dealing with hazardous wastes which are the responsibility of the producer even though he may contract out their disposal to a waste processor. Should the waste processor default or his disposal methods prove inadequate, the ultimate responsibility falls back on the producer.

A process flow diagram of the proposed project should be included to augment the descriptions. An example of this type of diagram is shown in Figure 3. This process flow chart is of a proposed coal gasification plant.

Resource Requirements

Land, raw materials and energy requirements of the project are to be described in detail. Industrial, commercial, domestic, agricultural and other land uses are outlined. A map or photograph of the designated area and a description of governmental jurisdictions, sewage lines, drainage basins, wetlands, river, stream and ground water proximity are included. This is to enable a proper land use assessment to be made. For example, property adjacent to the proposed project may derive its water supply solely from wells. Therefore, a spill of any kind could provide a source of leachate to the ground water resulting in contamination.

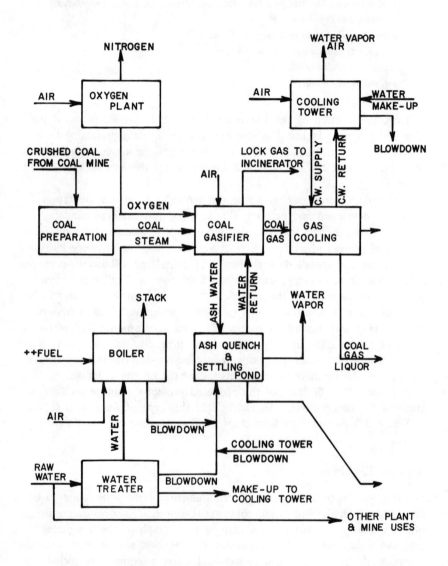

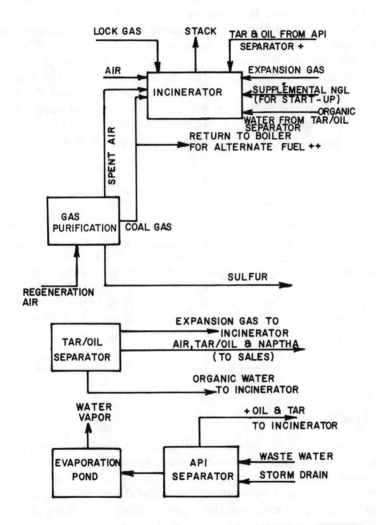

+Small batch about once a month, little or no environmental effect.
++Boiling will be capable of burning either natural gas liquids or s.g. from plant.

Figure 3. Process flow diagram of a proposed coal gasification plant.[14]

The project's power and energy requirements are then explained. This description will include the energy source and pollutants affected by this source. Pollutants generated by an energy source not as a part of the project are not to be included. Alternate, backup or substitute energy sources are to be listed, as are the current and future usage of these sources by all consumers in the region.

The type and quantity of raw materials necessary for the proposed project must be assessed. These materials, whether they be wood, oil, coal, water, iron ore, etc., are to be identified as to their source and availability and the rate at which they are being consumed. Other users of these materials should be recognized and accounted for, and alternate materials that may be used in the processes should be given consideration. Tables 7 and 8 are typical raw material analyses that might be used for a project utilizing coal, such as a coal gasification plant.

Table 7. Quality Data for Coal (the Raw Material)
to be used at the Coal Gasification Plant[15]

	Red	Green	Blue	Yellow	Average
Coal (Btu/lb)	8,627	8,627	9,006	8,589	8,664
Moisture (%)	16.48	16.96	17.02	15.25	16.25
Ash (%)	19.50	11.75	15.88	20.17	19.25
Volatile Material (%)	30.27	32.65	31.44	31.51	30.65
Fixed Carbon (%)	33.75	38.64	35.66	33.07	33.85
Sulfur (%)	0.85	0.53	0.53	.60	0.69

Process Facility

As we have mentioned, the unit operations are described and accompanied by a process flow diagram. Each unit operation description could be enhanced by a process schematic as shown in Figure 4.

Next, a description of the inputs and outputs of the raw materials versus their products, byproducts and wastes is given. Table 9 is a typical input/output analysis for two operating conditions of a coal gasification plant. The lower operating level (288 million cubic feet per day MMCFD) and peak capacity (785 MMCFD) conditions are identified along with the material balances for each individual unit operation.

Table 8.   An Analysis of Coal Showing Trace Element Concentration[15]

Element	Concentration in ppm by Weight	
	From	To
Antimony	0.3	1.2
Arsenic	0.1	3.0
Bismuth	0.0	0.2
Boron	60.0	150.0
Bromine	0.4	18.0
Cadmium	0.2	0.4
Fluorine	200	780
Gallium	0.5	8.0
Germanium	0.1	0.5
Lead	1.4	4.0
Mercury	0.2	0.3
Nickel	3.0	30.0
Selenium	0.1	0.2
Zinc	1.1	27.0

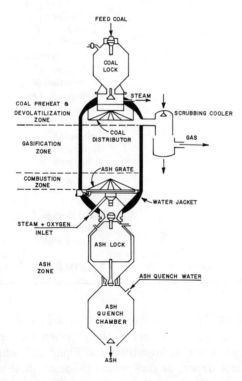

Figure 4. Lurgi gasifier, unit operation process schematic.[14]

Table 9.   Material Balance Analysis for Two Operating Conditions
of a Coal Gasification Plant[14]

	Material Balance in Thousand Pounds/Hour				
In	288-MMCFD Plant	785-MMCFD Plant	Out	288-MMCFD Plant	785-MMCFD Plant
Coal to gas			Boiler stacks	6,593	17,920
production	1,938	5,268	Superheater stacks	361	981
Coal to fuel			Fuel gas heaters	79	215
production	416	1,131	Waste nitrogen	1,587	4,314
Raw water	3,528	9,590	Cooling tower		
Air to O_2 plant	2,061	5,602	evaporation	1,403	3,813
Air to fuel			Wet ash	573	1,557
production	533	1,449	Sulfur plant vent	1,804	4,904
Air to turbine and			Incinerator stack	78	212
boilers	5,760	15,656	Product gas	514	1,397
Air to superheaters	315	850	Sulfur (byproduct)	15.6	42.4
Air to incinerator	31	84	Tar (byproduct)	88.8	241.4
Air to gas heaters	69	187	Tar oil (byproduct)	48.6	132.1
Air to sulfur plant			Naptha (byproduct)	20	54.4
oxidizer	168	457	NH_3 solution (by-		
			product)	107	291
			Phenols (byproduct)	11.3	30.7
			Wastewater to		
			evaporation ponds	450	1,223
			Cooling tower en-		
			trainment	80	217
			Evaporation from		
			ash handling	76	207
			Deaerator vents	32	87
			Steam losses to		
			atmosphere	163	443
			Water to offsite		
			users	644	1,751
			Water storage		
			evaporation	73	198
			Degasser vent	18	49
Total (M lb/hr)	14,819	40,280	Total (M lb/hr)	14,819	40,280

An analysis of the waste sources must be made. As shown in the
process flow diagram, the wastes of the processes are identified and an
inventory of air, water, radioactivity, and liquid and solid wastes is made.
Pollution control devices as they would influence these waste streams
should also be considered in this section.

Intermittent processes such as safety valve releases, start-ups, shut-downs, testing, and various unit-cleaning operations should be described. Emissions and discharges from these occurrences are to be presented.

Included in this section on the description of the process facility should be a timetable for project operation. Personnel training periods, start-up dates, date of full operation, and a brief description and date of future project additions should be tabulated and projected based on good management techniques.

Economic Benefits

The proposed new project should be assessed in relation to its economic, social and environmental benefits. Benefits should not be limited to material values, but must include the spiritual and recreational value of the project. These benefits are to be defined in terms of quantity and quality to the beneficiary. How a particular benefit derives from a specific value should also be included.[13]

THE STATUS-QUO ENVIRONMENT

The environment of the region of the proposed action should be described to better understand the impact of the action should it be implemented. This description of the environment without the proposed project would establish a control situation for assessing the socioeconomic environmental effects likely to occur. Further, the interrelationship of these effects can be better defined.

The control environment would be depicted as it presently exists and as it would exist in 10 or 15 years. In this manner the impacts of all project processes whether they be physical, chemical, biological, economical or social can be evaluated against a norm. The changes generated by these environmental effects (better known as cumulative impacts) can also be more easily understood.

An example of the cumulative and secondary impact would involve a highway being routed through a city neighborhood. The neighborhood might be a cohesive unit that most likely will continue for several years. However, a highway dividing the community has more impact than the removal of a row of houses. First there is a decline in property values for those houses bordering the highway because of location, noise and automobile exhaust. As these values drop, people begin to move out of, abandon or rent these residences along the periphery of the highway. Upkeep on these houses begins to taper off resulting in the spreading of deterioration. This deterioration is usually accompanied by an influx of crime to the area, which only completes the destruction of the neigh-borhood's environment and negates the ability of the inhabitant to use it.

Control Environment Description

The parameters now outlined should be considered when describing the characteristics of the control environment of the project, and how it may exist in 10 to 15 years. These considerations are made in light of the project not being implemented and, depending on the dominant effects of the project, specific parameters will be discussed in greater detail than others. That is, if the proposed action is primarily an air pollution source, much more detail on meteorology, climatology and air quality would be necessary than for, say, soil types. The status-quo environment is described by including in the EIS:

1. The social, economic, and environmental characteristics of a region and their interrelations.

2. Meteorology and climatology including temperature specifics; seasonal or monthly ranges and averages; rainfall; annual average, frequency and magnitude of storm conditions; other types of weather—hurricanes, snow, hail; the frequency of occurrence of wind direction and wind speeds; wind roses. Tables 10a through 10e and Figure 5 present data that are readily available from the National Climatological Center in North Carolina and are necessary to the EIS.

3. The air quality of the region expressed in terms of ambient concentrations of hydrocarbons, carbon monoxide, nitrogen oxides, sulfur oxides, photochemical oxidants, and particulates. Atmospheric quality may be improving or deteriorating and these trends should be reported. Any available information such as previous surveys and ambient concentrations are assessed. In the event data are unavailable, air quality surveys and/or air pollutant dispersion modeling may have to be initiated.

State and municipal departments of environmental protection should be consulted to determine current standards and regulations and their influences on the proposed action. Federal requirements for fuel combustion, fuel switching, individual pollutant ambient concentrations, and air quality degradation for the region of concern must be examined.

4. Topography of a region is readily shown with a USGS map. These maps are available in 7½-minute and 15-minute squares and show a great deal of information and detail about the region. Supplemental information may be provided in the forms of aerial photographs, map overlays, figure blow-ups, and written descriptions.[13] The major topographical features and land formations should be identified. These can include: plateaus, mountains, hills and valleys, lakes, rivers and drainage basins,

Table 10A. Temperature and Wind Speed—Relative Humidity Occurrences

Temperature (F)	0-4 mph						5-14 mph						15-24 mph						25 mph and Over						Total OBS
Relative Humidity	Under 30%	30-49%	50-69%	70-79%	80-89%	90-100%	Under 30%	30-49%	50-69%	70-79%	80-89%	90-100%	Under 30%	30-49%	50-69%	70-79%	80-89%	90-100%	Under 30%	30-49%	50-69%	70-79%	80-89%	90-100%	
99/95									1																1
94/90			11		+			3	44	2					10									+	61
89/85		1	8	7	1	+	1	11	305	230	27	1		2	118	121	15	+				2	5	+	849
84/80		1	18	42	34	6	+	15	196	570	443	60		3	48	151	148	18	1	1	1	2	5	4	1763
79/75	+	1	14	26	41	11	+	16	125	198	286	120		9	40	63	130	47	+	1	2	2	5	4	1144
74/70		1	6	14	23	20	1	20	85	105	217	220	3	15	37	52	88	83	1	4	3	2	3	4	1006
69/65		3	6	6	14	23	1	27	86	80	155	318	2	26	54	44	77	98	+	4	5	3	5	4	1042
64/60	+	3	10	9	11	30	1	30	85	66	143	308	3	27	63	45	57	84	+	8	7	4	4	4	1000
59/55		2	8	7	11	20	1	30	93	62	96	126	1	27	69	33	45	84	+	8	15	4	6	7	717
54/50		2	11	7	11	10	1	24	95	56	63	39	1	24	70	36	37	45	1	8	12	7	7	7	557
49/45		3	9	5	5	2	+	15	59	37	25	15	1	16	47	26	21	31	1	5	13	8	6	5	336
44/40		2	3	2	1	1	1	10	24	17	14	4	1	10	22	18	14	15	1	3	8	4	3	5	178
39/35			1	1	1			2	12	5	3	1	+	4	11	7	9	12	+	1	4	2	4	5	77
34/30					+			1	2	1	1	2		2	6	2	1	6			1	1	1	1	27
29/25				+	+				1	1	1	+			1	1	+	4				+	+		6
24/20															1	1	+	1				+	+		2
19/15																+						+	+		
Total	+	19	94	125	153	123	7	204	1213	1430	1474	1213	13	164	594	601	643	443	5	44	71	42	48	46	8767

Occurrences are for the average year (10-year total divided by 10). Values are rounded to the nearest whole number, but not adjusted to make their sums exactly equal to column or row totals. "+" indicates more than 0 but less than 0.5.

Table 10B. Percentage Frequencies of Wind Direction and Speed

Direction	Hourly Observations of Wind Speed (mph)									Total	Average Speed
	0-3	4-7	8-12	13-18	19-24	25-31	32-38	39-46	47 Over		
N	0.2	0.8	1.7	2.0	1.3	0.5	0.1	+	+	6.7	15.0
NNE	0.1	0.6	1.7	1.8	1.0	0.3	0.1	+	+	5.6	14.6
NE	0.2	0.8	2.2	1.9	0.6	0.2	+	+		5.9	12.9
ENE	0.1	0.6	1.4	1.3	0.5	0.1	+	+		4.0	12.9
E	0.2	0.9	2.2	1.9	0.6	0.1	+	+		5.9	12.3
ESE	0.1	1.1	3.7	3.3	0.6	0.1	+	+		8.9	12.2
SE	0.2	1.7	6.0	4.0	0.6	0.1	+	+	+	12.3	11.7
SSE	0.1	1.4	6.1	4.9	0.6	+		+	+	13.2	12.0
S	0.3	1.6	6.1	5.1	0.9	0.1	+	+	+	14.0	12.1
SSW	0.1	0.7	2.4	2.5	0.7	0.1	+			6.4	12.9
SW	0.2	0.7	1.5	1.1	0.4	0.1	+	+		4.1	11.8
WSW	0.1	0.5	0.8	0.3	0.1	0.1	+			1.8	9.7
W	0.1	0.5	0.7	0.3	0.1	+	+	+		1.8	10.1
WNW	0.1	0.4	0.8	0.5	0.2	0.1	+	+		2.1	12.4
NW	0.1	0.5	0.9	0.8	0.5	0.2	+	+	+	3.1	13.9
NNW	0.1	0.3	0.7	0.8	0.7	0.4	0.1	+	+	3.0	16.6
Calm	1.1									1.1	
Total	3.5	12.8	39.0	32.5	9.5	2.3	0.5	0.1	+	100	12.5

Table 10C. Occurrences of Precipitation Amounts

Intensities (in.)	Frequency of Occurrence for Each Hour of the Day																								No. of Days With
	A.M. Hour Ending At												P.M. Hour Ending At												
	1	2	3	4	5	6	7	8	9	10	11	Noon	1	2	3	4	5	6	7	8	9	10	11	Mid.	
Trace	10	11	10	12	12	16	17	19	25	22	22	21	18	17	15	18	18	16	16	14	15	13	12	11	41
0.01	4	3	3	4	4	4	4	4	4	5	4	3	4	4	4	4	4	3	3	4	3	4	4	4	9
0.02 to 0.09	5	4	5	6	8	7	9	10	8	8	9	8	6	7	6	5	5	5	5	4	4	4	6	6	26
0.10 to 0.24	2	2	3	3	3	3	2	4	3	3	3	3	3	2	3	2	2	1	1	2	2	2	1	2	18
0.25 to 0.49	1	1	1	1	1	1	2	1	1	2	1	1	1	1	1	1	1	1	1	1	1	1	+	1	15
0.50 to 0.99	+	+	1	+	+	1	+	1	1	1	1	1	1	+	+	1	+	+	+	+	+	+	+	+	13
1.00 to 1.99	+	+	+				+									+	+	+					+	+	7
2.00 and Over	+		+																						2
Total	22	22	23	26	29	32	35	39	43	41	38	37	33	32	29	31	27	26	24	24	24	24	23	21	128

Occurrences are for the average year (10-year total divided by 10). Values are rounded to the nearest whole number, but not adjusted to make their sums exactly equal to column or row totals. "+" indicates more than 0 but less than 0.5.

Table 10D. Percentage Frequencies of Ceiling-Visibility

(Miles)	Ceiling (ft)									
	0	100-200	300-400	500-900	1000-1900	2000-2900	3000-4900	5000-9500	Over 9500	Total
0 to 1/8	0.2	0.3	+	+	+	+	+	+	0.2	0.7
3/16 to 3/8	0.1	0.3	+	+	+	+	+	+	0.2	0.8
1/2 to 3/4	+	0.2	0.1	+	0.1	+	+	+	0.2	0.7
1 to 2-1/2	+	0.3	0.3	0.3	0.2	0.1	0.1	0.1	0.3	1.7
3 to 6		0.2	0.5	1.3	1.0	0.5	0.5	0.4	2.7	7.2
7 to 15		0.1	0.4	2.6	5.3	3.4	3.7	4.6	68.8	88.8
20 to 30										
35 or more									+	+
Total	0.3	1.5	1.3	4.3	6.6	4.1	4.4	5.2	72.3	100

Table 10E. Percentage Frequencies of Sky Cover, Wind and Relative Humidity

Hour of Day	Clouds Scale 0-10			Wind Speed (mph)				Relative Humidity (%)					
	0-3	4-7	8-10	0-3	4-12	13-24	25 & Over	0-29	30-49	50-69	70-79	80-89	90-100
00	49	18	33	4	57	36	3	+	2	14	25	31	28
01	48	19	33	5	56	36	3	+	2	14	23	32	29
02	47	19	34	6	56	35	3	+	2	13	23	32	30
03	45	19	36	6	58	34	3	+	1	13	22	33	31
04	45	19	36	6	58	33	3	+	1	12	22	33	33
05	40	22	38	6	59	32	3	+	1	11	19	35	34
06	35	23	42	6	58	33	3		1	11	20	35	34
07	33	24	44	6	55	36	3		1	12	25	33	29
08	33	24	44	4	51	42	3	+	2	18	28	30	23
09	31	25	44	2	48	46	3	+	4	25	29	25	17
10	31	26	44	2	45	50	3	+	6	31	29	22	12
11	32	25	43	2	43	53	3	+	7	35	28	20	9
12	33	24	43	1	41	55	3	1	9	36	27	19	8
13	34	24	42	1	40	56	3	1	10	37	26	18	8
14	35	23	42	1	40	56	3	1	11	38	24	19	7
15	37	20	43	1	42	55	3	1	12	36	25	19	8
16	38	19	43	1	45	51	3	1	10	35	24	19	10
17	38	18	44	2	51	45	2	1	9	31	25	20	14
18	37	19	44	2	54	41	2	+	7	27	27	21	17
19	39	20	41	4	57	37	3	+	5	22	27	25	21
20	42	20	38	4	57	36	2	+	4	20	26	28	22
21	45	20	34	4	58	35	2	+	3	18	26	29	24
22	47	20	33	4	57	37	3		3	16	26	29	26
23	48	19	33	5	56	36	3	+	3	16	25	30	27
Avg.	39	21	40	3	52	42	3	+	5	22	25	26	21

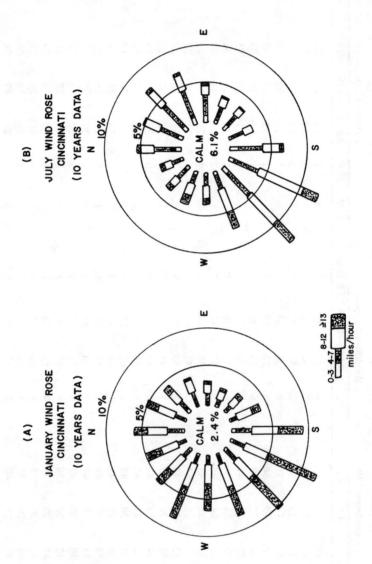

Figure 5   January and July monthly wind roses showing the relative frequency of occurrence of wind directions and wind speeds during these months. The data are averaged over a period of 10 years.[16]

forests, plains, etc.   Other important aspects would be: area size, elevation, slope, erosion, water velocity, wood types, and aesthetics.   These features may be represented pictorially as in Figure 6.   This sketch notes the various land forms found at a proposed dam site in New Jersey.

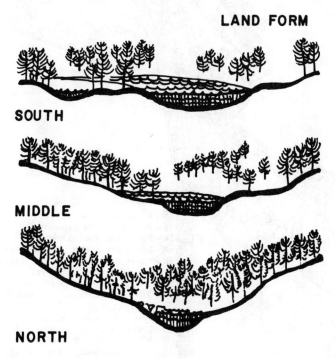

LAND FORM

SOUTH

MIDDLE

NORTH

Figure 6   Land forms available at the Lake Hackensack River damming project.[18]

5. The region's geology.   Basic geological formations are located and identified with respect to their historical setting.   In this manner Figure 7 defines the geology of the mid-Atlantic states.   Special attention should be given to those features that directly affect water resources.   Geological conditions should be described in relation to the activities of the pro-posed action.   Knowledge of the type of geological formation would be imperative if land disposal or chemical spills should occur above ground waters.   Regions of earth movement, landslides, volcanic activity, earthquakes, faults and fissures should be designated on survey maps and their past, present and future activity described.   Unique formations may be investigated more closely.   Figure 8 examines and illustrates the Baltimore Canyon Trough and the Georges Bank Trough of the eastern continental shelf.

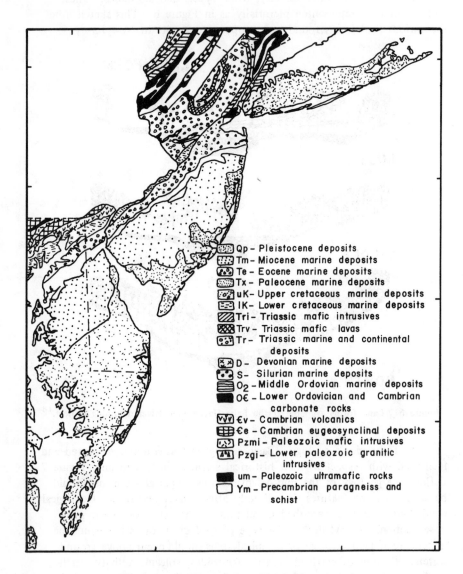

Qp – Pleistocene deposits
Tm – Miocene marine deposits
Te – Eocene marine deposits
Tx – Paleocene marine deposits
uK – Upper cretaceous marine deposits
lK – Lower cretaceous marine deposits
Tri – Triassic mafic intrusives
Trv – Triassic mafic lavas
Tr – Triassic marine and continental
    deposits
D – Devonian marine deposits
S – Silurian marine deposits
O₂ – Middle Ordovician marine deposits
O∈ – Lower Ordovician and Cambrian
    carbonate rocks
∈v – Cambrian volcanics
∈e – Cambrian eugeosynclinal deposits
Pzmi – Paleozoic mafic intrusives
Pzgi – Lower paleozoic granitic
    intrusives
um – Paleozoic ultramafic rocks
Ym – Precambrian paragneiss and
    schist

Figure 7 Geology of the mid-Atlantic states.

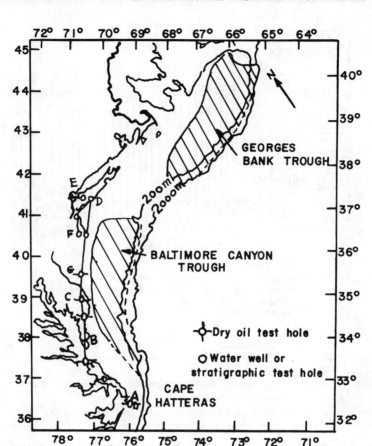

Figure 8 Utilizing a map such as this, the EIS writer can more closely examine and illustrate the unique geological formations (Baltimore Canyon and Georges Bank Troughs) of a region.[19]

6. A breakdown and analysis of a region's numerous soil types. The major types and/or categories should be discussed with respect to mineralogic and organic components, moisture content, soil temperature, slope characteristics, permeability, erodibility, expansion and compaction. A soils map such as Figure 9 should be included. Soils information is available from various atlases and the U.S. Soil Conservation Service.

7. A detailed hydrology statement of the region. Water quality of the various surface waters and ground waters is described in terms of physical, chemical and biological factors including detailed permeation and water transport information should land waste disposal techniques be used. Water quality may be expressed as in Table 11.

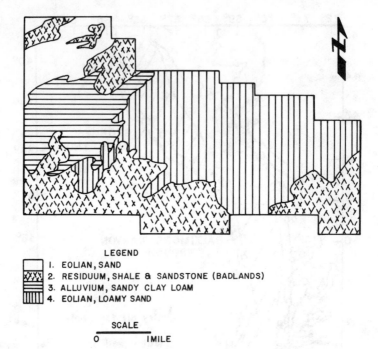

LEGEND
1. EOLIAN, SAND
2. RESIDUUM, SHALE & SANDSTONE (BADLANDS)
3. ALLUVIUM, SANDY CLAY LOAM
4. EOLIAN, LOAMY SAND

SCALE
0   1 MILE

Figure 9   Soil mapping units of the eastern portion of the El Paso coal base area for a proposed coal gasification project.[14]

Table 11. Typical Water Quality Data to be Included in the EIS[18]

Parameter	Rockland County[a]	Lower Basin
Median Dissolved Solids (mg/l)	170	192-1290
Total Hardness (mg/l)	112	192-1290
Chlorine		
Range	2-36	19-755
Median	8	
Alkalinity Cadox		
Range	17-162	
Median	93	
Specific Conductance (microohms)		579-3980
Iron	0.56	0.36
Sulfate		
Range	5.9-64	7-966
Median	21	
pH		
Range	5.4-8.4	7.3-7.7
Median	7.7	

[a]Corresponding to upper basin.

Water quantity of the various surface and ground waters should be related to current water uses (see below).   A map of the study area's drainage basin would illustrate streams and tributaries, water flow obstructions and diversions, and other structures such as dams, canals and tunnels (see Figure 10).

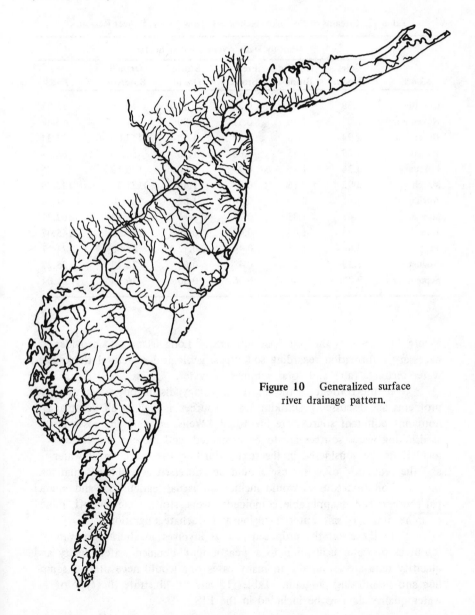

Figure 10   Generalized surface river drainage pattern.

This section should discuss volumes of surface (estuarine, marine, lake, pond, stream) waters and ground waters, stream flow rates as influenced by seasonal variations (see Table 12), ground water 7-day and 10-year low flows, the region's aquifers and their water exchange rates.

Table 12. Streams of the Lake Hackensack, New Jersey, Project Region.[18]

Month	Monthly Daily Average Flows in cfs					
	Coles Brook	Hirsfeld Brook	French Brook	Storm Drains	Oradell Reservoir	Total
October	2.38	1.45	0.37	0.72	10.95	15.87
November	2.87	1.70	0.51	0.90	67.50	73.48
December	2.94	1.75	0.57	1.08	117.00	123.34
January	3.29	1.85	0.68	1.20	56.60	66.62
February	3.36	1.90	0.70	1.50	130.10	137.56
March	3.92	2.15	0.97	1.44	135.30	143.78
April	3.85	2.10	0.96	1.62	183.50	191.02
May	3.43	1.93	0.71	1.14	95.10	102.31
June	2.73	1.60	0.46	0.78	139.50	145.07
July	2.66	1.55	0.44	0.72	65.80	71.17
August	2.59	1.55	0.43	0.60	20.10	25.27
September	2.38	1.40	0.36	0.60	76.30	81.04

Figure 11 identifies the principal aquifers of Long Island and provides necessary information regarding soil types, geological structural thicknesses, water recharge rates and total volumes of water.

In the content of the above hydrology discussion, water uses and problems are discussed. Drinking water sources as well as all point and nonpoint pollutant sources are identified. Wells, aquifers and reservoirs as drinking water sources are to be protected, and present and future regulations are considered in the text. Further, the recycling of water and the reduction of water use should be addressed as viable alternatives.

Pollution sources would include industrial, agricultural and municipal sources, and, if applicable, combined sewers, storm water runoff, mine or mine drainage, salt water intrusion and leachate migration.[13]

In the event the proposed project involves an altering of water resources or water utilization to a great extent, detailed water quality and quantity data are required. In many cases this would necessitate a sampling and monitoring program. Tables 13 and 14 illustrate the type of water quality data to be included in the EIS.

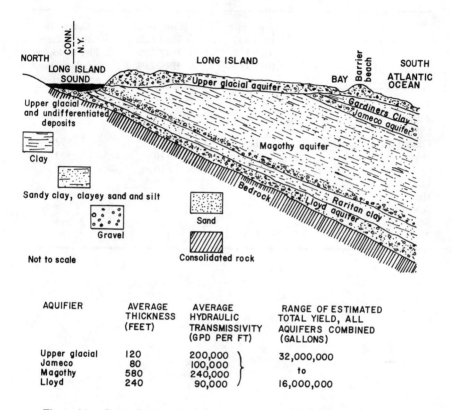

AQUIFER	AVERAGE THICKNESS (FEET)	AVERAGE HYDRAULIC TRANSMISSIVITY (GPD PER FT)	RANGE OF ESTIMATED TOTAL YIELD, ALL AQUIFERS COMBINED (GALLONS)
Upper glacial	120	200,000	32,000,000
Jameco	80	100,000	to
Magothy	580	240,000	to
Lloyd	240	90,000	16,000,000

Figure 11   Generalized stratigraphy and principal aquifers of Long Island. Aquifer characteristics are also noted.[20]

Regional management plans of water quality requirements and water usage are necessary to give the reader an understanding of the over-all water picture as affected by the proposed project. Regulations and standards as they apply to water quality and quantity and the pollution generated by the proposed action are discussed.

Areawide flood plans and levels are to be included. The Army Corps of Engineers should be consulted for data and information regarding their flood plan activities. Flood insurance plans should also be dealt with.

8. The biology of the region should be described in detail. However, like the other parameters considered in the EIS, highly technical information should be included in the appendix. The various ecosystems of the region are summarized.

Table 13. Water Analysis for Heavy Metals (Lake Hackensack, 1975)[18]

Parameter	Composite HR1 33% HR2 33% HR3 33% (ppb)	Composite HR4 50% HR5 25% HR6 25% (ppb)	Composite Equal Parts Five Biological Sample Stations (ppb)
Arsenic	1	1	9
Barium	35	35	104
Cadmium	1	1	68
Chromium (hexavalent)	12	6	592
Copper	70	110	342
Cyanide	10	10	10
Lead	25	75	675
Mercury	1	1	19
Selenium	1	1	7
Silver	44	20	544
Zinc	64	143	3720

Table 14. Sampling Program for Lake Hackensack Project—Water Quality Data
for Hackensack River Tributaries[18]

Parameter	Average Value–Tributaries
pH	7.2
Turbidity	18 JTU
Orthophosphate	0.20 mg/l
Dissolved Phosphate	0.42 mg/l
Total Phosphate	0.61 mg/l
NH_3-N	0.63 mg/l
NO_3-N	2.28 mg/l
NO_2-N	0.038 mg/l
Fe	0.68 mg/l
Mn	0.35 mg/l
BOD_5	3.8 mg/l

Aquatic life would include a discussion on the available animal and plant organisms, their habitats, and how they are affected by changes in their habitats. Distinctions between salt and fresh water, lake, stream and wetland life forms are made.

The terrestrial life discussion would be similar in content. The flora and fauna communities as they have been affected by human activities should be noted. Semiaquatic life must be considered.

Unique natural features are described and these may include forests, marshland, or tundra. The importance of the region's biology to man whether it be for food, recreation, or aesthetics must be assessed. Characteristic, rare and endangered flora and fauna species which inhabit the region are noted and possible effects on them are noted.

Information as to biological factors is available from the U.S. Fish and Wildlife Service, various state fish and game agencies, and local studies done by consulting companies and universities.

9. Probably the most important aspects to be considered are the land use requirements of the area in which the proposed project is to be built. Local ordinances, laws and administration procedures should be outlined. Jurisdictional boundaries, agency policies and governmental agreements can be defined.

The proposed project should be discussed in relation to master municipal land use plans. Maps of the affected region (as shown in Figure 12) detail the commercial, industrial and residential communities. Also pinpointed are highway, railroad and other transportation arteries, sanitary landfill sites, historical, agricultural and conservation sites, parks and open land. The effects of the proposed project may concur with current land use plans, but these plans may not be environmentally sound. Therefore, the environmental basis of the land use plan should be evaluated.

As previously noted, the projection of land uses should be about 10 to 15 years in the future. Governmental and industrial development trends including growth expectations are assessed. Much of the information needed for this section is available from local governments and agencies.

10. Other environmentally sensitive areas that have not been covered above should be discussed in this section. These may include proposed project affected waters, flood plains, land formations, silviculture and agriculture lands, rare and endangered species, archeological and historic sites, or recreational regions.

11. Of primary concern in an EIS is the proposed project's influence on society. Therefore, the population and economics of a region are

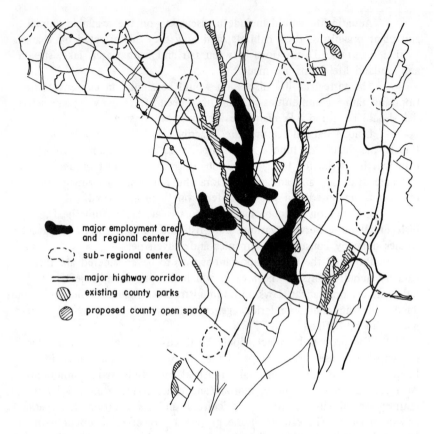

major employment area
and regional center

sub-regional center

major highway corridor

existing county parks

proposed county open space

Figure 12   Bergen County areawide development guide.[18]

assessed and projected into the near future. Trends and expectations are
outlined and regional growth rates established. Population density maps
could be provided for larger regions.[13]

    Information for this section is available from the U.S. Census
Bureau, Bureau of Economic Analysis, Department of Commerce, the Economic Research Service, and the Department of Agriculture. Forecasting
methods should be explained.

    12. In order to fulfill the EIS obligation to assess the cumulative
effects of several governmental actions, this section requires a brief description of the social, economic and environmental impacts of other local,
state and federal actions in the region of the proposed project. The cumulative effects of these actions should be evaluated.

ENVIRONMENTAL IMPACTS OF THE PROPOSED ACTION

In this section of the EIS the impacts of the proposed project on the environment are discussed in detail. These impacts are described in relation to the physical, biological and human environments. Figure 13 outlines the procedures that may be followed in the preparation of this portion of the EIS. First the impacts are identified. Evaluation and assessment of their interactions follow, and then comparisons are made with the control environment should no action take place.

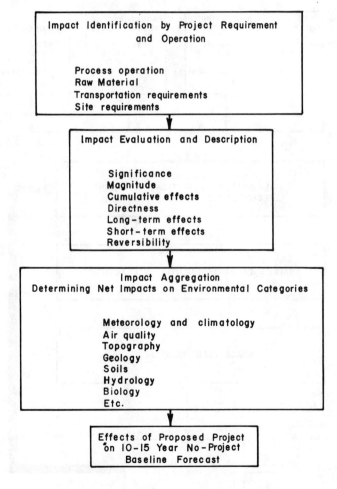

Figure 13   Environmental impact assessment outline.[13]

Identification of Impacts

The impacts on the environment of a project are determined through an examination of its process operations, its raw material requirements, its transportation requirements, and its site requirements. Obviously different projects will demand greater detail in different areas. A refinery or chemical plant would necessitate a large section on process impacts while a damming reservoir project would require detailed site impact analysis. Figure 14 illustrates the analysis operation necessary for final impact identification.

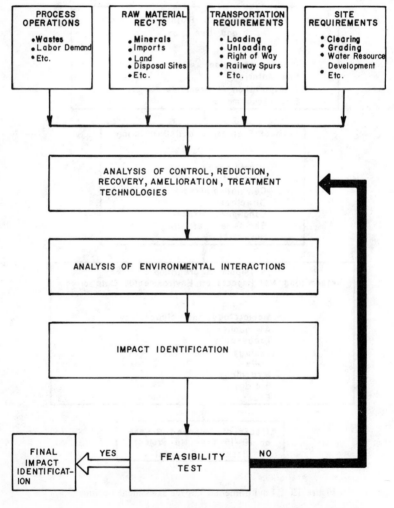

Figure 14 Environmental impact identification procedures.[13]

The impact of a project's process is determined through examination of all atmospheric, aquatic and terrestrial wastes. Compliance with governmental regulations must be shown. All pollution control techniques and devices (wastewater treatment plants, electrostatic precipitators, scrubbers, solid waste disposal) must be reviewed as should applicable process modifications. Water and waste recycling is encouraged and such processes should be described. New technology should be employed when feasible, with an explanation when necessary.

A flow diagram of the process' discharges is required to better visualize pollution control measures and waste production. Figure 15 is the type of pollutant flow diagram needed. Note that this diagram includes material balances and effluent descriptions. From this summary pollution control measures can be evaluated and categorized into normal (*e.g.,* gas scrubbers), intermittent *(e.g., *H_2S safety valve flares), and emergency *(e.g.,* oil spills).

The physical and economic requirements of these controls are described along with any process modifications for pollution reductions. Process control instrumentation and other monitoring devices and sampling systems should be discussed.

The impact identification must show the relationship between each process waste stream and governmental regulations. Figure 16 is the type of table and information required. This is the initial phase of impact evaluation; further evaluation is accomplished in the section on impact evaluation.

Any doubts concerning a waste stream's content, quantity and concentration are stated. All social and economic factors which can become environmental impact sources must be assessed.

The impacts of raw material usage at a project are defined. Raw material impact operations include handling and storage, minerals and other natural resources utilized, the derivation of these resources, and the disposal of wastes.

The impacts of transportation requirements of a project are defined. This involves the determination of the effects of highway and/or railway construction, obtaining right-of-ways, increased usage of highways or railways and increased air and noise pollution during construction and, finally, additional requirements as a result of increased use of facilities.

The impacts of site development and utilization of a project are defined. These might include clearing and grading, loss of agricultural land, effects of construction, loss of flora and fauna lands, reduction in property values, and aesthetic degradation.

Accident and spill possibilities are described. Previous statistical information derived from similar projects, modeling and actual data gathered

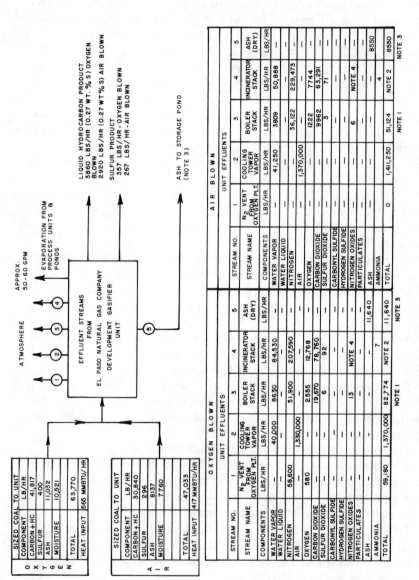

Figure 15   Typical pollutant process flow diagram needed for an EIS.[14]

TABULAR SUMMARY OF STANDARDS & REGULATIONS

SOURCE IDENTIFICATION	CONCENTRATION AVG. & MAX.	QUANTITY (FLOW RATE) AVG. & MAX.	METHOD OF DISPOSAL	STANDARD OR REGULATION		RESPONSIBLE REGULATORY AGENCY	OTHER
				REQUIRE-MENT	PROPOSED		

Figure 16.

if necessary should be used in determining the probability of an occurrence.

The identification of an accident or spill possibility is essential and cannot be understated. Figure 17 is a map of the Atlantic continental shelf oil-drilling platform area showing the probability of some oil spill material reaching Long Island in the winter. These accidents or spills must be identified as to toxicity, volatility, odor, flammability, treatability by conventional treatment methods, damage and upsets to flora and fauna and other biotic communities, persistence in food chains and collectibility.

The impacts to already defined environmentally sensitive areas must be identified. The impacts to already defined population and economic growth rates must be identified. The impacts as affected by other governmental projects must be identified. The impacts to already defined land use programs must be identified.[13]

THE EVALUATION OF THE ENVIRONMENTAL IMPACTS

After impacts are identified, they are explained and assessed in this portion of the EIS. This evaluation is based on each impact's significance, magnitude, cumulative effects, direct effects, long-term effects, short-term effects, and reversibility.

The methodology used to evaluate the environmental impact varies because proposed projects have different characteristics, *e.g.,* highway versus power plant, river dam versus chemical plant. Each project affects different waste streams and economic, social and aesthetic parameters. Each project requires a statement that demands more attention be given to one particular area than another area.

We present a methodology that is a straightforward system of environmental assessment in Chapter 4. The methodology chosen to write an EIS is left up to the writer of the EIS. The only criteria are that the method be accurate, reproducible, economical and easily understood.

In the environmental evaluation the magnitude of the action should be differentiated from its significance. The magnitude is a physical entity that can be described in physical terms, such as a 400,000 barrel/day refinery or SO_2 emissions of 1000 lb/hr. The significance of a project is determined by considering all the impacts and their effects on the environment and human life quality. The amount of harm or benefits a project may effect is an estimate of its significance.

In the environmental evaluation, consideration should be made of possible and inevitable impacts. The probability of an impact may be low but have a great significance on human life in the area. This would be

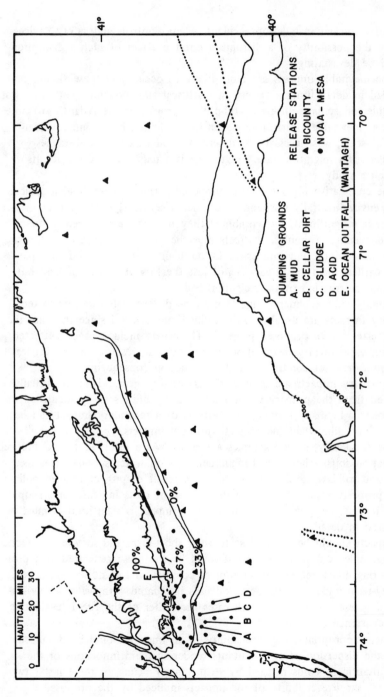

Figure 17 Atlantic offshore drilling platform area shows the percent probability of some oil spill material reaching Long Island in the winter within 10 days.[11]

true of a nuclear installation accident, where controls and safety devices reduce the possibility of a disruption but the effect of such a disruption would be devastating.

The inevitable impacts are more likely to occur, and these should be handled in detail using a variety of statistical and modeling methods. For example, oil spillage during an oil-drilling project of the Atlantic offshore size was inevitable. Therefore, extensive modeling, both mathematical and full-scale, was necessary to determine spill trajectories, distances travelled, and migration rate. Figure 17 is a map showing the results of such a study.

The cumulative impacts of the proposed action can result in a project being environmentally unsound. Considered separately each impact may be rather insignificant, but combined they may prove to be environmentally damaging. Cumulative effects must be assessed as to their ultimate degradation of the environment. In the highway construction example given earlier, the total impact is not just the loss of a strip of land but the final destruction of the neighborhood.

Primary and secondary impacts must be differentiated and evaluated. Primary impacts are usually easily defined and assessed since they are the direct effects of a proposed project. These may include process-effected air and water quality degradation, waste disposal, resource usage, process accidents or upset conditions, and construction consequences.

Secondary impacts are indirect effects of a project. They are usually directed from the primary impacts and are not always easily identified or controlled. In many instances the effects of a secondary impact can be more far-reaching than the project's primary impacts.

For familiarity, some examples are given. As a result of project-induced population shifts, there are changing demands on water supplies, treatment units and solid waste disposal sites. As a result of project-induced pollution, property values decrease, water treatment costs increase, and people move out of the area. When this type of impact is significant it must be discussed in detail.

Impacts should be assessed as to their short-term and long-term effects. For example, a short-term impact would be the noise generated at a project during construction. After project completion this noise is removed. A long-term impact would be woodlands inundated with water at a river-damming reservoir site. In both cases the writer of the EIS is evaluating the persistence of an impact.

Where appropriate, the reversibility of an impact should be discussed (e.g., the dispersion of air pollutants rendering them innocuous or the rehabilitation of lands ravaged by strip mines.) Thus, in these latter two sections, the irreversibility of the impacts induced by the proposed project are evaluated.

Summation of the Impacts

When the preceding work of impact analysis, identification and evaluation have been completed, their combined and interrelated impacts are discussed with respect to the status-quo environment previously described. These impacts are also discussed with regard to the future environment as it would exist in 10 to 15 years if no project were implemented (this condition was defined in the status-quo environment section).

ALTERNATIVES TO THE PROPOSED ACTION

The purpose of this section of the EIS is to identify and evaluate alternate plans and actions that may accomplish similar goals of the proposed action. The types of alternatives can include modifications to the proposed project, relocation of the proposed project, or the alternate that must always be considered: *no project.*

In order for the alternatives to a proposed project to be properly designated, its environmental assessment must be done early in the planning stages. In this manner, social, economic and environmental factors, against which each alternative is to be judged, can be set. Alternatives should not be limited to a cost/benefit analysis in deciding their attributes. Environmental and social benefits must also be weighed with each alternative.

Alternatives are based on the magnitude and significance of the impact of a proposed project. A project that is shown to have a rather small impact on a region would require fewer alternatives presented in the EIS. Table 15 defines the criteria on which project alternatives should be based.

The project and alternatives, as influenced by public opinion, should be evaluated. In this manner factors such as aesthetics, social settings and land use can be properly assessed.

Alternatives to a project and their impacts must be identified and evaluated in detail. Review of environmental assessment is made by the EPA, and then review of the draft EIS is made by various governmental agencies, technical experts and the public. Therefore, omitted alternatives that may have a lesser environmental impact will only result in delay of the project, possibly through court interventions.

Alternative Descriptions

The criteria for alternative description and comparison are given in Table 16. Details are not specified in these tables, but typical project-related characteristics are outlined with the judgment for alternative

Table 15. Criteria on which Proposed Project Alternatives are Based[13]

No.	Criteria—Process Related	Requirement
1	The proposed process is likely to be controversial, for environment or public health reasons.	Consider at least one other process option.
2	Process technology for industry in question is rapidly developing or expanding.	Consider postponement of the project.
3	Pollution control technology is rapidly expanding for some critical or costly facet of the industry.	Consider postponement of the project.
4	Renovation/expansion of existing facilities would eliminate the need to develop natural areas.	Consider renovation/ expansion.
5	The proposed project will rely upon relatively unproven technology.	Consider at least one other process option.
6	The proposed project utilizes scarce or rapidly diminishing resources (*e.g.,* natural gas).	Consider at least one other process using other resources.
7	The proposed project has several raw materials options.	Consider all raw material options, and determine one causing lowest pollution load.
8	Others, as defined by EPA Regional Administrator.	

requirements cited. For all EIS's the reasons for selecting the proposed project over the other alternatives must be described in detail.

Generally, the detail requirement for the environmental assessment of an alternative is less than the proposed project. For those alternatives that pose less environmental risk, the unacceptable economic, technical or other factors must be cited. In the discussion of an alternative, its general environmental characteristics and those parameters that effected its rejection in favor of the proposed project are stated.

Alternative environmental assessment is carried out in a manner similar to that discussed in the previous sections on *Environmental Impact of the Proposed Project.* A cost/benefit analysis is made between environmental and social and economic factors.

Finally, each alternative is listed in a table similar to Figure 18. In it, all the considerations, impacts and relative estimates of impact are summarized. Page and paragraph numbers are given for ease of reference with the text.

Table 16. Criteria for Alternative Considerations for Location Selection
of the Proposed Project[13]

No.	Criteria—Site Related	Requirement
1.	The proposed new source location is likely to be controversial.	Consider at least one additional site.
2	The proposed new source and/or associated facilities would infringe upon scientifically valuable areas, as determined by site uniqueness, primitiveness, amenability to study or observation. Such sites may be defined by local universities, colleges, research organizations, etc.	Consider at least one additional site.
3	The proposed new source and related facilities would directly or indirectly infringe upon recreational lands, park lands, wildlife refuge lands.	Consider at least one additional site.
4	The proposed new source and related facilities would either directly or indirectly accelerate change in rural, pristine or agricultural land areas.	Consider at least one additional site.
5	The proposed new source and related facilities would induce secondary residential, industrial and/or commercial growth in the community which which could not be supported by existing community services and financial capabilities.	Consider at least one additional site.
6	The proposed new source and related facilities would cause traffic congestion in the vicinity of the proposed site.	Consider at least one additional site.
7	The proposed site is prone to flooding, hurricane, earthquake, or other natural disasters.	Consider at least one additional site.
8	The proposed new source and related facilities would infringe directly or indirectly upon endangered species or their habitat, or upon wetlands (including freshwater wetlands), upon wild and scenic rivers, or sensitive or unique ecosystems.	Consider at least one additional site.
9	The proposed new source and related facilities would infringe directly or indirectly upon historical sites currently included or proposed for inclusion within the National Registery of Historical Landmarks. Archaelogically important sites are likewise covered by this criterion.	Consider at least one additional site.
10	Others to be specified by EPA Regional Administrator.	

GENERAL FORMAT FOR SUMMARIZING ALTERNATIVES

NEW SOURCE TITLE	ALTERNATIVE A (BRIEF DESCRIPTION)		ALTERNATIVE B		ALTERNATIVE C		ALTERNATIVE D		ALTERNATIVE E	
IMPACT DESCRIPTION	SIGNIFICANCE	PAGE	SIGNIFICANCE	PAGE	SIGNIFICANCE	PAGE	SIGNIFICANCE	PAGE	SIGNIFICANCE	PAGE
(EXAMPLES)										
1. PROCESS IMPACTS										
SOLID WASTE DISPOSAL EFFECTS										
WATER QUALITY EFFECTS										
AESTHETIC EFFECTS										
2. SITE IMPACTS										
LOSS OF AGRICULTURAL LAND										
EROSION										
LOSS OF HABITAT										
TRAFFIC CONGESTION										
3. ETC.										

Figure 18.

CHAPTER 4

ENVIRONMENTAL METHODOLOGY

To be an effective tool in the planning process the environmental assessment segment of the EIS must be a comprehensive and systematic analysis that encompasses an interdisciplinary approach to the environment and its many related aspects.[21]

In Figure 19 a plant (although it could be any living organism) is shown as it is acted upon by the many environmental factors. Each factor is shown to have a direct impact on the plant and also an impact on the other factors which, in turn, results in an indirect impact on the plant. Thus, the environment is a complex interaction of many factors and a change in one aspect of the environment will effect changes of the entire system. Therefore, a comprehensive analysis of the environmental impact of a proposed project is necessary to understand the physical and biological effects and their influences on the social, cultural and aesthetic concerns of man.

The analysis must also be systematic and interdisciplinary because an environmental assessment encompasses a variety of physical, biological and social sciences and is examined and evaluated by numerous special interest groups. It is subjected to a public hearing and expert testimony and must withstand careful scientific scrutinization.

ENVIRONMENTAL SEGMENTATION

The large amounts of data and information considered in an environmental assessment make it imperative that a system be created for analysis. One approach would be to categorize the environment into its physical and biological components. Table 17 lists the principal physical and biological factors of the natural environment. However, what must be stressed in the EIS is the concern with human life quality and any effects

75

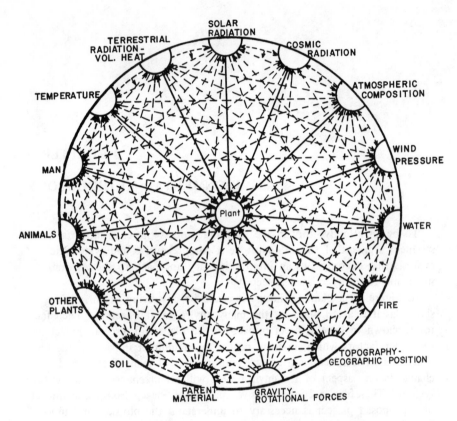

Figure 19   An illustrative view of the direct and indirect interactions of environmental factors on a single plant.[22]

Table 17.   Natural Environmental Components[23]

Physical Factors	Biological Factors
Energy	Green Plants
Radiation	Nongreen Plants
Temperature and Heat Flow	Decomposers
Water	Parasites
Atmospheric Gases and Wind	Symbionts
Fire	Animals
Gravity	Man
Topography	
Geologic Substratum	
Soil	

on it should be dealt with in detail. These would include cultural and social environments.

The system presented for environmental analysis divides the human environment into four major categories.[21] The first two are physical and biological, and the latter consider social and cultural factors.

1. Physical/chemical factors describe the physical and chemical effects of air, water and land pollution. Changes in this sector of the environment usually demonstrate their impacts on the other environmental sectors.

2. Ecological factors describe the flora and fauna of the environment and the impacts made on them. Each species and its habitat is considered as to population, growth rate, interactions with its own species, and interactions with other species and life cycles.

3. Aesthetic factors describe the land use impacts of the proposed project and other sensory effects, primarily visual.

4. Social factors describe the human life quality, health and welfare.

The above system of categories permits a more varied approach to the EIS. Each project will affect each of these environmental segments to a greater or lesser degree. More detail can then be given to the segment that requires it, and the project planner or decision-maker, without too much difficulty, can consider the various aspects and impacts of a project.

The four major components of the environment are broken down into subdivisions in Figure 20. A description of each subdivision and the parameters to consider in the environmental assessment follow. Further, the types of analysis that could be made are included.

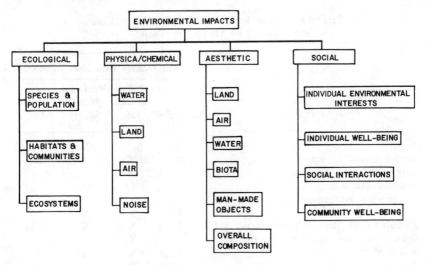

Figure 20. Environmental categories and components for environmental assessment.[21]

PHYSICAL/CHEMICAL CATEGORY

Water

Water is one of the necessities to life as we know it. It is used by all flora and fauna life cycles and as a sink for wastes. These wastes have now begun to inhibit the use of water for its intended purpose. Table 18 lists the major types of water pollution and the water uses most affected by them. Thus, water of a region must be assessed as to its quality, quantity and availability. Typical considerations might include:

- What are the nonpoint pollution sources, such as pesticides in agricultural runoff?
- Is a stream-flow reduction affected by diversions set up by the proposed project?
- Would BOD, COD and suspended solids be reduced to meet EPA standards with water treatment? How?
- What are the ground water effects due to land application waste disposal?

Land

Land has always been a cherished commodity for man. Its uses range from agricultural production to mineral ore supplier, from housing

Table 18.   Major Types of Water Pollution and the
Water Uses Most Affected by Them[24]

	Water Quality Uses Most Affected			
Types of Pollution	Recreation	Aquatic Life	Municipal Water Supply	Industrial Use
Bacteria	X		X	
Turbidity (muddy water)	X	X	X	X
Lack of Dissolved Oxygen		X		
Inorganic Materials				
(iron and manganese)		X	X	
(copper and lead)	X	X	X	
Phenols (smelly, bad taste substances)	X	X	X	
Organic materials (oil, pesticides, exotics)	X	X		
Radioactivity	X	X	X	X
Heat	X	X		X
Nutrients (phosphorus and nitrogen)	X	X	X	X
Dissolved Solids			X	X

foundations to industrial sites, from open space beauty to waste disposal sites. Proper land use management is imperative to reaching sound environmental goals. Typical considerations might include:

- What types of land weathering take place and what are the causes?
- Is the project suitable for local land use ordinances and regional planning criteria?
- What are the solid and liquid waste disposal practices and are they environmentally safe?

Air

The pollutants of our atmosphere are known to have detrimental effects on humans and vegetation; we now know that they contribute to the deterioration of the stratospheric ozone layer resulting in the admission of the cancer-causing ultraviolet rays of the sun. Anything that is emitted to the atmosphere that ordinarily would not be there is a contaminant. Typical considerations might include:

- What is the ambient air quality and how will it be affected by the proposed project?
- As a result of project-induced population shifts, are sulfur dioxide emissions increased due to increased power requirements?
- Will solid waste disposal sites produce objectionable odors?

Noise

Noise is usually considered to be unwanted sounds. It may consist of truck engines or screeching subways or many other undesirable sounds which may have a detrimental effect on man. Typical considerations might be:

- What will be the increase in noise intensity produced by construction blasting?
- Will increased truck and rail traffic to the proposed project increase the noise frequency to surrounding areas?
- Will the natural sounds of the environment (e.g., birds, flowing streams) be drowned out by increased noise levels of the project?
- Are sonic booms going to disrupt everyday life styles in adjacent areas?

ECOLOGICAL CATEGORY

Species and Populations

The local environment's ability to support the life of an organism makes it possible for that organism to live. Species and populations are the basis for the ecological world, and their existence can be greatly influenced by changes in their environments. Care should be taken to describe those species and populations that can be identified. However, many species cannot be identified and how they will be affected by environmental changes is unknown. This should be discussed.

Further, certain species are used as indicators of healthy and unhealthy human environments as well as the status of the general environments. These should be discussed in detail. Typical considerations might be:

- With the removal of natural feeding sites, is the animal life of the area adversely affected?
- Does the effluent of the proposed project suggest impairment to aquatic life?
- Are the flora and fauna of the region significantly affected?

Communities and Habitats

A group of species is a community. A habitat is the environment of a community. The existence of a variety of species and communities living together is the result of numerous and complex interactions. Major habitats include streams, lakes, estuaries, swamps, deserts, marshlands, forests, all components of some river basin. Figure 21 illustrates the variation in tree species with mountain altitude. Different temperatures and overall climates affect different habitats for the various trees. Typical considerations might be:

- Is the proposed action going to result in the destruction of rare and endangered species habitat (such as filling in a swamp)?
- Are project-induced population shifts eliminating important biological communities of the local ecosystem?
- Are soil communities destroyed due to leachate at waste disposal sites?

Ecosystems

An ecosystem is the entire natural life system of a local environment. River basins, forests and plains are typical natural systems in which ecosystems function. The fundamentals of most ecosystems are known, but data are lacking on the numerous processes and interactions that occur. Thus, in the ecosystem analysis, assumptions are made and the justifications for such assumptions are described.

Figure 22 illustrates an island ecosystem showing ocean to lagoon vegetation belts, soils, and fresh water lens. Locations and general environmental conditions for various types of terrestrial ecosystems are listed in

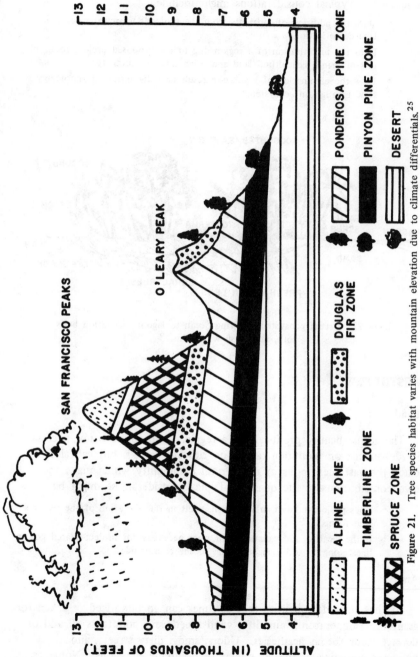

Figure 21. Tree species habitat varies with mountain elevation due to climate differentials.[25]

Table 19.[22]   Typical considerations might include:

- Does the addition of nutrients from the project's wastewater result in lake eutrofication?
- Does the urbanization of a region due to the proposed project result in loss of prime agricultural land and other land productivity?
- Does the elimination of vegetation result in a disruption of the energy flow through an ecosystem?

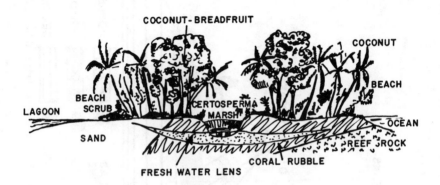

Figure 22.   Island ecosystem showing ocean to lagoon vegetation belts, soils, and fresh water lens.[26]

AESTHETICS

Land

The scenic beauty produced by a region's land forms is not necessarily dependent on mountain ranges or flowing brooks, but can also include the vast metropolis or rustic rural community which make their own contributions to life quality. Typical considerations might be:

- Does project-induced urbanization result in the alteration of the region's land forms?
- Is the clearing and excavation of trees necessary for project-related power lines going to add unsightly scars to the countryside?

Air

Air has aesthetic qualities that at times can go unnoticed or taken for granted. Temperature, humidity, wind speed, and pollutants all add or detract from the air aesthetics. Odors, smog, dirty streets, and buildings are some of the air effects of pollutant emissions. Noise, as an

Table 19. Various Types of Terrestrial Ecosystems, Their Locations, and General Environmental Conditions[22]

Climax Ecosystem Type	Principal Locations	Precipitation Range (in./yr)	Temperature Range (F) (Daily Max and Min)	Soils
Tropical Rain Forest	Central America (Atlantic Coast) Amazon Basin Brazilian Coast West African Coast Congo Basin Malaya East Indies Philippines New Guinea N.E. Australia Pacific Islands	50-500 Equatorial type frequent torrential thunderstorms Tradewind type: steady almost daily rains No dry period	Little annual variation Max 85-95 Min 65-80 No cold period	Mainly reddish latentes
Tropical Savanna	Central America (Pacific Coast) Orinoco Basin Brazil, S. of Amazon Basin N. Central Africa East Africa S. Central Africa Madagascar India S.E. Asia Northern Australia	10-75 Warm season thunderstorms Almost no rain in cool season Long dry period during low sun	Considerable annual variation, no really cold period Rainy season (high sun) Max 75-90 Min 65-80 Dry season (low sun) Max 70-90 Min 55-65 Dry season (higher sun) Max 85-105 Min 70-80	Some latentes, considerable variety
The Atoll	Principally in Tropical Pacific and Indian Oceans	15-150 Convectional but some tropical cyclones Droughts common	Little annual variation or range Max 80-100 Min 65-75 No cold period	Calcareous sand, gravel, and rubble Some atolls with phosphate "Jemo" hardpan soils

Table 19, continued

Climax Ecosystem Type	Principal Locations	Precipitation Range (in./yr)	Temperature Range (F) (Daily Max and Min)	Soils
Broad-Sclerophyll Vegetation	Mediterranean Region California Cape of Good Hope Central Chile S. W. Australia	10-35 Almost all rainfall in cool season Summer very dry	Winter Max 50-75 Min 35-50 Summer Max 65-105 Min 55-80	Terra rossa, noncalcic red soils, considerable variation
Temperate Grasslands	Central N. America Eastern Europe Central & Western Asia Argentina New Zealand	12-80 Evenly distributed through the year or with a peak in summer Snow in winter	Winter Max 0-65 Min -50-50 Summer Max 70-120 Min 30-60	Black prairie soils Chernozems Chestnut and brown soils Almost all have a lime layer
Warm Deserts	S.W. North America Peru & N. Chile North Africa Arabia S. W. Asia East Africa S. W. Africa Central Australia	0-10 Great irregularity Long dry season, up to several years in most severe deserts	Great diurnal variation Max 80-135 Min 35-75 Frosts rare	Reddish desert soils, often sandy or rocky Some saline soils
Cold Deserts	Intermountain W. North America Patagonia Transcaspian Asia Central Asia	2-8 Great irregularity Long dry season Most precipitation in winter, some snow	Great diurnal variation Winter Max 20-60 Min -40-25 Frosts common ½-¾ of yr Summer Max 75-110 Min 40-70	Gray desert soils, often sandy or rocky Some saline soils

Biome	Regions	Precipitation	Temperature	Soil
Temperate Deciduous Forest	Eastern N. America Western Europe Eastern Asia	25-90 Evenly distributed through year Droughts rare Some snow	Winter Max 10-70 Min -20-45 Summer Max 75-100 Min 60-80	Gray-brown podzolic Red and yellow podzolic
Temperate Rain Forest	N.W. Pacific coast, North America W. Coast, New Zealand Southern Chile Tasmania & S.E. Australia	50-350 Evenly distributed through year; wetter in winter Some snow	Winter Max 35-50 Min 25-45 Summer Max 55-70 Min 50-65	Podzolic, deep humus
Montane	Western N. America Appalachian N. America European mountains Asian mountains	15-100 Evenly distributed or with summer dry season Snow may be very deep in winter	Winter Max -20-60 Min -55-35 Summer Max 45-80 Min 20-60	Various podzolic, often shallow, rocky
Boreal Coniferous Forest	Northern N. America Northern Europe Northern Asia	15-40 Evenly distributed Much snow	Winter Max -35-30 Min -65-15 Summer Max 50-70 Min 20-55	True podzols Bog soils Some permafrost at depth, in places
Alpine Tundra	Western N. America N. Appalachian N. America European mountains Asian mountains Andes African volcanoes New Zealand	30-80 Much winter snow; long persisting snowbanks	Winter Max -35-30 Min -60-10 Summer Max 40-70 Min 15-35	Usually rocky Some turf and bog soils Polygons and stone nets Some permafrost
Artic Tundra	Northern N. America Greenland Northern Eurasia	4-20 Shallow snowdrifts, but many bare and dry areas in "High Artic"	Winter Max -40-20 Min -70-0 Summer Max 35-60 Min 30-45	Rocky or boggy Much patterned ground Permafrost

air medium, can affect the tranquility of an area. Typical considerations might be:

- What is the smog potential in reducing the visibility of a skyline due to stack emissions of the proposed project?
- Are any project chemical processes likely to release odor emissions?
- Will project-related sounds drown out bird calls?

Water

The beauty of water is unceasing. In its many forms—rain, lakes, streams, geysers, waterfalls, oceans, ponds, estuaries—it creates an almost incomparable beauty. Typical considerations might be:

- Does the diversion of a river by a dam eliminate any natural sites?
- Will the project's effluent introduce any pollutants to a stream, thereby changing its color?
- Will project-related sludge dumped offshore result in unsightly and unusable beaches?

Biota

The flora and fauna of a region contribute greatly to its aesthetic qualities. The elimination of vegetation not only removes its beauty, but also any animal life it may support. Typical considerations might be:

- Does the clearing of vegetation for a highway eliminate a natural animal habitat?
- Will vegetation types differ as a result of project-induced land development?

Man-Made Objects

Man-made objects such as art figures, skyscrapers, bridges, monuments and other buildings and structures create their own aesthetics through combination of beauty, uniqueness, age and history. The visual effect of the Manhattan skyline is at times awe-inspiring as would be a lonely bridge connecting opposite ridges above a steep ravine. Typical considerations would be:

- Does the design of the proposed project conform to the visual landscaping as it now exists?
- Would further development induced by the project conform to the present environment?

SOCIAL

Individual Environmental Interests

Individual aspirations must be dealt with here. They can add new dimensions and qualities to individual life styles. These factors may include historical settings, recreational features, or university facilities. Typical considerations might include:

* Does the proposed project infringe upon a historical setting disrupting its continuity?
* Are recreation facilities endangered by accidents such as oil spills from a tanker?

Individual Well-Being

People are greatly affected both physically and psychologically by their environment. Like other living things, changes in their environment can cause adverse effects on many individuals. Typical considerations might include:

* What are the psychological effects of frequent sonic booms, *e.g.*, the SST?
* What health problems will result from short and long term exposure to project effected emissions?

Social Interactions

The organization of individuals into various groups and communities is a social function. Of primary concern is the environment of the society and changes that may occur in the environment. This environment includes not only physical parameters, but job opportunities, adequate housing, social services, and recreational and educational opportunities. Typical considerations might be:

* Will the proposed project result in large population shifts?
* What are the economic and social opportunities created by this project?
* What will be the changes in the existing population's life styles?

Community Well-Being

The individual and social interactions that make up a community are in part governed by its environment. Any changes in the environment will produce changes in the community. Typical considerations might be:

* In what manner is the community disrupted by the proposed project?
* Is the overall environment of the community, as a result of the project, improved, deteriorated or unchanged?

The environmental aspects to be considered in an environmental assessment have been summarized. Each of these aspects should be considered fully with regard to: (1) the existing environment; (2) the impacts of the proposed project on the environment; (3) each alternative to the proposed project; and, (4) the irreversible and irrevocable uses of the environment.

INTERACTIONS OF THE ENVIRONMENTAL COMPONENTS

We have shown the interactions of the natural environment to be highly complex. The natural environment interacts with man and his social settings, and Figure 23 illustrates these intricate interrelationships. In most cases, a proposed action's impact is in the physical/chemical sector of the environment—such as some type of air, water or noise pollution. These, in turn, have their impacts in the various ecological, social and aesthetic sectors of the environment. Interactions occur among sectors and between sectors. Thus, in the making of an environmental assessment, the writer must be cognizant of these interrelationships and be able to analyze each component as it exists separately and as it fits into the whole environmental picture.

Environmental Factors

Once the environment has been segmented into the previously identified components, each component can be subdivided into actual parameters to be measured and assessed. These parameters are listed in Table 20 and, depending on the proposed action, the list under each environmental component may expand or decrease.

The environment is systematically evaluated through the segmenting of the whole into major categories. A further subdivision into components establishes the ecological impacts conceptually. Then the paramaters of each environmental component are identified. This identification process for each parameter would be followed again for: (1) the existing environment; (2) the impacts of the proposed project on the environment; (3) each alternative to the proposed project; and (4) the irreversible and irrevocable impacts. This parameter assessment would include:

* Data gathering or other measurements to determine the quality of each parameter and the existing environment.

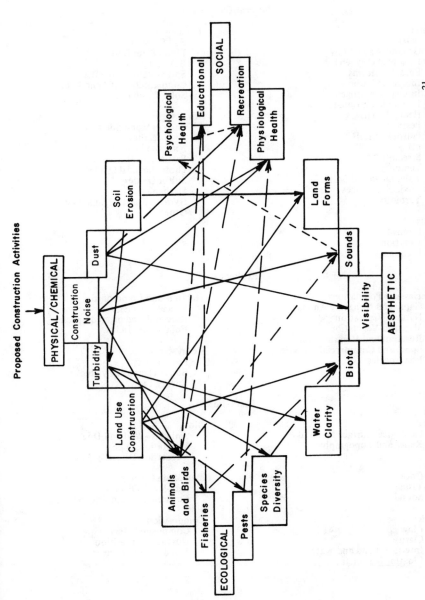

Figure 23. The interrelationships of the environmental impacts of a proposed project.[21]

Table 20. Environmental Parameter Assessment List[27]

Physical/Chemical

Water
BOD
Ground Water Flow
Dissolved Oxygen
Fecal Coliforms
Inorganic Carbon
Inorganic Nitrogen
Inorganic Phosphate
Heavy Metals
Pesticides
Petrochemicals
pH
Stream Flow
Temperature
Total Dissolved Solids
Toxic Substances
Turbidity

Noise
Intensity
Duration
Frequency

Land
Soil Erosion
Flood Plain Usage
Buffer Zones
Soil Suitability for Use
Compatibility of Land Uses
Solid Waste Disposal

Air
Carbon Monoxide
Hydrocarbons
Nitrogen Oxides
Particulate Matter
Photochemical Oxidants
Sulfur Oxides
Methane
Hydrogen and Organic Sulfides
Other

Ecological

Species and Populations
Game and Nongame Animals
Natural Vegetation
Managed Vegetation
Resident and Migratory Birds
Sport and Commercial Fisheries
Pest Species

Habitats and Communities
Species Diversity
Rare and Endangered Species
Food Chain Index

Ecosystems
Productivity
Biogeochemical Cycling
Energy Flow

Aesthetic

Land
Geological Surface Material
Relief and Topography

Air
Odor
Visual
Sounds

Water
Flow
Clarity
Interface Land and Water
Floating Materials

Biota
Animals—Wild and Domestic
Vegetation Type
Vegetation Diversity

Man-Made Objects
Man-Made Objects
Consonance with Environment

Composition
Composite Effect
Unique Composition
Mood Atmosphere

Table 20, continued.

Social

Individual Environmental Interests	Individual Well-Being
Educational/Scientific	Physiological Health
Cultural	Psychological Health
Historical	Safety
Leisure/Recreation	Hygiene
Social Interactions	Community Well-Being
Political	Community Well-Being
Socialization	
Religious	
Family	
Economic	

- Extrapolation of each parameter and/or group of parameters to estimate the future environmental status with and without the project.
- Differentiate between each parameter and/or group of parameters and their extrapolation to determine the beneficial and adverse environmental impacts.[21]

Thus, each parameter is analyzed. Each is compared with respect to changes in itself should the proposed project or any of its alternatives be implemented. If these changes are negligible then the environmental impact due to this parameter is nonexistent. If these changes are evident, then their size and significance must be determined, since an impact would occur.

The significance of an impact due to each parameter is assessed as to its contribution to the entire environmental impact of the proposed project. Alternative parameters are compared to the original project parameters, and an evaluation is made as to the desirability of the alternative. Alternatives can then be ranked with respect to the magnitude and significance of their environmental impact.

CHAPTER 5

AIR QUALITY IMPACT

National air quality standards have been established for total suspended particulates (TSP), sulfur dioxide, carbon monoxide, photochemical oxidants, and nitrogen dioxide. These standards have been established for the short-term and long-term averaging periods by the 1970 Clean Air Act and its amendments. Table 21 identifies the ambient concentrations allowable for the various pollutants.

Further, nondegradation of air quality has become a politically passionate issue. No significant air quality deterioration regulations establish classes of air quality for various regions of the country. Regulations would not allow further degradation of the air quality within these regions beyond the limits set. Environmentalists believe these laws are necessary to prohibit air quality degradation beyond the national standards and to preserve the pure and pristine atmospheres of the national parks and similar areas. On the other hand, many feel this type of law would be too restrictive to industrial expansion and economic growth.

At the time of this writing Congress was still haggling over the issue. However, present pollution control technology can be applied in most cases to allow expansion into the pristine environments. The problem is economics.

ATMOSPHERIC PROCESSES

The atmosphere has always been used as a sink for pollution. It has many natural pollutant removal mechanisms including: foliar absorption, soil absorption, absorption by natural water bodies, absorption by natural rock, rainout and washout (scavenging), and ambient chemical reactions.[1] These processes are dependent on the conditions of the atmosphere (meteorology). Pollutant transfer and reactions are a function of the

Table 21. National Air Quality Standards for Pollutant Ambient Concentrations

Pollutant	Description	Pollutant Standard	
		Primary	Secondary
Total Suspended Particulates	Solid and liquid particles in the atmosphere including dust, smoke, mists, fumes and spray from many sources.	75 $\mu g/m^3$, annual geometric mean; 260 $\mu g/m$, maximum 24-hour average	60 $\mu g/m^3$, annual geometric mean; 180 $\mu g/m$, maximum 24-hour average
Sulfur Dioxide	Heavy, pungent, colorless gas formed from combustion of coal, oil and other.	80 $\mu g/m^3$ (0.03 ppm), annual arithmetic mean; 365 $\mu g/m^3$ (0.14 ppm), maximum 24-hour average	1300 $\mu g/m^3$ maximum 8-hour average
Carbon Monoxide	Invisible, odorless gas formed from combustion of gasoline, coal, and other; largest man-made fraction comes from automobiles.	10 mg/m^3 (9 ppm), maximum 8-hour average; 40 mg/m (35 ppm) maximum 1-hour average	Same as primary
Photochemical Oxidants (as O$_1$)	Pungent, colorless toxic gas; one component of photochemical smog.	160 $\mu g/m^3$ (0.08 ppm), maximum 1-hour average	Same as primary
Nitrogen Dioxide	Brown, toxic gas formed from fuel combustion. Under certain conditions, it may be associated with ozone production.	100 $\mu g/m^3$ (0.05 ppm), annual arithmetic mean	Same as primary

the atmospheric dispersion process because in most instances it is the prime pollutant transport mechanism enabling the above removal processes to occur.

Atmospheric dispersion of a pollutant is primarily dependent on meteorological conditions and pollutant stack emission parameters such as gas velocity, temperature and molecular weight. Thus, the air quality of a region is greatly influenced by meteorological conditions. Weather parameters such as ambient temperature, wind speed, cloud cover, insolation, and inclement conditions (rain, snow, hail, etc.) can determine the impact severity of atmospheric pollutants. Obviously, the weather of most areas is well defined over long periods of time, but the daily or even monthly forecasts prove to be very difficult.

Therefore, in the selection of a plant site the planner usually considers the climatology of the area. The atmospheric emission rates of the proposed plant are assessed as influenced by this climatology, but taking into account the worst possible cases. In this manner stacks can be designed and control devices considered and implemented, if necessary, to lessen the air quality impact of the plant.

The location of the plant within an area can be dependent on the local wind directions. For example, residential areas may lie downwind of the plant, in line with the prevailing wind direction. A more suitable site might have to be considered to reduce the air impact of the plant. Figure 24 illustrates this point. Further, in the planning of the plant, consideration should be given to local, state and federal air pollution authorities which may require a shut-down or curtailment of plant emission activities during times of extreme air pollution problems. These backup procedures should be discussed in detail in the EIS.

In the description of meteorological conditions for the EIS the data used are usually obtained from the local airport weather station. However, localized weather conditions can in some instances be very different from those of weather stations just 5 miles away. This condition can be caused by variations in topography, buildings, urban areas (heat islands), and land forms.

METEOROLOGICAL CONDITIONS OF AIR DISPERSION

The wind is the primary atmospheric transport mechanism. The winds of the earth are the result of the pressure differences induced by the heating and cooling of the atmosphere by the sun. The rotation of the earth also imparts a motion on the atmosphere combining to produce a localized wind rose. The wind rose of a region is its characteristic wind patterns with respect to wind speed and wind direction.

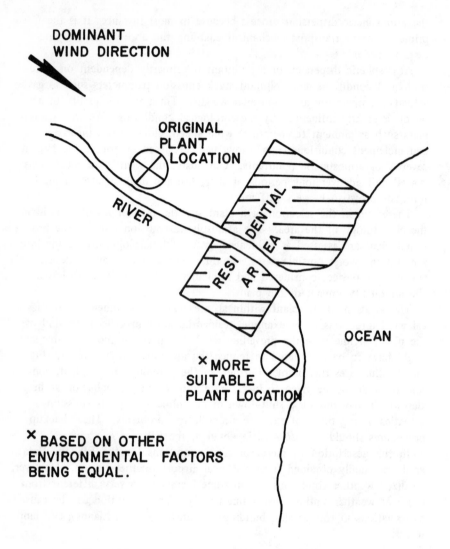

Figure 24. Plant site selection within a localized region can be influenced by meteorological conditions.

The wind roses of a region can be for monthly, seasonal or annual weather conditions. Figure 25 is a typical monthly wind rose of Cincinnati. Note the south to southwest wind is the dominant wind direction, the "prevailing wind direction."

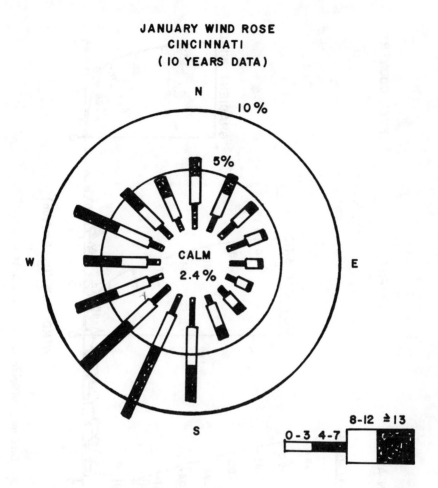

Figure 25. Typical wind rose.

The wind speed also varies with height, and this is known as the wind shear. Figure 26 illustrates the wind speed variation with increased elevation. It further shows a gradient wind variation with topography and population densities. Mountains, hills, trees, buildings and other obstructions can divert wind patterns, increase atmospheric turbulence, and affect general atmospheric stability.

The atmospheric stability is related to the rising and falling of volumes of air. It is a function of temperature gradient, atmospheric turbulence, wind speed, insolation, cloud cover, and other weather conditions (rain, hail, snow).

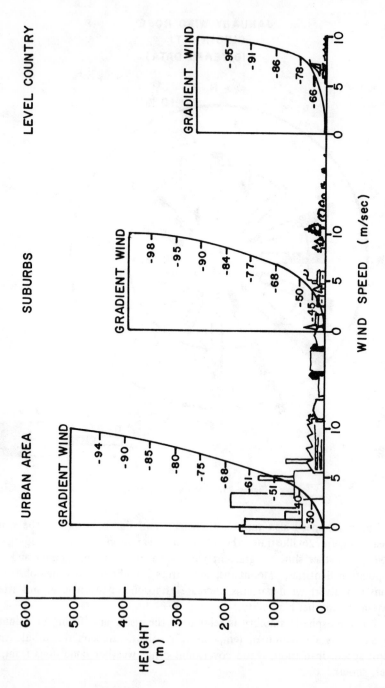

Figure 26.   Gradient wind speed variations with location types.[28]

In general, the atmospheric stability is determined by the atmospheric thermal gradient. The dry adiabatic lapse rate, or neutral stability has a temperature gradient of -1°C/100 meters, or temperatures decrease 1°C for every 100 meters. This condition is such that a volume of pollutant in air would neither gain nor lose buoyancy upon emission. Unstable conditions with lapse rates greater than -1°C/100 meters add to the buoyancy of an emission, and stable conditions or inversions with lapse rates less than -1°C/100 meters tend to inhibit vertical motion of the pollutant gases (plume). Typical environmental lapse rates as defined above are shown in Figure 27.

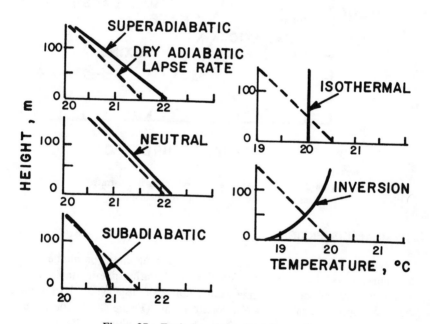

Figure 27. Typical environmental lapse rates.

The actual region of the atmosphere where pollutant emissions and ambient parameters can intermingle is confined to the mixing layer. The mixing layer is the region of the atmosphere capped by a layer of warm air which inhibits any movement past it in the upward direction by cooler air. Figure 28 illustrates this lid effect with a temperature versus elevation diagram. The height of this mixing layer can greatly affect the dispersion process. Low mixing layer heights, confining dispersion and trapping pollutants can result in air pollution emergencies and in many cases sickness and death.

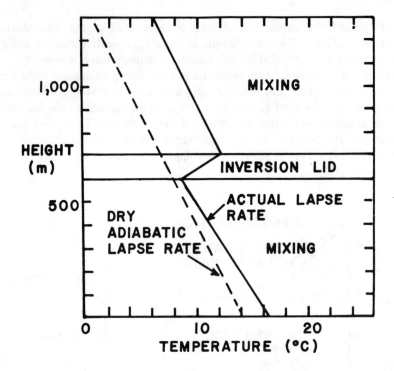

Figure 28. The mixing layer mechanism is illustrated as a function of temperature gradient.

PHYSICAL DISPERSION

Pollutants exit a stack in the form of a flowing cloud or plume. The configuration of the plume is a function of the atmospheric stability which, in turn, affects the manner and amount of dispersion. Atmospheres that are nearly neutral produce a coning plume (Figure 29). These conditions do not greatly influence the motion of the plume. Unstable atmospheres tend to cause the plume to be buoyed up and down and break apart as in Figure 30. This is known as a looping plume. A fanning plume is characteristic of an extremely stable or inversion condition inhibiting vertical motion, resulting in a "pencil-like" plume that can be continuous for over 50 miles (Figure 31). When the mixing layer is very low and/or stack exiting conditions right (high velocity and temperature), the mixing layer may be penetrated. No pollutants will penetrate back down through this layer, thus a lofting plume is formed (Figure 32). On the other hand, a low mixing height could result in no upward penetration and a build-up of pollutants beneath this layer. This condition is shown in Figure 33.

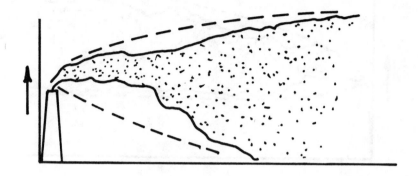

Figure 29. Near neutral (coning).[29]

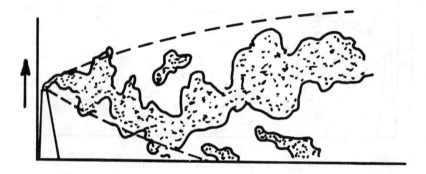

Figure 30. Unstable conditions (looping).[29]

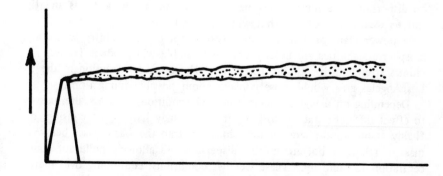

Figure 31. Surface inversion (fanning).[29]

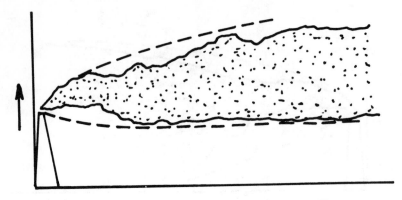

Figure 32.   Surface inversion below stack (lofting).[29]

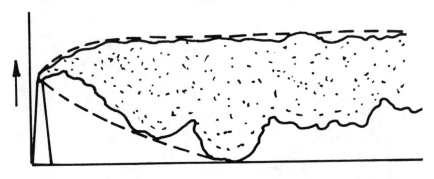

Figure 33.   Inversion aloft above stack.[29]

The writer of an EIS must be aware of the proposed project's regional as well as local meteorological conditions and the effects they would have on dispersion. He must be cognizant of the various plume types and the various weather conditions that will produce them.

A power plant built in the southwest sector of the country where atmospheric stabilities are usually extremely stable will produce fanning plumes stretching for almost 100 miles. However, construction in the Los Angeles area would cause concern about possible fumigations.

Depending on a region's meteorological conditions, process alternatives to effect different stack emission parameters may have to be considered. Highly stable atmospheres or inversions can trap the waste gases below a mass of relatively hot air limiting dispersion and allowing pollutant concentration to build up. These weather conditions, continuing over metropolitan areas for a period of days, have caused many illnesses and deaths.

However, of usual concern to governmental agencies is the maximum ground level concentrations (glc) produced by an emission source, and for the short term these normally occur under unstable atmospheres.

Under unstable atmospheres, it is more likely for atmospheric turbulence and crosswinds to carry the plume to the ground. Critical glc's can usually be predicted based on an unstable atmosphere allowing the EIS writer to determine the proposed project's worst cases and the short-term effects. For any given day the atmospheric stability can usually be obtained from the local weather bureau or estimated from Pasquill's commonly used chart given as Table 22. This chart can be used in the air quality prediction methods we present.

Table 22. The Pasquill Chart for Determining the Atmospheric Stability Class[30]

Atmospheric Stability Class	Class Description
A	Extremely unstable
B	Unstable
C	Slightly unstable
D^a	Neutral
E	Slightly stable
F	Stable to extremely stable

Surface Wind Speed (at 10 m), m sec^{-1}	Day Incoming Solar Radiation			Night	
	Strong	Moderate	Slight	Thinly Overcast or ≥4/8 Low Cloud	≤3/8 Cloud
< 2	A	A-B	B		
2-3	A-B	B	C	E	F
3-5	B	B-C	C	D	E
5-6	C	C-D	D	D	D
> 6	C	D	D	D	D

aThe neutral class, D, should be assumed for overcast conditions during day or night.

Since the EIS must assess the impact on air quality of a proposed project, a determination of the effects of stack emissions must be made. This is accomplished through the use of mathematical models of the dispersion process. The dispersion process will be discussed briefly to give the reader a better understanding of the many parameters and interactions involved, and to get a general feel for the limits of these models.

THE DISPERSION PROCESS

Dispersion from an elevated source (stack) is produced by the mixing and dilution of waste gases with the atmosphere. This is generally accomplished by the turbulent action of the exiting gases, and the crosswind, eddy currents, wind shear, etc.

At the effective stack height, pollutant gases are diluted further by increased wind speeds. Higher wind speeds make available more volumes of air to be mixed with the plume in a shorter time period. However, higher wind speeds also tend to "bend" a plume, retarding its vertical motion and increasing downwind pollutant concentrations. Ground level concentrations are greater for higher wind speeds since the plume is forced to ground level before the pollutants can be dispersed over a much broader region and atmospheric volume. Thus, in order to reduce the environmental impact of an emission, there is the need to increase the area over which a pollutant is dispersed as well as to keep the emissions from harming surrounding structures—resulting in tall stacks and the desirability for large plume rises.

The calculation methods to predict ambient pollutant concentrations which we will offer are based on a two-step process for dispersion. First, the pollutant gases from a stack rise as a result of their own conditions of release, and then they are dispersed in accordance with a Gaussian or normal distribution. Figure 34 shows the Gaussian distribution as the commonly known bell-shaped curve.

Meteorology plays an important role in determining dispersion and the height to which pollutants rise. Wind speed, wind shear and eddy current influence the interaction between plume and atmosphere. Ambient temperatures affect the buoyancy of a plume. However, in order to make equations of a mathematical model solvable, the plume rise is assumed to be a function of the emission conditions of release, and many other effects are considered insignificant.

Short-Term Effects

The calculation methods presented first are for the prediction of short-term effects. Short-term pollutant glc's are predicted for time periods of 24 hours or less. Under these conditions meteorology is assumed to remain constant for the calculated time period.

Plume Rise

The vertical motion of the plume to the height where it becomes horizontal is known as the plume rise. The plume rise is assumed to be a

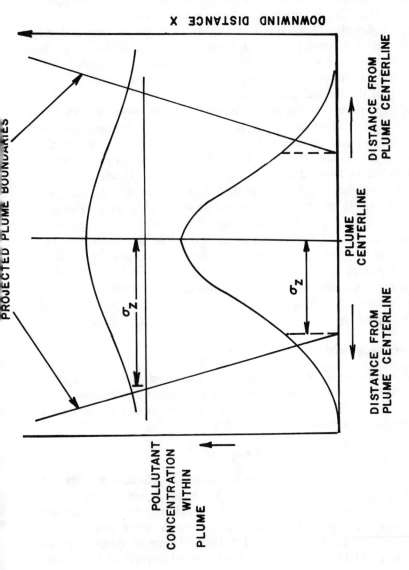

Figure 34. Gaussian or normal distribution bell-shaped curve. Largest pollutant concentrations lie within one standard deviation σ_z, σ_y of the centerline (see Figure 36).[31]

function primarily of the emission conditions of release (*i.e.,* velocity and temperature characteristics). A velocity in the vertical plane gives the gases an upward momentum causing the plume to rise until atmospheric turbulence disrupts the integrity of the plume. At this point the plume ceases to rise. This is known as the momentum plume rise.

Stack gas exiting temperatures are usually much greater than ambient making them less dense than the surrounding air. This difference in densities gives the gases a buoyancy allowing the plume to rise until it is cooled by the atmosphere, reducing the density differential to zero. This is known as the thermal plume rise.[28]

The momentum and thermal plume rises combine to produce the plume rise of an emission. These effects are not independent: gases with a high exit velocity are cooled faster as a result of normal atmospheric mixing of the plume. The thermal buoyancy contribution to plume rise can therefore be lessened by increased exit velocities. Low exit velocities can cause the plume to become trapped in the turbulent wake along the side of the stack, and fall rapidly to the ground (fumigation). Fumigation can usually be prevented by keeping the emission velocity greater than 10 meters/second. An emission velocity that is one and one-half times greater than the atmospheric crosswind is generally accepted as a safety factor to prevent fumigation.

In many calculation methods, the momentum contributions to plume rise are considered negligible as compared to the thermal plume rise and are disregarded.

Effective Stack Height

The importance of plume rise is that it determines the effective stack height, or the height at which most calculation procedures assume dispersion to begin. The plume rise added to the actual height of the stack is the effective stack height (see Figure 35).

$$h = h_s + \Delta h$$

At the effective stack height the dispersion of pollutants are assumed spread out as a Gaussian distribution. The basic dispersion equation considers a continuously emitting point source emanating through a coordinate system with its origin at the base of the source as shown in Figure 36.

Emissions of gases or particles less than 20 microns (larger particles settle more quickly due to gravitational effects) disperse with an origin and plume centerline at the effective stack height. They then spread out in the horizontal and vertical planes. Pollutant concentrations are greatest

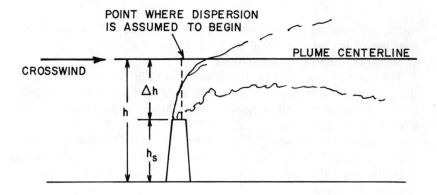

h = effective stack height

Δh = plume rise

h_s = actual stack height

Figure 35. At the effective stack height most calculation procedures assume dispersion to begin.

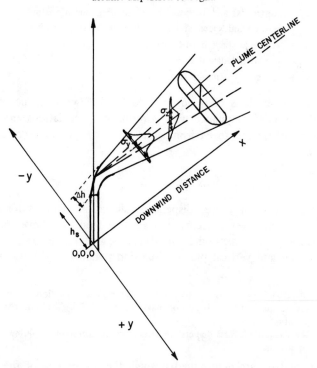

Figure 36. Coordinate system of a Gaussian distribution plume dispersal.[31]

within one standard deviation of the plume centerline. Thus, the determination of the value of these standard deviations is an important factor in calculating ambient concentrations.

The standard deviations in the vertical σ_z and in the horizontal σ_y of the dispersing plume along the centerline are functions of meteorological conditions and downwind distances. These dispersion parameters and the effective stack height can be calculated in a number of ways, using various empirical constants. Each manner of calculating these parameters, σ_y, σ_z and Δh defines a different calculation method without disrupting the basic Gaussian calculation procedures. These methods will now be discussed.[31]

The basic dispersion equation to calculate glc's directly downwind from a point source is as follows:

$$\chi = \frac{Q}{\pi \, \bar{u}_s \, \sigma_y \, \sigma_z} \, \exp - \left[\frac{h^2}{2\sigma_z} \right]$$

where
χ = ground level concentration (gm/m^3)
Q = pollutant exit rate (gm/sec)
σ_y, σ_z = horizontal and vertical plume standard deviations/m
$\bar{u}_s$ = mean wind speed at height of stack (m/sec)
h = effective stack height (m)
x = downwind distance (m)
y = crosswind distance (m)

The maximum glc is of importance to the EIS writer in determining the proposed project's compliance with governmental regulations and the worst case with regard to environmental impacts. It can be calculated as follows:

$$\chi_{max} = \frac{2Q}{e \, \pi \, \bar{u}_s h^2} \, \frac{\sigma_z}{\sigma_y}$$

The three commonly used dispersion calculation methods for the prediction of ground level concentration are based on the above equation. The variance in each method is the calculation of plume rise, Δh, and the horizontal and vertical plume dispersion parameters. These methods are:

1. the ASME plume rise equation and the ASME dispersion parameters;
2. the Pasquill-Gifford dispersion parameters and Brigg's plume rise equations; and
3. the Pasquill-Gifford dispersion parameters and Holland's plume rise equations.

Δh, σ_z and σ_y are determined through the use of one of the above equations to obtain glc's.

THE ASME DISPERSION CALCULATION METHOD

ASME Plume Rise Equations[38]

The ASME method is one of the few calculation methods to consider emissions having relatively low exit temperatures and relatively high exit velocities. Under these conditions of release, the momentum effects of the plume dominate over the thermal, thus the momentum plume rise should be used in the glc calculations. If

$$V_s \geqslant 10 \text{ m/sec}; \ T_s < 50°K + T_a$$

$$\Delta h = D \ (V_s/\overline{u}_s)^{1.4}$$

where Δh =   the plume rise (m)
  D =   stack diameter (m)
  V_s =   stack gas exit velocity (m/s)
  $\overline{u}_s$ =   mean wind speed at height of stack (m/s)
  T_s =   stack gas exit temperature °K
  T_a =   ambient temperature °K

The thermal plume rise is based on the relative buoyancy of a plume to the surrounding atmosphere. When the stack gas exit temperatures are 50°K greater than ambient temperature and relatively large volumes of gases are being discharged, the following set of equations should be used to determine plume rise.

For stable atmospheric conditions:

$$\Delta h \ = \ 2.9 \ \left(\frac{F}{\overline{u}_s \, G} \right)^{1/3}$$

where F = $2.45 \ V_s D^2 \ [(T_s\text{-}T_a)/T_s]$
  G = $(9.8/T_a)[(\Delta T/\Delta Z) + 0.98]$
  $\Delta T/\Delta Z$ =   the atmospheric temperature gradient °K/100 m

For neutral or unstable atmospheric conditions:

$$\Delta h = \frac{7.4 \ h_s^{2/3} \ F^{1/3}}{\overline{u}_s}$$

The ASME Dispersion Parameters

The horizontal and vertical dispersion parameters are represented by the following empirical power law equation:

$$\sigma_z, \ \sigma_y = aX^b$$

where   a,b   =   the empirical ASME dispersion constants given below
  x   =   downwind distance from the source

Atmospheric Stability	σ_y		σ_z	
	a	b	a	b
very unstable	0.40	0.91	0.40	0.91
unstable	0.36	0.86	0.33	0.86
neutral	0.32	0.78	0.22	0.78
stable	0.31	0.71	0.06	0.71

THE PASQUILL-GIFFORD-BRIGGS DISPERSION METHOD

Brigg's Plume Rise Equation

For

$$x \leqslant 3.5 \ W$$

$$\Delta h = \frac{1.6F^{1/3} \ x^{2/3}}{\overline{u}_s}$$

and

$$x > 3.5 \ W$$

$$\Delta h = \frac{1.6F^{1/3} \ (3.5W)^{2/3}}{\overline{u}_s}$$

$$W = 14F^{5/8} \ \text{if} \ F \leqslant 55$$

$$W = 34F^{2/5} \ \text{if} \ F > 55$$

$$F = 2.45 \ V_sD^2 \left(\frac{T_s\text{-}T_a}{T_s}\right)$$

x = distance between source and receptor

The Pasquill-Gifford Dispersion Parameters

The Pasquill-Gifford dispersion parameters are functions of downwind distance and meteorological conditions. The parameters σ_z and σ_y may be obtained from Figure 37 and 38, respectively. The user must know the atmospheric stability as well as downwind distance from the source to select the appropriate dispersion parameter. Atmopsheric stability can be estimated with use of Table 22.

The Pasquill-Gifford-Holland Dispersion Method[31]

Holland's Plume Rise Equation:

for

$$x > 300 \ \text{meters}$$

$$\Delta h = \frac{V_sDK_s}{\overline{u}_s} \ 1.5 + (2.71)(D)\left(\frac{(T_s\text{-}T_a)}{T_s}\right)$$

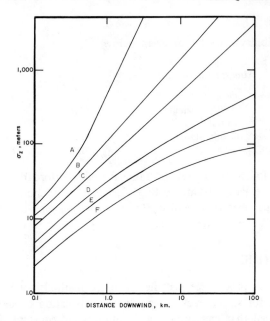

Figure 37. σ_z the Pasquill-Gifford vertical dispersion parameter is a function of atmospheric stability and downwind distance from the source.[31]

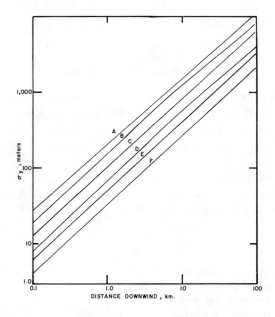

Figure 38. σ_y, the Pasquill-Gifford horizontal dispersion parameter is a function of atmospheric stability and downwind distance from the source.[31]

where K_s = stability class factor given below

Atmospheric Stability	K_s
unstable	1.2
slightly unstable	1.1
neutral	1.0
slightly stable	0.9
stable	0.8

This calculation method uses the same Pasquill-Gifford-dispersion parameters as the Pasquill-Gifford-Briggs calculation method. Therefore, the dispersion parameters σ_z and σ_y for this method may be obtained from Figures 37 and 38, respectively.

AVERAGING TIME

The calculation procedures just described are methods for predicting short-term glc's for single stack cases. In the calculation of short-term glc's it is assumed that meteorological conditions are constant throughout the measuring time period. Short-term periods are usually considered to be 24 hours or less. The averaging times for the various methods discussed are as follows:

Method	Averaging Time
ASME	60 minutes
Pasquill-Gifford-Briggs	10 minutes
Pasquill-Gifford-Holland	10 minutes

The problem the environmental assessor faces is that governmental regulations may specify 10-, 15-, 30- or 60-minute averaging times and one of the above calculation procedures. It is his job to interpret known monitoring data of a different averaging time and relate it to his calculated impact.

A simple means of converting from one averaging time to another in order to compare calculation methods, regulations, or calculated and measured glc's of different averaging times is given below.[32]

$$\frac{X_1}{X_2} = \left(\frac{t_1}{t_2}\right)^{-\alpha}$$

where X_1 = glc of averaging time period t_1
X_2 = glc of averaging time period t_2
t_1 = averaging time period 1
t_2 = averaging time period 2
α = an empirical constant equal to 0.17

These calculation procedures can be used to assess a number of stack emission problems including:

1. glc's to be affected by the proposed plant;
2. glc's to be affected by process modifications (such as reduced emission rates, velocity or temperature); and
3. plant design requirements to meet governmental regulations and other criteria.

The limiting factor of the calculation methods as presented is their use only with a single point source. The impact from multiple sources and area sources can be determined only through the use of a computer program. Computer analysis of these source types is necessary because of the numerous calculations involved.

COMPUTER ANALYSIS

A computer program for the prediction of glc's would enlist the use of the previously presented calculation methods. Its input requirements would include:

1. a coordinate system to identify stack and receptor site positions;
2. stack coordinates—north and east of origin;
3. stack emission parameters (exit gas velocity, temperature, molecular weight, pollutant rate);
4. desired meteorological data (wind speed, direction, atmospheric stability);
5. receptor site coordinates.

Figure 39 shows typical input forms for the prediction of short-term glc's. The output of a short-term glc point source computer program lists the specific meteorological data input and the pollutant glc's at each receptor site. The output might also include the computer-found maximum glc for any receptor site, print glc's for a specified grid system, and/or give glc's at a receptor site due to the contributions of all sources or just a single source. This latter feature might be used to compare the effects of each source of a proposed project to other sources and to sources of the whole area. Output may be of the form shown in Figure 40.

LONG-TERM AVERAGE GROUND LEVEL CONCENTRATIONS

Long-term averaging periods are normally considered to be 24 hours or greater which is too long a time interval to assume constant meteorology. Therefore, long-term average glc's are calculated with the use of actual meteorological data to predict monthly, seasonal or annual average ambient concentrations. This method of determining the effects of a proposed

CARD #1 - EMISSION PARAMETERS

STACK NUMBER	STACK COORDINATES X	STACK COORDINATES Y	POLLUTANT EMISSION (gms/s) SI	STACK HEIGHT (m) SH	STACK DIAMETER (m) D	EXIT GAS VELOCITY (m/s) VS	EXIT GAS TEMPERATURE (°C) T	EXIT GAS MOLECULAR WEIGHT
01	1000	1000	1500.	25.	2.5	6.5	20.	28.8
.	.	.	.	.	.	.	.	.
08	2430	1512	1000.	10.	1.0	5.0	70.	28.8

CARD #2 - METEOROLOGICAL PARAMETERS

WIND SPEED (m/s)	WIND DIRECTION N=0°	ATMOSPHERIC STABILITY+	AVERAGING TIME (min.)
10	130	3	15

CARD #3 - RECEPTORS

RECEPTOR NUMBER	RECEPTOR COORDINATES X	RECEPTOR COORDINATES Y
01	500.0	125.0
02	10.0	0.0

+ATMOSPHERIC STABILITY

INPUT CODE	CLASS DESCRIPTION
1	Extremely unstable
2	Unstable
3	Slightly unstable
4	Neutral
5	Slightly stable
6	Stable to extremely stable

Figure 39. Input forms for the calculation of glc's due to point sources based on a short-term analysis.

POINT SOURCE OUTPUT $\mu g/m^3$

WIND SPEED - 10 m/s
WIND DIRECTION - 130° FROM THE NORTH

RECEPTOR NUMBER	RECEPTOR COORDINATES X Y	POLLUTANT GROUND LEVEL CONCENTRATIONS
01	500.0 125.0	1243.
02	10.0 0.0	462.

Figure 40. Typical output form for short-term glc's predictions.

project can illustrate the long-term impacts of its pollutant emissions and also its compliance with annual governmental pollutant standards.

The weather data are based on thousands of observations of wind speed, wind direction, and atmospheric stability taken over the desired averaging interval at local weather bureau stations. The data can then be compiled, processed and made available in the form of a joint frequency function. The joint frequency function is a distribution of the relative occurrence of wind speed, wind direction, and atmospheric stability averaged over the time period considered. A joint frequency function for each weather station in the U.S. is available or can be made available to the public for various monthly, seasonal or annual time intervals from the National Climatological Center (NCC) in North Carolina. It is in a form readily acceptable as input to a computer program.

One computer program for the estimation of long-term concentrations of pollutants due to emission from area and point sources is the Climatological Dispersion Model (CDM).[32] It is available from the EPA and uses a joint frequency function for its meteorological data. It uses the Pasquill-Gifford-Brigg's calculation method in its computations, but the other methods can easily be added for comparison.

Aside from the meteorological data, input requirements are similar to the previously illustrated input forms. Meteorological and other data necessary for the CDM include:

1. the joint frequency function of the proposed project region (usually from the closest airport weather bureau), a three-dimensional matrix consisting of six wind speed classes, sixteen wind directions, and six stability classes;
2. average mixing layer heights for afternoon and morning (averaged over the calculation time period);

3. average temperature (averaged over the calculation time period);
4. average emission parameters (averaged over the calculation time period).

The computer program then takes the data and performs hundreds of computations to generate the long-term average ambient concentration at various receptor sites due to one or more continuously emitting sources. Output from the CDM is similar to the form shown. The output can be calibrated based on actual monitoring data if available. Further, the output can be used as input to another program that can generate isopleths (lines of constant glc's) over the proposed project region. Figure 41 illustrates a new source region overlayed with isopleths.

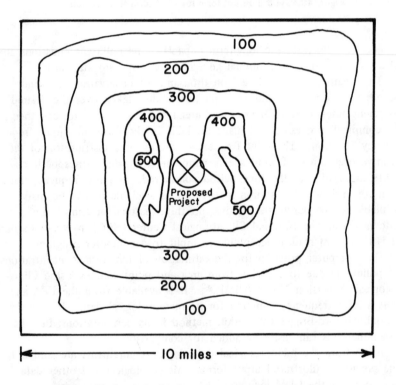

Figure 41. Isopleths for the region of the proposed project, noting the impact of the project's emission ($\mu g/m^3$).

From this map the impact of the proposed project can be evaluated. Long-term effects are assessed, the overall region that may be affected is determined and the extent of the impacts can be projected. The EIS would include these data and subsequent estimates of future air quality.

By obtaining an emission inventory of the region without the new source and estimating glc's, a comparison can be made with glc estimations based on an emission inventory including the new source.

AIR SURVEY

An air survey over the region of the proposed project may be necessary to supplement available data, and to determine the baseline environment. Several steps are to be followed in order to obtain the optimum results from work performed:

1. Establish the need for the air survey.
2. Establish the region to be surveyed. A large region probably would require a van to aid in the conduction of the survey.
3. Determine the pollutants to be sampled.
4. Utilize the most effective sampling instrumentation. This may have to be done through a literature search and a general assessment of the various measurement devices available.
5. Determine the sampling program — selection of receptor sites, frequency of sampling region, duration of the sampling probe, and time of day of the sample.
6. Take the data.
7. Analyze the data.
8. Make conclusions concerning the impact of the proposed project and the baseline environment. References should be made to populated areas and the impact of the new source on the air quality in this region.

CHAPTER 6

IMPACT ON WATER RESOURCES

A general misconception is that water pollution abatement is costly and no useful products are derived from it. However, an expanding economy and water population need an ample water supply. The water resources of the nation must be protected. The reckless dumping of wastes into our waterways is not only poor environmental planning, but also poor management.

Economic growth is related to the availability of water and other land resources. Water demand is dependent on the requirements of the various sectors of the economy. Industrial and municipal users have become competitive for the waters in their regions, since most are located in or near metropolitan centers with water supply, flood control, and waste disposal problems.

Recycling and water conditioning processes have become reality in those areas where water is scarce or of poor quality. The current trends are for more governmental planning in water usage. Flood control plans, 208 water studies, and the EIS among others provide the data and the surveying information necessary to effect restraints on water usage or inducements for recycling. In this manner a stabilized economic growth with adequate water supplies can result.

WATER USES

Water uses fall into two categories: (1) withdrawal uses that remove the water from its natural course, and (2) instream uses that utilize the water while it runs its natural course.

Typical withdrawal uses are for domestic, industrial, utility and agricultural purposes. Table 23 projects the quantities of water required in

119

Table 23. Estimated Water Use and Projected Requirements by Type of Use in the U.S.[33]

(Million Gallons Daily)

Type of Use	Withdrawals				Consumptive Use			
	Used 1965	Projected Requirements			Used 1965	Projected Requirements		
		1980	2000	2020		1980	2000	2020
Rural Domestic	2,351	2,474	2,852	3,334	1,636	1,792	2,102	2,481
Municipal (public-supplied)	23,745	33,596	50,724	74,256	5,244	10,581	16,478	24,643
Industrial (self-supplied)	46,405	75,026	127,365	210,767	3,764	6,126	10,011	15,619
Steam-Electric Power								
Fresh	62,738	133,963	259,208	410,553	659	1,685	4,552	8,002
Saline	21,800	59,840	211,240	503,540	157	498	2,022	5,183
Agriculture								
Irrigation	110,852	135,852	149,824	160,978	64,696	81,559	89,964	96,919
Livestock	1,726	2,375	3,397	4,660	1,626	2,177	3,077	4,238
Total	269,617	442,626	804,610	1,368,088	77,782	104,418	128,206	157,085

the U.S. by the various withdrawal uses through 2020. Obviously, these projections are made in conjunction with projections of national growth rate, recycling and reuse, and other factors. In the EIS a description should be made of the present water uses with projected data 10 to 15 years into the future.

Municipal water uses include all of the water that passes through the system. Such uses are for typical household consumption, fire protection, watering lawns, cleaning streets, and industrial and commercial concerns that tap the municipal water main. These water systems may be publicly or privately owned, and a description in the EIS with respect to size and expansion possibilities is necessary. Water consumption varies between rural and urban communities and also between urban communities of the same size. Affluence, type of industry within the city, climate, water costs, local economy and water distribution facilities, among other factors, can influence water consumption.

The per capita water use is calculated by dividing the total water passing through a municipal water system (all domestic, commercial and industrial) by the population served by the system. Per capita water use has continually increased and is expected to continue to rise through the year 2020. Table 24 lists past and projected water per capita consumption rates for the U.S. from private and public systems.

Table 24. Past and Future Per Capita Water Requirements of Public and Private Facilities[33]

	(Gallons per capita per day)					
	Public Systems					Individual
Year	Domestic	Public	Commercial	Industrial	Total	Systems
1965	73	20	28	36	157	51
1980	77	18	28	40	163	58
2000	81	16	28	43	163	71
2020	83	14	28	45	170	83

We have discussed projections of water use and, although these data are often available from other sources (*i.e.,* Public Health Service, USGS, Economic Research Service), the EIS writer should be familiar with their preparation.

Domestic use (*i.e.,* drinking, cooking, washing, etc.) water requirement projections are usually worked up with the use of Census Bureau population projections and an estimation of per capita domestic water use. The previously listed factors help to establish the per capita use rate. Domestic use is also dependent on type and size of industrial, commercial and municipal facilities.

Water supplies are often not available in those places that have the greatest demand. Industrial and municipal pollution has resulted in water shortages in some areas. Further, population growths and shifts can cause an uneven distribution of water supplies and demands. The EIS should take into account these influences on water usage.

Also considered in the EIS are descriptions of the legal and political structures and organizations that hover over the water supply and distribution facilities. Other institutional problems may involve water jurisdictional conflicts, multistate river allotments, and competition among governmental agencies.

Industrial activities are divided into four categories: manufacturing, mining and mineral processing, ordnance, and construction. The latter two categories require relatively small amounts of water. Further, defense installations and materials are normally exempt from NEPA requirements. Construction of highways, housing and other structures usually meet their water requirements with municipal supplies and these should be accounted for in that section.

Industrial water consumption is dependent on an expanding population and its ability to purchase goods. Industrial expansion and the construction of facilities are stimulated by a demand for goods. These events increase the water requirements of industry. Table 25 lists the quantities of water used by industry in the U.S. during 1965.

Table 25. Industrial Water Use in the United States[33]

Type of Use	Withdrawal bgd	Gross Use bgd	Recirculation Ratio	Consumption bgd	Discharge bgd
Manufacturing					
Processing	10.6	–	–	2.0	8.6
Process Cooling	18.5	–	–	0.2	18.3
Power Cooling	8.0	–	–	0.1	7.9
Other	2.9	–	–	0.3	2.6
Total	40.0	90	2.25	2.6	37.4
Mineral Industries					
Mining Operations	0.19	–	–	–	–
Processing	2.07	–	–	–	–
Cooling	0.87	–	–	–	–
Other	0.11	–	–	–	–
Total	3.24	11.1	3.43	0.76	2.48

Projections for manufacturing-type water requirements are based primarily on production and the value of goods. Water consumption in the mineral and mining industry is estimated based on several factors including usage, recycling, humidity and temperature. Table 26 gives the projected manufacturing and mineral industry water requirements through 2020. Depending on the scope of the EIS under consideration a localized estimation of industry's water requirements is necessary.

Table 26. Manufacturing and Mining Industry Water Requirements[33]

Type of Use	Unit	1965	1980	2000	2020
Manufacturing					
Gross	bgd	90.0	164.3	349	763
Recirculation Ratio	–	2.25	3.0	4.4	6.3
Withdrawal	bgd	40.0	54.8	80	121
Discharge	bgd	37.4	50.2	70	100
Consumption	bgd	2.6	4.6	10	21
Mineral Industry					
Gross	bgd	11.1	20.0	33.0	88.0
Recirculation Ratio	–	3.43	4.9	7	10
Withdrawal	bgd	3.24	4.1	4.7	8.8
Discharge	bgd	2.48	2.9	3.3	7.0
Consumption	bgd	0.76	1.2	1.4	1.8

Electric power generation in most cases requires the use of some form of water. Steam plants and hydropower supply most of the electric power needs of the nation. A proposed project that promotes population and industry shifts—and thus energy needs—results in increased water demands.

Hydropower plants, except for the first filling of a reservoir, are instream uses of water. The water is not consumed, but merely redirected on its course downstream. The introduction of a hydropower plant can result in adverse and beneficial environmental impacts. The latter may include continuous irrigation of farm lands, enhancement of fish and wildlife, and many water-oriented recreation opportunities.

For steam-electric power plants water consumption is also kept to a minimum. Water is converted into steam to drive the electricity-producing turbogenerators. Another water source then cools the steam in a condenser. The only water not returned to its original stream, lake or reservoir are small amounts due to losses in the system.

The generation of electricity, while not a great consumer of water, can affect the quality of waters through: (1) thermal pollution, (2) land removed for reservoirs, (3) varying power plant water withdrawal and return rates, and (4) the return of dissolved oxygen deficient waters. Effects on flora and fauna can also be significant.

Projected use of water to meet the power requirements of future demand are based on accurately predicting that demand. This demand is based on economic growth rate, industrial growth rate, governmental regulation, population growth rates, fuel costs and availability, and other factors.

The industry using the greatest quantity of water is agriculture. Precipitation is the source of most of the agriculturally used water. It is used for crops, pastures, livestock and woodlands. Irrigation contributes to the production of nearly 20% of the crops in the U.S. Irrigation waters are derived from diverted streams, tapped lakes or storage ponds, and underground aquifers. Irrigation uses of water are those of present concern since they are withdrawing and/or consuming surface or ground waters.

The EIS should consider the many federal, state and local government laws, policies and administrative regulations that influence water quality, quantity and usage. Water is so critical to agriculture and other industries, there has always been a constant battle over rights to it. A typical example would be the water rights laws of eastern states as opposed to their western counterparts. In the East, water law has been based on the riparian doctrine. This doctrine gives water in undiminished quality and quantity to the lands bordering streams. On the other hand, in the West, the appropriation doctrine is in effect, giving water rights to the earliest claim.

Agricultural usage of water has been constantly increasing and is expected to continue to grow. Tables 27, 28 and 29 show the increasing levels of water usage for irrigation and livestock on the national level. Similar figures are available on the local level and should be used when discussing current and future water trends. These trends will be influenced on a national/worldwide scale and a more localized regional scale. On the large scale, agricultural water usage will be influenced by population shifts, national and worldwide population growth rates, worldwide climate conditions, technological advances, and changes in social and economic conditions. On the more local, farm-level scale, water usage is affected by climate, type of crop grown, pattern of crop growth, type and quantities of livestock, the amount of land irrigated, irrigation requirements, and irrigation efficiencies.

Table 27. Regional Projections of Stream, Lake and Aquifer Water Withdrawals
for Agricultural Irrigation[33]

Region	(Million Gallons Daily)			
	1965	1980	2000	2020
North Atlantic	151	230	330	420
South Atlantic Gulf	3,270	3,900	6,000	8,200
Great Lakes	75	110	170	230
Ohio	24	40	80	115
Tennessee	8	18	23	29
Upper Mississippi	95	110	200	280
Lower Mississippi	.1,320	3,030	4,400	6,000
Souris-Red-Rainy	24	200	562	576
Missouri	16,039	19,300	21,600	23,000
Arkansas White-Red	6,960	9,400	10,700	11,500
Texas Gulf	7,450	9,400	9,000	8,500
Rio Grande	6,670	6,840	6,840	6,840
Upper Colorado	3,880	5,800	5,850	4,900
Lower Colorado	6,400	7,700	7,000	6,500
Great Basin	4,575	6,200	6,100	5,800
Columbia-North Pacific	26,400	31,400	37,500	42,500
California	26,200	30,950	31,700	32,600
Subtotal	109,542	134,128	147,555	157,990
Alaska	_[a]	4	9	18
Hawaii	1,060	1,420	1,910	2,570
Puerto Rico	250	300	350	400
Total	110,852	135,852	149,824	160,978

[a]Insignificant.

We have presented the major water users and the relative magnitude of their withdrawals and consumptions. Several factors to be considered when projecting current and future industrial water requirements as affected by a proposed action have been noted. These parameters are all interrelated, and the final water demand of any of the above users is greatly influenced by the other users.

THE WATER RESOURCE

The water resource has always been considered renewable. The hydrological cycle, illustrated in Figure 42, is based on evaporation from oceans and evaporation and transpiration from lakes, streams and vegetation. In a continuous cycle, the warm moist air rises and passes inland, the

Table 28. Regional Projections of Irrigation Consumption, Those Waters Used
and Not Returned Directly to Their Original Water Body[33]

	(Million Gallons Daily)			
Region	1965	1980	2000	2020
North Atlantic	150	230	330	420
South Atlantic-Gulf	1,400	1,600	2,450	3,350
Great Lakes	68	95	140	190
Ohio	24	40	80	115
Tennessee	8	16	21	26
Upper Mississippi	83	95	170	240
Lower Mississippi	890	2,180	3,170	4,320
Souris-Red-Rainy	24	150	402	416
Missouri	9,798	12,100	13,500	14,400
Arkansas-White-Red	5,030	6,800	7,800	8,300
Texas-Gulf	5,810	7,100	7,100	7,100
Rio Grande	4,165	4,270	4,270	4,270
Upper Colorado	1,934	2,600	2,880	2,880
Lower Colorado	3,170	3,630	3,760	3,760
Great Basin	2,100	3,040	3,110	3,110
Columbia-North Pacific	10,050	12,900	15,900	18,700
California	19,290	23,800	23,700	23,800
Subtotal	63,994	80,646	88,783	95,397
Alaska	_a	3	6	12
Hawaii	477	640	860	1,150
Puerto Rico	225	270	315	360
Total	64,696	81,559	89,964	96,919

aInsignificant.

moisture in the air condenses upon impact with a cold air mass (precipitation), and then runoff enters rivers, lakes, oceans and vegetation.

However, the cycle is not consistent. Precipitation not only varies
from region to region, but from season to season and year to year. Runoff to streams and aquifers also varies depending on annual precipitation,
land characteristics (dry, arid regions versus woodlands), ground water and
reservoir storage capacities, and other meteorological conditions. On the
average, runoff constitutes about 30% of the water in the cycle. The
remaining 70% is absorbed by vegetation or soils to be reevaporated and
retranspired.

During the cycle, the water quality is affected by biological and hydrological factors. Natural water quality is primarily affected by sediments
and dissolved minerals. This water quality, then, varies from region to

Table 29. Livestock are a Major Consumer of Agricultural Waters. Past and Projected Water Requirements are Given[33]

(Million Gallons Daily)

Region	Withdrawal				Consumption			
	1965	1980	2000	2020	1965	1980	2000	2020
North Atlantic	81	90	150	170	69	80	110	140
South Atlantic-Gulf	146	200	300	380	139	190	285	360
Great Lakes	79	96	132	183	72	87	120	167
Ohio	134	129	194	258	132	129	194	258
Tennessee	37	48	67	93	36	47	64	89
Upper Mississippi	314	477	695	956	305	392	563	775
Lower Mississippi	39	59	79	114	38	58	75	110
Souris-Red-Rainy	19	21	29	40	19	21	29	40
Missouri	368	521	726	1,015	355	502	701	980
Arkansas-White-Red	150	228	318	443	146	223	311	433
Texas-Gulf	100	180	300	440	100	170	280	410
Rio Grande	69	70	70	70	68	69	69	69
Upper Colorado	11	15	20	25	10	14	20	20
Lower Colorado	15	20	30	40	12	15	25	35
Great Basin	16	20	30	40	11	15	21	29
Columbia-North Pacific	59	77	107	147	55	71	100	137
California	80	110	150	220	50	80	90	160
Alaska	—	—	—	—	—	—	—	—
Hawaii	3	4	5	6	3	4	5	6
Puerto Rico	6	10	15	20	6	10	15	20
Total	1,726	2,375	3,397	4,660	1,626	2,177	3,077	4,238

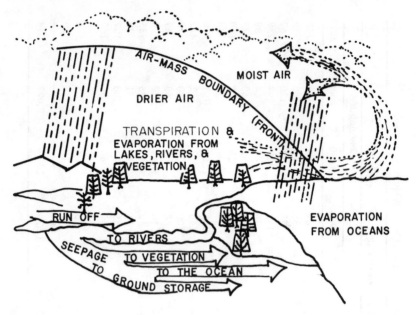

Figure 42. The hydrologic cycle.

region and in its naturally occurring state can be unsuitable for human consumption or other uses.

Man's pollution of water has been very disruptive to its availability for use over large regions. Entire industries have come to a halt as a result of water pollution. The recent banning of fishing in Lake Ontario because of toxic materials can attest to this. Continued dumping of untreated or slightly treated wastes has resulted in the closing of beaches in New York and New Jersey. These effects have eliminated the recreational use of water on the short term and may permanently disrupt its utilization.

Domestic, industrial and agricultural wastes are the primary water pollutants generated by man. These not only include the initial waste discharge, but also those wastes picked up in runoff through urban, industrial, mining and agricultural lands. Stream diversions such as dams, dredging operations or the removal of large quantities of water can also affect water quality.

The EIS considers the water quality as it presently exists and how it would be in the future with and without the proposed project. Often data are available from local, state or federal agencies or private organizations on the water quality in a particular area. However, many locations have an insufficient data base to complete the EIS—thus the need for a sampling and monitoring program.

SAMPLING PROGRAM FOR WATER QUALITY

Figure 43 is a flow diagram of the planning and implementation process of a typical sampling program for a proposed project. A monitoring program is established for pre- and post-project implementation. After the project is completed the monitoring program is continuous to show continued compliance with regulations and the estimates established in the EIS.

Although the sampling program to be outlined is for an industrial plant in operation, similar but modified procedures would be followed in establishing a sampling program for assessment of stream, lake, estuarine, marine and other waters.

Decision

After the initial assessment of an environmental data base, a decision is made with regard to its adequacy. If it is adequate, it is used in any permit, license or compliance schedule. If it is inadequate, or if a continuing sampling program is required, then the following procedures can be used as a basis for a monitoring system. Table 30 is a questionnaire for potential new sources of wastewater discharge. This is the type of information required by regulatory agencies for permit, license and grantee applicants.

Personnel

Depending on the scope and magnitude of the sampling program desired or required, the selection of personnel will be made. A plant's personnel, their expertise, its facilities, and analytical capabilities all contribute to the decision of whether to use inhouse people or to engage consultants and/or laboratories. Outside assistance may be required to do the entire survey or just portions of it such as analyzing grab samples.

WASTE SURVEY

The reason for the sampling program is to identify qualitatively and quantitatively the water contaminants of a plant effluent. As a result, regulations can be met as well as data obtained to complete permit and license applications. Further, an effective program can warn the plant operator of a wastewater problem in time for it to be corrected or diverted before its introduction to receiving waters.

The waste survey locates the points of waste generation, establishes their quality and quantity, and identifies the discharge facilities. The

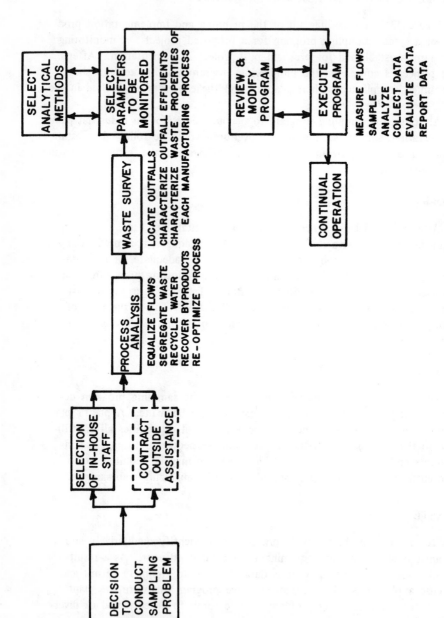

Figure 43. Flow diagram of the planning and implementation process of a typical sampling program.[34]

**Table 30. National Pollutant Discharge Elimination System
Questionnaire for Potential New Sources of Wastewater Discharges**

Complete the following items and return this form to the U.S. Environmental Protection
Agency (EPA) regional office having jurisdiction over the state in which the proposed
source of discharge will be located (hereinafter "the appropriate EPA regional office,"
see list of EPA regional offices on the last page of this form.)

I. Identification

   1. Applicant's name: _____

   2. Mailing address: _____

                                                                         Zip _____

   3. Location of proposed source:

      Street address: _____

      City or town: _____

      County: _____

      State: _____

      Range and township, if grid system is used: _____

      Assessor's parcel number, if known: _____

   4. Name of person to be contacted by EPA if necessary to discuss any of the information provided on this form: _____

                                 Phone: _____ Area code: _____

   5. Standard Industrial Classification (SIC) code(s) _____
(If unknown, consult SIC Manual, 1972 edition, or contact the appropriate EPA
regional office. If more than one code applies to the potential new source, give all
applicable codes.)

   6. a) Is the potential new source an addition or alteration to an existing facility,
or a completely new facility? _____

      b) If an addition or alteration to an existing facility current operating under
National Pollutant Discharge Elimination System (NPDES) permit, give
permit number. _____

      c) If an addition or alteration to an existing facility,

         1) Will there be an increase in productive capacity? _____
How much? _____

         2) Will there be a change in product? _____
From what to what? _____

         3) Will there be a change in production process? _____
From what to what? _____

   7. Date discharge is expected to begin _____

II. Status of Construction

On separate sheets of paper, briefly answer each of the following questions concerning
the progress that has been made in the construction of the potential new source. Be
sure to supply the dates that are requested. You may be required to provide documentation to support these answers.

Table 30, Continued.

---

1. Has land been purchased as a site for the potential new source? If so, on what date?

2. Have any options been taken on alternate site or sites? If so, on what date does the option(s) terminate?

3. Have contractual obligations been made for purchase of facilities or equipment for the potential new source? If so, for each obligation give:

   a) The facility or equipment that is the object of the contractual obligation;
   b) The total amount of the contract;
   c) The amount of money which has been irrevocably obligated to date under the contract;
   d) The terms by which any such obligation is irrevocable (e.g., penalty clause, payment schedule, etc.);
   e) The date on which such obligation became irrevocable.

4. Are any contractual obligations expected to be made in the future for facilities or equipment for the potential new source? If so, for each obligation give:

   a) The facility or equipment that is the object of the future obligation;
   b) The date on which such obligation will become irrevocable;

5. Has clearing of land begun? If so, on what date? If not, when is clearing expected to begin?

6. Has excavation begun? If so, on what date? If not, when is excavation expected to begin?

7. Has any other type of site preparation work begun? If so, specify and give date.

8. Has erection of buildings or other structures begun?

   a) If so: Specify the type of building or structure (e.g., manufacturing plant, storage tank, dock, treatment works);
   Give the date erection of such building or structure began;
   Give the cost of erecting such building or structure.
   b) If not: When is erection of any building or structure scheduled to begin?

9. Has any equipment been placed, assembled or installed at or in the potential new source:

   a) If so: Specify the type of equipment;
   Give the date of such assembly or installation;
   Give the cost of such assembly or installation.
   b) If not: When is any assembly or installation scheduled to begin?

10. What is the expected total cost of the facility, excluding cost of land?

Table 30, Continued.

I. Identification

   A. Name _____

   B. Mailing Address:
      1. Street address or P.O. No. _____
      2. City_____ 3. State _____
      4. County _____ 5. Zip _____

   C. Location

      1. Street or other description _____

      _____

      2. City _____ 3. County _____
             (if applicable)

      4. State _____

   D. Telephone No. _____
              Area Code

   E. Name of person to be contacted by EPA to discuss any of the information requested
      by this form _____

II. Facility Description

   A. What is the proposed source of wastewater discharge? (check one or more of the
      following blocks, as appropriate. Where necessary to explain, attach additional
      sheets.)

      1. Type of Facility

         ☐ a) Manufacturing or materials processing plant. If checked, what is
             principal product?
         ☐ b) Animal feedlot
         ☐ c) Irrigation project
         ☐ d) Fish farm
         ☐ e) Other (explain)

      2. Specific source of wastewater (check one or more blocks, as appropriate).

         ☐ a) Construction of facility
         ☐ b) Manufacturing or processing operation
         ☐ c) Sanitary facilities
         ☐ d) Storm drainage
         ☐ e) Other (explain)

   B. Location

      1. Please supply a map showing the location of the proposed source. A U.S.
      Geological Survey map is preferred; however, a roadmap will suffice.

      2. Check the box which best describes the development in the area in which the
      proposed source will be located.

         ☐ a) Urban   ☐ e) Shopping center
         ☐ b) Suburban   ☐ f) Commercial strip
         ☐ c) Small town   ☐ g) Housing development
         ☐ d) Rural   ☐ h) Other (explain

Table 30, Continued.

3. Briefly describe the natural features of the area in which the proposed source will be located, such as: level ground, valley, mountainous, desert, wooded, flood plain, etc.

4. Give the name of the navigable waterway to which the proposed source will discharge and indicate the location of the discharge on the map in II.B.1.

5. Is the proposed source located on any government designated recreational or wildlife area or near any historical or archeological site listed or eligible for listing in the *Federal Register* pursuant to the Historical Preservation Act of 1966? If yes, explain.

C. Size

1. Give the size of the site in acres or square feet, whichever is more appropriate.

   _____

2. Give the size of any buildings in number of stories and square feet per story.

   _____

3. Give the expected number of employees when operating at capacity.

   _____

4. a) If you checked "a" under II.A.1. above, give the production capacity of the manufacturing or processing plant. _____

   b) If you checked "b" under II.A.1. above, give the largest number of animals that will be confined at any one time for a period of 30 days or more during a 12-month period. _____

   c) If you checked "c" under II.A.1. above, give the number of acres that will be irrigated. _____

   d) If you checked "d" under II.A.1. above, give the number of pounds of fish that will be produced per year. _____

D. Cost

1. What is the expected total cost of constructing the source? $ _____

2. What is the expected cost of constructing any buildings involved? $_____

3. What is the expected cost of machinery, equipment, etc.? $ _____

4. What is the expected cost of any wastewater treatment system? $_____
   (If no treatment system is planned, state "none.")

III. Water Impacts

A. Give the volume of wastewater that will be discharged per day. _____ gal/day

B. Give the frequency of discharge, or if continuous, indicate "continuous."

   _____

C. Give the contents of the discharge, using the most specific name available. For example, if chemical compounds are unknown, use a more general description, such as "soap suds," "oil," "food waste." _____

   _____

Table 30, Continued.

D. Are any of the normal expected pollutant concentrations in the discharge known to you to adversely affect human or other plant or animal health?
   ☐ Yes   ☐ No

   1. If yes, explain, indicating what steps will be taken to reduce or eliminate such potential damage. _____

E. Will the proposed source withdraw water from a river, stream, ground water aquifer, etc.?   ☐ Yes   ☐ No

   1. If yes: Give the name of the waterway _____
      Give the volume to be withdrawn per day _____ gpd.

IV. Air Impacts

A. Will construction or operation of the source result in any emissions to the air from any of the following:

Yes No
☐ ☐ a) manufacturing or materials processing
☐ ☐ b) heating facility
☐ ☐ c) cooling facility
☐ ☐ d) vehicle exhausts, in significantly greater volume than would be expected if the source were not constructed
☐ ☐ e) waste treatment or incineration
☐ ☐ f) other (explain)

B. Give type and quantity of emissions, using the most exact names and data available. _____

C. Are any of these emissions subject to federal, state or local regulation? If so, explain, indicating what control measures will be taken as a result of such regulation.

   _____

D. Are any of these emissions known to you to have an offensive odor?
   ☐ Yes   ☐ No

   1. If yes, explain, indicating what steps will be taken to reduce or eliminate potential offense to the public. _____

E. Are any of the emissions known to you to adversely affect plant or animal health?
   ☐ Yes   ☐ No

V. Noise

A. Will construction or operation of the new source result in any violation of ambient noise standards beyond the property line of the applicant?   ☐ Yes   ☐ No

   1. If yes, explain, indicating what steps will be taken to reduce or eliminate any potential annoyance or injury. _____

Table 30, Continued.

VI. Solid Wastes

A. Briefly describe any plans made for disposal of solid wastes resulting from construction of the source, such as cleared trees, excavated soil, left-over construction materials, sanitary wastes. _____

B. Will operation of the source generate any of the following types of waste?

Yes No
□ □ a) Residuals from manufacturing or processing
□ □ b) Containers from raw materials or supplies
□ □ c) Garbage or waste paper
□ □ d) Sanitary or sewage wastes
□ □ e) Radioactive materials
□ □ f) Other (explain)

C. If you answered "yes" to any part of B above, briefly answer the following questions for each type of waste that will be generated:

1. How will these wastes be collected? _____

2. How will these wastes be stored, and for how long? _____

3. How will these wastes be disposed of, and how often?

VII. Land Use Impacts

A. Briefly describe any anticipated effect of construction or operation of the source of the surrounding area, including, but not limited to: Any increase in residential or commercial development; any increase or change in traffic volume; any change in land use; any foreclosure of alternative future uses of the land. _____

_____

B. Are any zoning restrictions or any currently approved land use plans applicable to this site?   □ Yes   □ No

1. If yes, will this source be in accordance with such restrictions or plans?
□ Yes   □ No

If no, explain what corrective steps must be taken. _____

VIII. Miscellaneous

A. Have you considered any alternative sites for the proposed source?
□ Yes   □ No

1. If yes, briefly explain why the chosen site was selected. _____

_____

B. Briefly describe any social or economic benefits expected to result from construction or operation of the proposed source. _____

_____

C. Are you aware of any public objections to the construction or operation of the proposed source?   □ Yes   □ No

Table 30, Continued.

1. If yes, explain, indicating what steps will be taken to eliminate or reduce such public objections. _____

D. List all other environmentally related permits, licenses and approvals that will be required to construct and operate this source, giving the name of the issuing or approving authority and the status (*e.g.,* applied for, issued, etc.) of such licenses, permits or approvals.

Type of License	Issuing Authority	Status

initial step is to provide a waste flow diagram outlining the plant operations. In this manner, detailed information is obtained for each process, raw materials, final products, by-products, and liquid and solid wastes. Also considered are the periods of discharge, whether they are continuous or intermittent and the times of occurrence. Wastewater characteristics (*i.e.,* temperature, pH, flow rate) are necessary.

A mass balance of the plant's operations is useful in determining the types and amounts of wastes generated. Additionally, estimates of the waste survey parameters to be measured can be made. All water, wastewater, sanitary, storm and drain line locations are identified.

Sampling Site Location

A preliminary sampling survey is done to establish those sites where monitoring should take place. All significant sources of pollution are considered. Sampling sites should be located where:

1. wastestream flow rate is available,
2. the wastewater is well mixed, and
3. the monitoring station is safely oriented and easily accessible.

The final factor may require construction or facility modifications to include the sampling station.

Parameters to be Measured

One of the most important aspects of the sampling program is deciding which parameters to monitor. Obviously, measurement for heavy metals at a grain milling operation is useless. From the mass balance and the preliminary survey the proper parameters to be measured can be established. Table 31 lists those parameters of various industries which can cause pollution problems and should be monitored.

Table 31. Significant Wastewater Parameters for Selected Industrial Classifications[34]

Group I[a]	Group II[b]
A. Aluminum Industry[c]	
Suspended Solids	Total Dissolved Solids
Free Chlorine	Phenol
Fluoride	Aluminum
Phosphorus	
Oil and Grease	
pH	
B. Automobile Industry[c]	
Suspended Solids	COD
Oil and Grease	Chlorides
BOD_5	Nitrate
Chromium	Ammonia
Phosphorus	Sulfate
Cyanide	Tin
Copper	Lead
Nickel	Cadmium
Iron	Total Dissolved Solids
Zinc	
Phenols	
C. Beet Sugar Processing Industry	
BOD_5	Alkalinity
pH	Total Nitrogen
Suspended Solids	Temperature
Settleable Solids	Total Dissolved Solids
Total Coliforms	Color
Oil and Grease	Turbidity
Toxic Materials	Foam
D. Beverage Industry	
BOD_5	Nitrogen
pH	Phosphorus
Suspended Solids	Temperature
Settleable Solids	Total Dissolved Solids
Total Coliforms	Color
Oil and Grease	Turbidity
Toxic Materials	Foam
E. Canned and Preserved Fruits and Vegetables Industry[c]	
BOD_5	Color
COD	Fecal Coliforms
pH	Total Phosphorus
Suspended Solids	Temperature
	TOC
	Total Dissolved Solids
F. Confined Livestock Feeding Industry[a]	
BOD_5	Fecal Coliforms
COD	Nitrogen
Total Solids	Phosphate
pH	TOC

Table 31, Continued.

Group I[a]	Group II[b]

G. Dairy Industry[c]

BOD_5	Chlorides
COD	Color
pH	Nitrogen
Suspended Solids	Phosphorus
	Temperature
	Total Organic Carbon
	Toxicity
	Turbidity

H. Fertilizer Industry[c]

Nitrogen Fertilizer Industry

Ammonia	Calcium
Chloride	COD
Total Chromium	Gas Purification Chemicals
Dissolved Solids	Total Iron
Nitrate	Oil and Grease
Sulfate	pH
Suspended Solids	Phosphate
Urea and Other Organic Nitrogen	Sodium
Compounds	Temperature
Zinc	

Phosphate Fertilizer Industry

Calcium	Acidity
Dissolved Solids	Aluminum
Fluoride	Arsenic
pH	Iron
Phosphorus	Mercury
Suspended Solids	Nitrogen
Temperature	Sulfate
	Uranium

I. Flatglass, Cement, Lime, Gypsum and Asbestos Industries

Flat Glass

COD	BOD_5
pH	Chromates
Phosphorus	Zinc
Sulfate	Copper
Suspended Solids	Chromium
Temperature	Iron
	Tin
	Silver
	Nitrates
	Organic and Inorganic Waterbreaking Chemicals
	Synthetic Resins
	Total Dissolved Solids

Table 31, Continued.

Group I[a]	Group II[b]
Cement, Concrete, Lime and Gypsum	
COD	Alkalinity
pH	Chromates
Suspended Solids	Phosphates
Temperature	Zinc
	Sulfite
	Total Dissolved Solids
Asbestos	
BOD_5	Chromates
COD	Phosphates
pH	Zinc
Suspended Solids	Sulfite
	Total Dissolved Solids
J. Grain Milling Industry[c]	
BOD_5	COD
Suspended Solids	pH
Temperature	TOC
	Total Dissolved Solids
K. Inorganic Chemicals, Alkalies and Chlorine Industry[c]	
Acidity/Alkalinity	BOD_5
Total Solids	COD
Total Suspended Solids	TOC
Total Dissolved Solids	Chlorinated Benzenoids and Polynuclear
Chlorides	Aromatics
Sulfates	Phenol
	Fluoride
	Silicates
	Total Phosphorus
	Cyanide
	Mercury
	Chromium
	Lead
	Titanium
	Iron
	Aluminum
	Boron
	Arsenic
	Temperature
L. Leather Tanning and Finishing Industry[c]	
BOD_5	Alkalinity
COD	Color
Total Chromium	Hardness
Grease	Nitrogen
pH	Sodium Chloride
Suspended Solids	Temperature
Total Solids	Toxicity

Table 31, Continued.

Group I[a]	Group II[b]

M. Meat Products Industry

BOD_5	Ammonia
pH	Turbidity
Suspended Solids	Total Dissolved Solids
Settleable Solids	Phosphate
Oil and Grease	Color
Total Coliforms	
Toxic Materials	

N. Metal Finishing Industry

COD
Oil and Grease
Heavy Metals
Suspended Solids
Cyanide

O. Organic Chemicals Industry[c]

BOD_5	TOC
COD	Organic Chloride
pH	Total Phosphorus
Total Suspended Solids	Heavy Metals
Total Dissolved Solids	Phenol
Free-Floating Oil	Cyanides
	Total Nitrogen
	Other Pollutants

P. Petroleum Refining Industry[c]

Ammonia	Chloride
BOD_5	Color
Chromium	Copper
COD	Cyanide
Total Oil	Iron
pH	Lead
Phenol	Mercaptans
Sulfide	Nitrogen
Suspended Solids	Odor
Temperature	Total Phosphorus
Total Dissolved Solids	Sulfate
	TOC
	Toxicity
	Turbidity
	Volatile Suspended Solids
	Zinc

Q. Plastic Materials and Synthetics Industry

BOD_5	Total Dissolved Solids
COD	Sulfates
pH	Phosphorus
Total Suspended Solids	Nitrate
Oil and Grease	Organic Nitrogen
Phenols	Ammonia
	Cyanides
	Toxic Additives and Materials
	Chlorinated Benzenoids and Polynuclear Aromatics
	Zinc
	Mercaptans

Table 31, Continued.

Group I[a]	Group II[b]
R. Pulp and Paper Industry	
BOD$_5$ COD TOC pH Total Suspended Solids Total and Fecal Coliforms Color Heavy Metals Toxic Materials Turbidity Ammonia Oil and Grease Phenols Sulfite	Nutrients (Nitrogen and Phosphorus) Total Dissolved Solids
S. Steam Generation and Steam Electric Power Generation	
BOD$_5$ Chlorine Chromate Oil pH Phosphate Suspended Solids Temperature	Boron Copper Iron Nondegradable Organics Total Dissolved Solids Zinc
T. Steel Industry	
Oil and Grease pH Chloride Sulfate Ammonia Cyanide Phenol Suspended Solids Iron Tin Temperature Chromium Zinc	
U. Textile Mill Products Industry	
BOD$_5$ COD pH Suspended Solids Chromium Phenolics Sulfide Alkalinity	Heavy Metals Color Oil and Grease Total Dissolved Solids Sulfides Temperature Toxic Materials

[a]Group I consists of the most significant parameters for which effluent limits will most often be set.
[b]Group II consists of some additional parameters for which effluent limits can be set on an individual basis.
[c]Guidelines for these industries not currently available at time of publication.

WATER POLLUTION EFFECTS OF UNDESIRABLE
WASTEWATER CHARACTERISTICS

Soluble organics, through the activity of aerobic bacteria, use up the dissolved oxygen in wastewater. They are present in nearly all industrial and municipal wastewaters, such as paper and pulp mill discharges, brewery wastes, meat packing wastes, dairy wastes, canning wastes, domestic wastes, and effluents from textile and dye manufacturers. Measurements to quantify the soluble organic portion of the waste stream include Biological Oxygen Demand (BOD), Chemical Oxygen Demand (COD), Total Organic Carbon (TOC), and Total Oxygen Demand (TOD).

Soluble constituents affecting tastes and odors are particular nuisances to waste streams and water resources. They can result from: metallic substances (iron, manganese), decaying organic matter, algae and other organisms containing or liberating odorous compounds, organic chemicals (mercaptans), chlorine or other disinfectants, nonbiodegradable synthetic compounds and surfactants.

Toxic materials and heavy metal ions in waste streams are not only hazardous to the environment's health and welfare, but can destroy the workings of a biological waste treatment plant. Maximum concentrations in wastewater discharges for many of these toxicants have been established to reduce the chance of a "bug kill" rendering a treatment facility ineffective. Examples of these contaminants are ammonia nitrogen, sulfide, copper, zinc, lead and mercury.

Color and turbidity are problems associated with aesthetics. Materials in low concentrations such as lignins, dyes, suspended solids and colloids can cause discoloration of natural waters.

Nitrogen and phosphorus are nutrients which promote eutrophication and algae growth. Thus, their discharge into surface waters is undesirable. Agricultural fertilizers provide the major source of these nutrients; they reach streams and lakes via precipitation runoff. Nitrogen may be in the form of ammonia, nitrates, nitrites and organic nitrogen. Organic nitrogen is present as a result of proteins, urea and amino acids, which are common constituents of human wastes, food processing plants and tanneries. Phosphorus is available as orthophosphates and organic phosphorus.

Refractory materials are nonbiodegradable and remain in waterways indefinite lengths of time. Common refractory materials are long-chain nonbiodegradable surfactants or detergents that cause foaming of waters.

Oil, grease and immiscible liquids are highly detrimental to water quality. Effluents containing them are regulated with regard to permissible concentrations. Immiscible liquids from the inking and printing industry (toluene, xylene) are health hazards and when introduced to others, such as naphtheone and ether, may create explosive atmospheres in sewer and drain lines.

Grease materials (hydrocarbons, oils, fats, waxes) at high concentrations commonly cause problems for aerobic treatment plants. Oils are defined as the hydrocarbons of mineral origin (*e.g.,* gasoline, fuel oil, lubricating oils) and are usually removed with an API separator. Other similar materials of plant and animal origin can be degraded biologically; therefore, depending on the concentration, these substances can be introduced to biological treatment plants.

Acid and alkali presence in a waste stream can be extremely troublesome since the pH of a biological and a chemical system is many times a governing factor in the reaction. Variations in pH can destroy biological organisms, change the rate of biodegradation in a treatment plant, and affect the assimilation of wastes in receiving waters.

The acidity of a wastewater can be due to mineral acids and hydrolizing salts. In some cases it can be neutralized biologically, but more often its removal is achieved through the addition of a neutralizing agent. The alkalinity of a water can be due to salts of weak acids, such as borates, silicates and phosphates, and weak and strong bases. The major contributor to the alkalinity of waters is the bicarbonate, usually formed by carbon dioxide activity on basic materials in the soil.[35]

Odorous substances resulting in unpleasant atmospheres can be caused by anaerobic activity releasing hydrogen sulfide or through the introduction of volatile wastes to the waste or water stream.

Suspended solids will adversely affect waterborne ecosystems. Just their deposition in a stream can cause problems. Organic solids deplete the dissolved oxygen content of waters and in the process can release odorous or toxic gases to aquatic life. Suspended solids also add to the turbidity of waters. There are two types of suspended solids: (1) organic or volatile solids—those materials that are oxidized at $550°C$, and (2) inorganic solids.

Dissolved solids, such as salts and sulfates, are common to many industrial wastes, and their introduction to waterways can render the water useless to the public. Potassium, sodium, calcium and magnesium in water (hard water) can cause deposition problems in piping and other transport systems. Sulfates can corrode concrete as well as metal pipelines. Further, salt concentrations above 10,000 mg/l can cause havoc at biological treatment plants.

Temperatures of a waterway are highly critical to aquatic ecosystems. Large fish kills have been associated with sudden variations in temperatures. Temperature variations affect the dissolved oxygen content of surface waters and the natural ability to neutralize wastes. Microbial activity increases at greater temperatures. This condition would be beneficial in biological treatment plants because bacterial waste degradation would be faster. However, higher temperatures in a surface water would promote bacteria and algae growths enhancing the eutrophication process.

Radioactive material can be highly hazardous to all forms of life. These materials can enter surface waters via laboratory waste discharges, nuclear reactor actions, uranium mines and uranium processing. Radioactive laboratory wastes are normally isotopes of carbon (C^{14}), iodine (I^{25}), radioiodine and radiophosphorus. These are commonly used in tracer studies for medical, biological and chemical research.

Alpha (α) and beta (β) rays do not have much penetration power; the former can be stopped with a piece of paper, and the latter with a thin sheet of aluminum. Gamma (γ) rays are much more powerful. They require several inches of lead or several feet of concrete to limit their release.

Pathogenic wastes can originate from a wide variety of municipal, industrial and agricultural sources. Chlorination of wastewaters is the common treatment for pathogen destruction. Ozone, though a much more costly method, can also kill pathogens.

Fecal coliform bacteria count of a wastewater is the measurement used to determine the presence of pathogens. Although fecal coliform bacteria are not pathogenic, they are indicators of salmonella, shigella, leptospira and vibrio and other enteric viruses and parasites.[34]

PREDICTION METHODS FOR THE ENVIRONMENTAL IMPACT OF WATER RESOURCES

The prediction of the environmental impact of wastes on water resources is often accomplished through use of computer programs. Mathematical models have been developed to simulate a wide range of water impacts. They are made into computer programs because of the large amounts of input data and computations required to generate results.

The computer modeling of effluents to ascertain their effects can often result in viable answers. However, in order for such models to be useful they should be employed by someone with expertise in the area of the simulation. This individual would be better qualified to interpret the computer output, based on the mathematical assumptions made and the inherent limitations of the model. Several models for the prediction of various water impacts are presented:[13]

Storm Water Management Models

These models simulate storm and combined sewer systems quantifying and qualifying storm and sanitary runoff.

Storm Sewer Design Models

The storm sewer design models generate the information implied by their name. A sewer system network is input to the computer. The program calculates runoff hydrographs at sewer line inlets and also sizes individual sewer lines from known invert elevations.

Sanitary Sewer Design Models

The design of sanitary sewer systems can be aided with these models. The model input requirements include system geometry, flow rates (minimum and maximum), minimum depth of cover, infiltration rate, and minimum sewer slope. Calculations are performed on the data to gather design information with regard to pipe sizes, invert elevations, slopes, system capacity and flow rates.

Plume Models

Plume models simulate point source pollutant discharges into stagnant, density-stratified surface waters. These waters would include lakes, reservoirs, estuaries and oceans. The models can be used for most municipal, industrial and thermal pipeline effluents.

Stream Assimilative Capacity Models

These models predict DO levels in a stream as it is affected by several point and/or BOD discharges. Further, they can, in order to maintain a predetermined DO standard, specify reductions in waste discharges and compute the necessary treatment efficiencies of these waste streams.

DOSAGE 1 and QUAL 1

DOSAGE 1 and QUAL 1 are computer programs used to assess stream water quality due to pollution discharges. QUAL 1 evaluates such stream parameters as multiple waste discharges, water withdrawals, tributary flows, and runoff with respect to varying BOD, DO, temperature and mineral content. This program is used for a detailed study of a stream. For a less detailed study encompassing a broader range of stream conditions the similar program DOSAGE 1 is utilized.

Water Surface Profiles

Water surface profiles for stream channels are calculated and plotted by this program. Stream flow conditions can be varied, and the effects of bridges, dams and embankments can be determined. Water profiles and flooding areas of a watershed can be computed for floods of various return frequencies.

System Design Optimization

This model is used to evaluate treatment plant process designs with regard to desired effluent characteristics. Alternative treatment methods to provide similar effluent characteristics can be evaluated and compared with respect to effectiveness, costs and reliability.

Thermal Plume Prediction Model

This model simulates the behavior of heated surface jets for varying ambient and initial discharge conditions. Surface jets are employed for the introduction of heated water into streams because, at the surface, heat transfer to the atmosphere is greater due to normally higher surface water temperatures. Input to the programs requires a description of a large body of water, either motionless or moving at a uniform velocity. The output describes the surface plume trajectory, width, temperature, depth, surface area and time of travel along the plume centerline.

Dispersion of Barged Wastes in Ocean Waters

This is a simple model used to predict liquid waste dispersion in ocean waters. Surface waste concentrations along the centerline of the dumping barge's wake are predicted for various times after discharge. As a result, specific concentration levels can be met through the estimation of pollutant dumping rates. The input requirements include barge capacity and speed, disposal time, and a coefficient for horizontal dispersion.

CHAPTER 7

SOLID WASTES–NOISE–ECONOMICS

SOLID WASTES

Domestic wastes will, of course, be increased as a result of population migration to a proposed project's area, so present municipal disposal systems should be assessed as to capability of handling additional wastes. Existing systems should also be examined to determine whether they will remain environmentally sound. Modifications or complete system redesign may be necessary (*i.e.,* alternate disposal methods used) to meet the demands of increased wastes and to meet regulatory standards. In areas with totally inadequate disposal systems, completely new systems may be required. Commercial wastes due to project-affected business activity must be quantified as well as quality assessed and the disposal methods identified. This type of waste is usually disposed of along with other urban solid wastes. Table 32 lists the urban sources of solid wastes and their composition.

Any mining and mining operation solid wastes produced as a result of a proposed project must be assessed and disposed of in an environmentally safe manner. Additionally, the wastes generated during construction of a project are determined and any excavation and demolition wastes must be eliminated. Typical mineral solid wastes and quantities produced are listed in Table 33.

As an industrial project, the proposed action can generate wastes as varied in scope as industry itself. Table 34 lists a number of industries and their resultant solid wastes. These wastes can vary from innocuous to highly hazardous with far-reaching environmental effects if not disposed of properly.[37]

149

Table 32. Urban Sources of Solid Wastes and Their Composition[36]

Urban Sources	Waste	Composition
Domestic, household	Garbage	Wastes from preparation, cooking and serving of food; market wastes from handling, storage and sale of food.
	Rubbish, trash	Paper, cartons, boxes, barrels, wood, excelsior, tree branches, yard trimmings, metals, tin cans, dirt, glass, crockery, minerals.
	Ashes	Residue from fuel and combustion of solid wastes.
	Bulky wastes	Wood furniture, bedding, dunnage, metal furniture, refrigerators, ranges, rubber tires.
Commercial, institutional, hospital, hotel, restaurant, stores, offices, markets	Garbage	Same as domestic.
	Rubbish, trash	Same as domestic.
	Ashes	Same as domestic.
	Demolition wastes, urban renewal, expressways	Lumber, pipes, brick masonry, asphaltic material and other construction materials from razed buildings and structures.
	Construction wastes, remodeling	Scrap lumber, pipe, concrete, other construction materials.
	Special wastes	Hazardous solids and semiliquids, explosives, pathologic wastes, radioactive wastes.
Municipal, streets, sidewalks, alleys, vacant lots, incinerators, power plants, sewage treatment plants, lagoons, septic tanks	Street refuse	Sweepings, dirt, leaves, catch basin dirt, contents of litter receptacles, etc.
	Dead animals	Cats, dogs, horses, cows, marine animals, etc.
	Abandoned vehicles	Unwanted cars and trucks left on public property.
	Fly ash, incinerator residue, boiler slag	Boiler house cinders, metal scraps, shavings, minerals, organic materials, charcoal, plastic residues.
	Sewage treatment residue	Solids from coarse screening and grit chambers, and sludge from settling tanks.

Table 33. Mineral and Fossil Fuel Industry Solid Wastes Generation[36]

Industry	Mine Waste	Mill Tailings	Washing Plant Rejects	Slag	Processing Plant Wastes	Total (thousands of tons)
Copper	286,600	170,500	—	5,200	—	466,700
Iron and Steel	117,599	100,589	—	14,689	1,000	233,877
Bituminous Coal	12,800	—	86,800	—	—	99,600
Phosphate Rock	72	—	54,823	4,030	9,383	63,308
Lead-Zinc	2,500	17,811	970	—	—	20,311
Aluminum	—	—	—	—	5,350	5,350
Anthracite Coal	—	—	2,000	—	—	2,000
Coal Ash	—	—	—	—	24,500	24,500
Other[a]	—	—	—	—	—	229,284
Totals	419,571	288,900	114,593	23,919	40,233	1,146,500

aEstimated waste generated by remaining mineral mining and processing industries.

Table 34. Types of Industries and Industrial Processes Generating Specific Wastes[36]

Industrial Classification	Waste Generating Processes	Expected Specific Wastes
Plumbing, heating, air conditioning Special trade contractors	Manufacturing and installation in homes, buildings and factories	Scrap metal from piping and duct work; rubber, paper and insulating materials, construction and demolition debris
Ordnance and accessories	Manufacturing and assembling	Metals, plastic, rubber, paper, wood, cloth and chemical residues
Food and kindred products	Processing, packaging and shipping	Meats, fats, oils, bones, offal vegetables, fruits, nuts and shells, and cereals
Textile mill products	Weaving, processing, dyeing and shipping	Cloth and fiber residues
Apparel and other finished products	Cutting, sewing, sizing and pressing	Cloth and fibers, metals, plastics and rubber
Lumber and wood products	Sawmills, mill work plants, wooden container, miscellaneous wood products, manufacturing	Scrap wood, shavings, sawdust; in some instances metals, plastics, fibers, glues, sealers, paints and solvents
Furniture, wood	Manufacture of household and office furniture, partitions, office and store fixtures, and mattresses	Those listed under Code 24, and cloth and padding residues
Furniture, metal	Manufacture of household and office furniture, lockers, bedsprings and frames	Metals, plastics, resins, glass, wood, rubber, adhesives, cloth and paper
Paper and allied products	Paper manufacture, conversion of paper and paperboard, manufacture of paperboard boxes and containers	Paper and fiber residues, chemicals, paper coatings and fillers, inks, glues and fasteners
Printing and publishing	Newspaper publishing, printing, lithography, engraving and bookbinding	Paper, newsprint, cardboard, metals, chemicals, cloth, inks and glues
Chemicals and related products	Manufacture and preparation of inorganic chemicals (ranges from drugs and soaps to paints and varnishes, and explosives)	Organic and inorganic chemicals, metals, plastics, rubber, glass, oils, paints, solvents and pigments
Petroleum refining and related industries	Manufacture of paving and roofing materials	Asphalt and tars, felts, asbestos, paper, cloth and fiber

Industry	Operations	Solid wastes
Rubber and miscellaneous plastic products	Manufacture of fabricated rubber and plastic products	Scrap rubber and plastics, lampblack, curing compounds and dyes
Leather and leather products	Leather tanning and finishing; manufacture of leather belting and packing	Scrap leather, thread, dyes, oils, processing and curing compounds
Stone, clay and glass products	Manufacture of flat glass, fabrication or forming of glass; manufacture of concrete, gypsum and plaster products; forming and processing of stone and stone products, abrasives, asbestos, and miscellaneous nonmineral products	Glass, cement, clay, ceramics, gypsum, asbestos, stone, paper and abrasives
Primary metal industries	Melting, casting, forging, drawing, rolling, forming and extruding operations	Ferrous and nonferrous metals scrap, slag, sand, cores, patterns, bonding agents
Fabricated metal products	Manufacture of metal cans, hand tools, general hardware, nonelectric heating apparatus, plumbing fixtures, fabricated structural products, wire, farm machinery and equipment, coating and engraving of metal	Metals, ceramics, sand, slag, scale, coatings, solvents, lubricants, pickling liquors
Machinery (except electrical)	Manufacture of equipment for construction, mining, elevators, moving stairways, conveyors, industrial trucks, trailers, stackers, machine tools, etc.	Slag, sand, cores, metal scrap, wood, plastics, resins, rubber, cloth, paints, solvents, petroleum products
Electrical	Manufacture of electric equipment, appliances, and communication apparatus, machining, drawing, forming, welding, stamping, winding, painting, plating, baking, and firing operations	Metal scrap, carbon, glass, exotic metals, rubber, plastics, resins, fibers, cloth residues
Transportation equipment	Manufacture of motor vehicles, truck and bus bodies, motor vehicle parts and accessories, aircraft and parts, ship and boat building and repairing motorcycles, bicycles and parts, etc.	Metal scrap, glass, fiber, wood, rubber, plastics, cloth, paints, solvents, petroleum products
Professional, scientific controlling instruments	Manufacture of engineering, laboratory, and research instruments and associated equipment	Metals, plastics, resins, glass, wood, rubber fibers, and abrasives
Miscellaneous manufacturing	Manufacture of jewelry, silverware, plated ware, toys, amusement, sporting and athletic goods, costume novelties, buttons, brooms, brushes, signs and advertising displays	Metals, glass, plastics, resins, leather, rubber, composition, bone, cloth, straw, adhesives, paints, solvents

Solid Waste Management

The proper disposal of solid wastes derived from any source is dependent on management practices. A management system must be developed and described that incorporates many diverse factors. Those factors considered may include economics, engineering, land use ordinances, environmental regulations, geography and sociology. A solid waste management system that would optimize these parameters would be designed based on Figure 44.

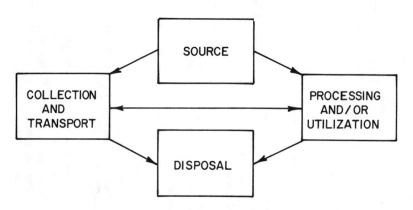

Figure 44. A solid waste management system.

After generation, solid wastes can be merely collected and transported to the disposal site. They might be collected and transported to a processing or recycling plant with the residual materials disposed. Depending on the economics and the availability of resources, solid wastes might be processed or recycled at the generation site. Onsite disposal might follow, or the residues may be trucked to municipal disposal facilities.[38]

Collection and transportation operations involve several steps that are necessary for proper disposal of solid wastes. The wastes are first transferred to the collection vehicle with a loading operation. This loading could be manual or automatic (conveyors, pneumatic chutes) depending on waste type and location. At this point the wastes are in in-transit storage and will be during the transportation operation.

The wastes are then hauled to the disposal facilities, or to a processing plant, or to a transfer point. The ultimate destination of these latter two operations is the final disposal site. At this site, the wastes are unloaded and the collection vehicles return to the generation site for refilling.[36-39]

Sanitary Landfill Disposal

The final step in the solid waste management system is disposal. Disposal may be in the form of incineration, open burning, ocean disposal, land application, animal feed or sanitary landfill.

In most areas, open dumping and open burning have been banned as unacceptable disposal practices. Ocean dumping of wastes has come under increased attack as a result of beach closings and the banning of fishing in certain areas.

The method of ultimate solid waste disposal which is in practice the most sound is the sanitary landfill. Sanitary landfilling is a land disposal method that utilizes engineering principles to ensure:

1. no public health or safety hazards and nuisances are created;
2. the refuse is compacted to the smallest practical volume;
3. the area of disposal is kept to a minimum, and
4. at least once a day (more if necessary) the refuse is covered with a layer of soil.

The sanitary landfill is designed, located and operated to minimize environmental impact. These impacts are possible because within the disposal site various physical, biological and chemical processes are taking place producing gases and compounds. The gases can escape to the atmosphere producing strong odors (hydrogen sulfide) or explosive conditions (methane). Water, due to precipitation or flooding, percolates through the solid waste dissolving compounds and picking up other contaminants. The waters contaminated in this process are known as leachate. Typical sanitary landfill leachate composition is given in Table 35. Depending on the site's soils and geology this leachate may or may not pollute surface and ground waters. Provisions may have to be made to collect and treat the leachate. This would require a sloping bottom, some sort of impermeable bottom liner, a collection area, and pumping and treatment facilities.

The utilization of sanitary landfills has been declining in recent years as alternate disposal and recycling methods are sought. The reasons for this trend include: increased transportation and handling costs, increased dumping costs, the unavailability of landfill sites, and the need to preserve present sites as long as possible.[38, 40]

Incineration and Energy Recovery

Incineration or the combustion of solid wastes has been used for centuries as a means of disposal. Highly efficient incinerators can reduce organic matter volumes by 90%, but many of the older incinerators can expect a volume reduction of only 40%. Combustion products of fly ash,

Table 35. Typical Sanitary Landfill Leachate Composition[40]

Analysis	Range of Values (μg/l) Low	High
pH	3.7	8.5
Hardness (carbonate)	35	8,120
Alkalinity (carbonate)	310	9,500
Calcium	240	2,570
Magnesium	64	410
Sodium	85	3,800
Potassium	28	1,860
Iron (total)	6	1,640
Chloride	96	2,350
Sulfate	40	1,220
Phosphate	1.5	130
Organic Nitrogen	2.4	550
Ammonia Nitrogen	0.2	845
Conductivity	100	1,200
BOD	7,050	32,400
COD	800	50,700
Suspended Solids	13	26,500

particulates and gases must be controlled. Pollution equipment often reduces the economic feasibility of such a system to nonimplementation. However, at the modern incineration plant, the residues of combustion are rather innocuous and can be landfilled without environmental harm.

In order to increase the cost/benefits of disposal and reduce natural resource consumption at a project, solid wastes can be burned as the primary or a supplementary fuel of the plant. There are numerous systems in operation utilizing solid wastes directly or processing them and deriving fuel gases and solids to run plant furnaces and boilers. One such system is illustrated in Figure 45. This is the PUROX process of Union Carbide and uses a pyrolysis furnace as the basis of its operations. Much background information on the various systems and processes is available in the literature.[38, 41]

Recycling

The recycling of materials is encouraged by the EIS. Not only is it an environmentally prudent procedure, but in many cases the processing of the recycled material is much less costly than the virgin ore. The processing of recycled aluminum requires just 5% of the energy necessary for the original process. Figure 46 is a flow diagram of a recycling system.

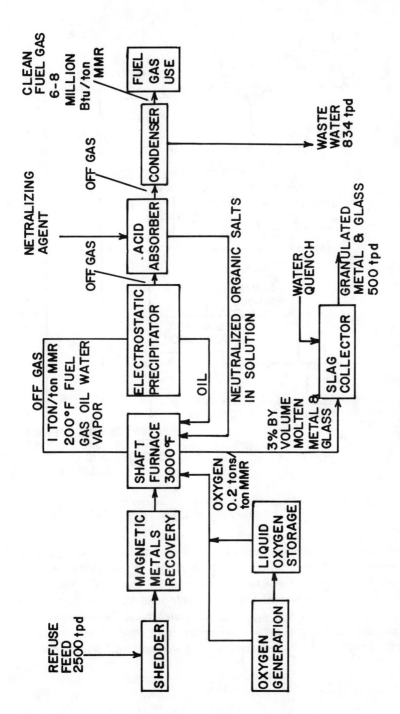

Figure 45. The PUROX process of Union Carbide for the recovery of energy from solid wastes.[38]

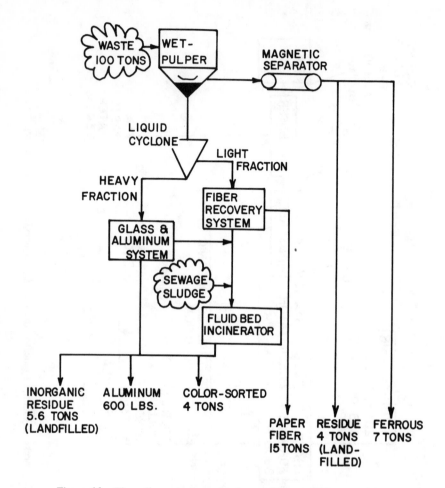

Figure 46. Flow diagram of process to extract recyclable materials from solid wastes.[42]

Glass, aluminum and other metals are removed and sold to scrap dealers for reprocessing. Paper fiber is also reprocessed. Final residues which amount to less than 10% of the original mass are landfilled. Sewage sludge is used to supplement the combustible solid wastes in the incinerator. The PUROX system previously illustrated takes the recovery process one step further and utilizes the energy from combustion. In this manner complete recycling is achieved and environmental impact minimized.[38, 42]

NOISE IMPACT

A noise is any unwanted, undesired or valueless sound. It can cause physiological and psychological damage to human beings and has an adverse effect on activities such as work, recreation, sleep, communication and rest.

Depending on exposure, noise levels and noise frequencies, hearing losses can occur. Noise also produces mental stress, fatigue, dizziness and loss of balance. It affects sleep and the recuperatory powers during illnesses.[43] Noise is a major environmental pollutant that has come under governmental regulation. Aside from local ordinances, in many cases noise impacts must be assessed béfore federal permits or licenses can be granted.

Noise Measurement

Sound is the localized fluctuation in atmospheric pressure to which the human ear responds. The fluctuations vary in size and in frequency of occurrence diversifying the range of sounds heard. The magnitude of the pressure fluctuations is expressed as a decibel (dB). The decibel scale is a logarithmic scale used because of the large range of sounds that are experienced. The louder the sound, the higher the dB reading. Zero dB is near the threshold of human hearing capabilities.

The frequency of a sound wave is the number of occurrences per second. Meters are used to analyze the sound energy over the audible range. They measure sound pressure levels and frequencies. Meters and other measurement devices can be used to determine the levels of noise-generating equipment, the levels of noise produced at a plant, and the general ambient noise levels of a community.

The impact of a proposed project's noise levels on a region should be evaluated with respect to local ordinances, present ambient noise levels, and federal and state noise guidelines. Local ordinances specify the allowable noise levels in a given region. Therefore, the noise levels of the proposed project should be known. Noise level criteria are usually set for residential, commercial, retail and industrial zones. Many areas also differentiate between daytime and nighttime noise levels. Table 36 lists some typical noise level criteria for day and night, and several zones of public occupation.

All the noises generated in a region constitute its ambient noise level. A proposed project's noise should be evaluated with regard to these levels to determine its acoustical impact on the community. Figure 47 gives the noise levels of various human and human/mechanized activities. In assessing the impact of noise due to the proposed project on the surrounding area, several factors should be considered including:

Table 36. Typical Noise Level Ordinance[44]

Area	Sound Level (dBA)		
	Day[a]	Night	Other
Rural (residential)	50	40	45
Suburban (residential–also hospital, church, and similar zones)	55	45	50
Urban (residential–also apartment)	60	50	55
Urban (residential–with some commercial, retail or light industry)	65	55	60
Predominantly industrial	70	60	65
Heavy industrial, few dwellings	75	65	70

[a]*Day* represents the period 0700-1700 hours on weekdays, and *night*, 2200-0700 hours. *Other* includes 1700-2200 hours and, on weekends and holidays, 0700-1700 hours. Where local custom differs, the period should be adjusted to conform. Each time period during which the noise is present should be considered.

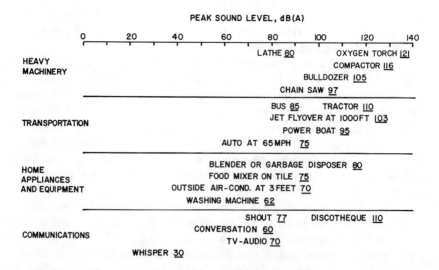

Figure 47. Typical noise levels of some common human activities.[44]

1. the social and economic benefits of the project;
2. the necessity of the noise in relation to the functions performed;
3. the adequacy of the control measures instituted;
4. the damaging health effects due to the noise; and
5. the aesthetics, such as the drowning out of bird calls or crickets.

Types of Noise

There are two major categories of noise: steady and nonsteady. Steady noises are continuous with a rather stable level. Changes in noise levels of about 6 dB are noticeable. (The human ear can normally detect differences of about 3 dB.) Nonsteady noises are any fluctuating, intermittent or impulsive sounds. These noises, such as impact machine or train noises, occur in varying degrees of frequency. They are sometimes allowed to exceed ordinance levels because of their infrequency and lack of persistence.[44]

EIS Noise Impact

A noise analysis of the region for the proposed project should be conducted according to the following:

1. Identify existing activities or land uses which may be affected by noise from the proposed project.
2. Prepare the noise levels for each alternate considered in the EIS (including the no project alternatives).
3. Measure the existing noise levels for existing activities or developed land uses.
4. Compare the predicted noise levels for each alternative in the EIS with the existing noise levels and with the proposed project's noise levels.
5. Examine and evaluate alternate noise abatement measures for reducing or eliminating the noise impact on existing activities.[45]

Noise Level Predictions

Computer programs are available for the prediction of noise levels to be generated by a proposed project. They also can be used to determine the noise impact on surrounding areas. These programs (hand calculations are possible for simple cases) can determine the noise levels at some distances away from the project. The levels can be determined for various atmospheric effects, barriers and topographical features. Equations for these prediction methods are well documented.[43, 44]

Noise Survey

A noise survey may be necessary to determine the present noise levels and to define a baseline for noise level analysis. Further, after implementation a noise survey may be necessary to determine the project's compliance with local ordinances, and federal and state laws. The survey could be conducted as follows:

1. Determine time of day measurements are to be taken.
2. Determine the type of acoustical data to be taken.
3. Determine the location of sample points and the significant sound sources.
4. Utilize a map to identify significant topographical features and other structures and sound barriers.
5. Meteorological conditions can greatly affect noise levels. These should include relative humidity, wind speed and direction, barometric pressure and temperature.
6. Determine and correct measured values due to calibration requirements, cable length and temperature.
7. Map significant information such as residential, commercial or industrial zones, population densities, special areas (hospitals), and areas of unique noise characteristics.

Figure 48 is a map of a region surrounding a source on which lines of constant noise levels (isopleths) are drawn. These isopleths are used to illustrate the impact of a noise source on an area. From this type of map a planner can determine the size of the buffer zone necessary to reduce or eliminate the noise source impact on the community. Further, barriers and other noise reduction techniques can be assessed with respect to their impact on the environment.

FORECASTING ECONOMICS

The forecasting of economics is very difficult and all too often impossible. Usually the condition of the economy is determined through the use of indicators. Gross national product (GNP), wholesale price index, retail price index, unemployment rate, imports versus exports, stock market trends, capital investment, interest rates and other factors contribute to the economic stability of a region, the nation or the world.

However, recently economics have become quite unpredictable. Spiraling inflation and large unemployment rates have at least rendered most indicating techniques unreliable. Yet, the economic impact of the proposed project must be assessed.

There are several methods of predicting economically induced changes in a region due to a proposed project. These methods are estimates of

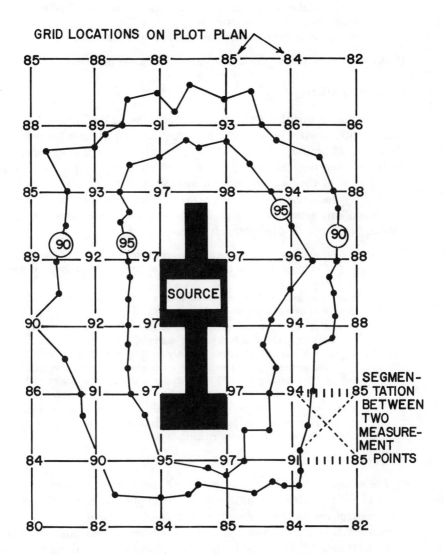

Figure 48. Grid plot of noise source and surrounding areas.
Isopleths of noise levels are plotted.[44]

the secondary impacts that can occur. The most important of these impacts are:

1. the imports and exports of commodities in the region as influenced by the
   project;

2. the stimulation of investment by other related industries to the project, known as the multiplier effect; and

3. the stimulation and level of stimulation in other industries.

Forecasting methods for the above impacts are now presented.

Location Quotients

Using readily available data on employment levels, the location quotient analysis differentiates between the exporting and importing industries of the region of project influence. A basic industry is one that exports, while the importing industries are designated nonbasic.

The basis of the location quotient analysis is that generally the basic industries support the nonbasic industries and result in the greatest economic and environmental impact. This influence of basic industries on nonbasic industries is determined as follows:

1. A comparison is made between the total employment level within the region of influence and the local employment level for the industry including the new industry.

2. A comparison is made between the total national employment level and the national employment levels of the industry.

3. A ratio of the above two comparisons estimates the proposed project's role as a basic or nonbasic industry. For example, a ratio greater than one would be considered as a basic industry in the project's region of influence. Thus, the proposed project would be a net exporter out of this region, reflecting the relative effects of this industry with regard to the rest of the nation. A ratio less than one would show a nonbasic industry.

4. A ratio similar to the previous one, but based on state or other appropriate regional employment data instead of national data, might reflect the local economic impacts more realistically.

Multiplier Effects

The estimate of nonbasic industry investment induced by the investment in basic industries is made with the multiplier concept. The multiplier concept considers the many and continued effects of the exports of a region such as employment and regional income. It estimates the total income affected by the proposed project. The ratio of total employment in the region of influence to the total employment for all basic industries is a measure of the multiplier effect. Table 37 gives a sample calculation of the multiplier effect for a hypothetical community and project.

The limiting factor of the multiplier effect is that it assumes the proposed industry will act in a fashion similar to those basic industries already in the area. Obviously this can introduce large errors into the estimates because the new industry may be better automated, produce a product

Table 37. Sample Calculation of Employment, Employment Coefficient, and
Location Quotient for Hypothetical Community

	(1) Employment Hypothetical Community	(2) Coefficient U.S.	(3) Coefficient Local	(4) Location Quotient (3) ÷ (2) −
Agriculture	6,106[a]	0.03710	0.58335	15,73720[a]
Chemical and Allied	11	0.01290	0.00105	0.08140
Construction	59	0.05972	0.00564	0.09444
Utilities and Sanitary Services	95	0.01677	0.00906	0.54085
Food and Kindred	2,603[a]	0.01816	0.24868	13,69380[a]
Textile Mills and Fabricated Textiles	5	0.02853	0.00047	0.01647
Trucking and Warehousing	80	0.01414	0.00764	0.54031
Communications	16[a]	0.01402	0.01530	1.09130[a]
Railroads	23	0.00831	0.00219	0.26353
Printing and Publishing	42	0.01556	0.00401	0.25777
Machinery, except Electrical	98	0.02600	0.00930	0.36000
Furniture, Lumber and Wood Products	169[a]	0.01278	0.01614	1.26291[a]
Eating Establishments	304	0.03003	0.02904	0.96703
Education	97	0.08030	0.00926	0.11530
Nonprofit	42	0.01519	0.00401	0.26398
Professional	151	0.02552	0.01443	0.56543
Governmental	566	0.05488	0.05407	0.98524
Total	10,467			
Total Basic Industries	8,894			
Multiplier: 10,467/8,894 = 1.18				

[a]Indicates a basic industry.

that does not stimulate the local economy, or have other unique characteristics. Therefore, the proposed industry must be assessed by the EIS
writer to determine its conduciveness to the economics of the region of
influence. Modifications to the multiplier concept may then be required
to get a better forecast of the economics of the proposed project.

Industries to be Stimulated

An estimate of the type of industry and the magnitude of stimulation
by the proposed industry can be determined from Bureau of Economic
Analysis data for the "Input/Output" (I/O) structure of the U.S. economy.
These data relate the effect that investment in one industry will have on
other industries. Using these data the economic impact of a proposed
project can be estimated. Table 38 is an example of this type of estimate. The dollar investment of a new petroleum refinery is projected

over other industries, and the amount of economic growth in these industries is estimated.

Like the multiplier effect these estimates are affected by the uniqueness of the proposed industry such as one that is dependent on rapid technological changes. Modification of the final estimates, based on the assessor's evaluation of the industrial impact, may then be necessary.

Table 38. The Relation of the Investment of One Dollar in a New Petroleum Refinery
to the Economic Growth of Other Industries

Industry		Per Dollar Investment
1.	(New) Petroleum Refining	1.086
2.	Chemicals and Selected Chemical Products	0.114
3.	Plastics and Synthetic Materials	0.062
4.	Paints and Allied Products	0.054
5.	Other Agricultural Products	0.052
6.	Transportation and Warehousing	0.050
7.	Stone and Clay Mining	0.043
8.	Maintenance and Repair	0.037
9.	Livestock	0.032
10.	Business Travel and Entertainment	0.032
11.	New Construction	0.030
12.	Forestry	0.018
13.	Broad and Narrow Fabrics and Yarns	0.026
14.	Stone and Clay Products	0.026
15-82	All Others	0.068

CHAPTER 8

OIL SPILL IMPACT

Any oil exploration and recovery operation will usually require environmental assessment. Economically and socially, on the national scale, petroleum resources demonstrate their significance—from the gasoline to run cars, to plastic furniture and apparel, from the heat for our homes, to the synthetic fertilizers that help keep us well-fed.

Petroleum-related localized impacts can be felt, for example, in the primary and secondary employment sectors. Further, large sums of money can be infused into local economies by way of taxation on mineral withdrawals, royalty percentages, or payments for exploration leases.

On the other hand, petroleum and petrochemical projects can have environmentally unsatisfactory impacts. Yet, because of the rewards petroleum resources bring to an area, undesirable impacts are often accepted as trade-offs. Environmentalists now consider these trade-offs imbalanced and advocate further controls to keep petroleum industry impacts to a minimum.

The 1969 oil spill off Santa Barbara and the Alaskan pipeline controversy have stimulated a public awareness of the inherent environmental (physical/chemical, ecological and aesthetic) problems associated with the petroleum industry. The impact of offshore drilling has received particular interest because of:

1. the North Sea oil discoveries and the selling of leases for exploration on the mid-Atlantic continental shelf; and

2. the environmental damage that can be caused by an oil spill on a water body. The water provides the medium for oil transportation over great distances where an oil spill affects not only the ecosystems of the immediate area, but also those of surrounding regions.

Oil spills on water result during the handling of the product during offshore drilling and withdrawal operations—through pipelines and tankers,

167

marine loading and unloading terminals, and refinery processing. Most of these operations would require that an environmental assessment be written because their environmental significance is related to their initiation through one of several governmental avenues, e.g.:

- the federal government granting oil exploration and development leases,
- the federal government granting pipeline and offshore drilling permits,
- the granting of funds for tanker construction,
- legislation for deep-water port licensing.[46]

Such an assessment should consider the impact of oil spills on the physical/chemical, ecological, aesthetic, and social and economic environments; the magnitude of the spill; the inevitability of the spill; prevention techniques, cleanup techniques, and restoration programs.

EFFECTS OF AN OIL SPILL

An oil spill can affect the environment in numerous ways. The magnitude of the impact would be dependent on the type of accident (blowouts, explosions, pipeline ruptures), the region of the spill, and the cleanup and control techniques. Usually, the study can describe the impacts of each of the above using past history of similar occurrences and relating them to the region of concern. Research studies may be required for any unique aspects of the region.

Of primary interest is the possibility of an oil spill reaching shore, resulting in adverse economic and social impacts. Recreational features of shorelines, swimming, boating and fishing can be disrupted for long periods of time. Docks, boats and other personal property can be severely damaged. Monetary losses could be astronomical, not only as a result of property damage, but also due to the disruption of businesses. Recreation facilities and commercial fisheries as well as their supportive industries could be closed down until cleanup measures have been completed. Upon completion of cleanup, these industries may still have trouble in approaching normalcy because of the effects on local ecosystems.

Initially, the water quality deteriorates as the percentage of dissolved hydrocarbons and trace metals increases. Particles on or near the surface adsorb the oil and upon sinking contaminate deeper water and bottom sediments. These effects persist until the surface oil is removed. Table 39 compares the concentrations of certain heavy metals in various media.

Under natural conditions marine organisms contain and synthesize hydrocarbons some of which are not unlike petroleum hydrocarbons. However, petroleum hydrocarbons (PHC) are also composed of the toxic aromatic hydrocarbons and cycloparaffins. Petroleum can enter the marine food chain in three manners:

Table 39. Comparison of Certain Heavy Metals in Various Media[11] (in ppm)

	Cr	Cu	Ni	V	Zn
Ocean Water	0.0005	0.003	0.007	0.002	0.01
Water Column of New York Bight – Northern and Southern Areas	0.003-0.015	0.00023-0.018	0.00072-0.001	NA	0.0018-0.038
Water Column of New York Bight Apex in Vicinity of Sewage Sludge and Dredged Materials Dump Sites	NA	0.0006-0.047	NA	NA	0.0021-0.190
Weighted Average Concentration of Ocean-Dumped Sludge, EPA Region II, 5/1/73-12/31/73	63	60	9.5	0.75	160
Median Concentrations in Produced Waters	ND-0.01	ND-<1	ND-0.015	–	ND-<0.01
Concentrations in Six Crude Oils	<0.09-10.6	0.19-1.25	1.28-116.8	4.9-112	–

NA = not available; ND = not detectable.

1. adsorption in organisms and particles;
2. the ingestion of oil-containing organisms by other organisms; and
3. the intake of dissolved or dispersed petroleum through the gills and body surface.

These hydrocarbons can accumulate in marine tissues and concentrate in certain organs such as the liver, gall bladder, and the nervous system. Heavy metals are taken up by organisms in any of the manners of petroleum ingestion. They can build up in tissues and progress through the food chain.

Water fowl or seabirds are highly vulnerable to oil spills. During the Torrey Canyon spill between 40,000 and 100,000 birds were killed, at Santa Barbara over 3,600, and about 7,000 birds killed during the San Francisco Bay spill.[47]   Oil clings to the birds' feathers destroying their buoyancy and insulation properties.   Oil ingested by the birds can poison or at least disrupt their feeding habits.

Other marine mammals—whales, dolphins, seals and walruses—are not as adversely affected as far as numbers are concerned. Animal species dominant in the deeper waters are less likely to be affected by the toxicity of the spills because of the dilution associated with depth. However, these mammals can suffocate when oil coats their outer surfaces, or become seriously injured should they ingest appreciable amounts.

Clams, abalones, starfish, sea urchins, lobsters, oysters, and other benthic invertebrates can become tainted by hydrocarbons and, depending on the concentration, these hydrocarbons can be toxic. Studies have shown that the impact on the benthic ecosystem can range from total destruction to minor effects. Shoreline areas are affected most severely by an oil spill. Most of the damage to benthic community life forms results from the oil smothering organisms rather than its toxicity. However, organisms near open shorelines are less likely to be suffocated by the oil.[11, 46, 47]

Biological effects on some marine life forms can be expected to continue for up to two years. Hydrocarbons are metabolized by some sea organisms, but this phenomenon is not entirely understood. Certain marine fishes and animals can metabolize hydrocarbons, thus minimizing the effects of the crude oil spill after a relatively short period of time. Some hydrocarbons are not easily oxidized by certain hydrocarbon-metabolizing systems. In many shellfish, the hydrocarbons persist in the muscle tissue and other organs for several weeks and possibly months. Their consumption by humans is therefore limited until these hydrocarbons have been eliminated.

Oil spills result in fish kills of varying degrees. In some respects fish are less vulnerable to oil spills than other marine life because they can vacate the area of the spill. Further, fish surfaces are coated with an oil-repellent mucus material which tends to inhibit oil problems. Fish which inhabit the ocean close to the surface are more likely to be adversely affected by oil toxicity and suffocation impacts.

Crude and other oils have been shown to harm or kill fish eggs. Crude oil concentrations of 1 ppm are hazardous to anchovy, scorpion fish, and sea parrot eggs. Fish respiration is adversely influenced by gill tissue damage and clogged gills. Other physiological activities are also disrupted.

Potentially, the most dangerous oil spills to fish are the sublethal discharges. These discharges are relatively small and can possibly result in widespread damage to fish populations over a long time period. Very low concentrations of hydrocarbons (*i.e.,* 10-100 ppb) have been shown to disrupt the feeding, reproduction, and social activities of fish. These low concentrations may interfere with fish taste receptors and possibly with the natural chemical messengers which attract predators to their prey. Further, inherent migratory and homing detection traits can be misaligned.

Fish whose ecosystems involve estuarine or other fresh water systems can be especially affected by an oil spill. Spills during the spawning season could result in the elimination of entire species through the prevention of egg hatching and the destruction of fry.

Marine flora are affected by oil spills to varying degrees. Marine algae are somewhat resistant to oil, especially those with a mucilage coating. Algae, even after a large initial kill, can be expected to recolonize impacted areas rapidly after the disturbance. Certain green algae forms can regenerate more quickly than brown algae forms. Severe and long-lasting damage to sea grasses can also result from an oil spill.

Many of the biological impacts of an oil spill are unknown. Subtle long- and short-term effects can be potentially devastating to ecosystems. However, the recoverability of marine life after the disturbance of an oil spill has been documented although several years may be required.[11]

OIL SPILL PREVENTION

Obviously, the best way of minimizing the impact of an oil spill on marine life is to prevent its occurrence. The Spill Prevention Control and Countermeasure Plans (SPCC) are required by federal law for all state, local and federal installations where oil handling and storage processes take place. Spill prevention methods as specified in these plans (including dikes and other preventive techniques) must be certified as adequate by a professional engineer and will be inspected by the EPA which has the power to assess fines for inadequate facilities.

Estimates show that up to 10 million tons of oil enter the oceans annually. However, the major impacts of oil are generated when large amounts of oil enter a relatively small area. Various sources of oil to the oceans are given in Figure 49 and Table 40. Tankers, offshore wells, and pipelines are of primary interest with regard to oil spills. Evidence has shown that most of the spills (60-70%) are the result of human error, but that the technical-error spills are much more devastating. Technical-error spills result in almost 75% of the quantities of oil introduced to the oceans. Improved technology can therefore eliminate much of the oil pollution.

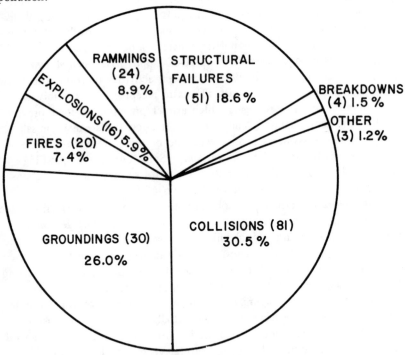

Figure 49. Oil pollution sources of the oceans.[38]

Human failure can be minimized through the introduction of instrumentation monitoring systems. For example, many of our tanker-filling operations are controlled by an operator looking into the tanker allage (gas vent) to see when the oil level has reached the top. This is an absurdity especially when considering that over 40% of the spills at marine terminals are the result of neglected tank overflows.

Table 40. Annual Ocean Oil Pollution Estimates[48]

Marine Operations	Metric Tons	Percent
Tankers		
LOT tank-cleaning operations	265,000	5.41
Non-LOT tank-cleaning operations	702,000	14.34
Discharge due to bilge pumping,		
leaks and bunkering spills	100,000	2.04
Vessel casualties	250,000	5.11
Terminal operations	70,000	1.42
Tank Barges		
Discharge due to leaks	20,000	0.41
Barge casualties	32,000	0.65
Terminal operations	18,000	0.38
All Other Vessels		
Discharge due to bilge pumping,		
leaks and bunkering spills	600,000	12.25
Vessel casualties	250,000	5.11
Offshore Operations	100,000	2.04
Nonmarine Operations		
Refineries and petrochemical plants	300,000	6.12
Industrial machinery	750,000	15.31
Highway motor vehicles	1,440,000	29.41
Total	4,897,000	100.00

TANKERS

As a result of accidents, carelessness or mismanagement, almost 36.5 million barrels of oil are annually discharged into the oceans during tanker operations. Figure 50 gives the present and projected quantities of oil transported by tankers. Oil transported by vessels is expected to continuously increase, thus making oil spill preventive measures imperative.

Approximately 98% of tanker oil spills involve the discharging of 1000 barrels of more.[47] Further, major tanker spills usually occur within 50 miles of the shoreline as the result of groundings, collisions and rammings. Figure 51 gives the types of tanker casualties and the frequency of their occurrence. Explosions, fires, tanker breakdowns, structural failures, and other accidents also contribute significantly to oil pollution of the oceans.

The fact that most of the major spills from tankers occur within a short distance of shore is related to the difficult maneuverability in

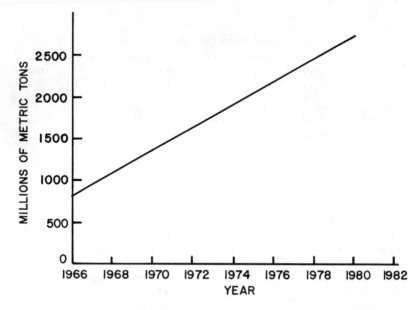

Figure 50.   Tanker oil quantities transported.[38]

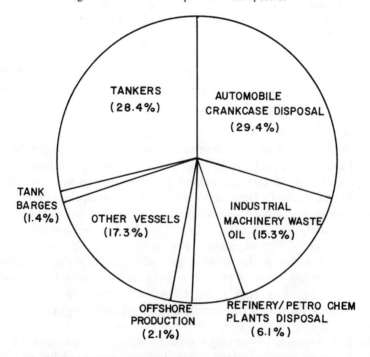

Figure 51.   Tanker oil discharges as a result of casualties.[38]

restricted and congested waterways. Potential oil spill preventive systems
have been designed which incorporate the concept of deep-water offshore
mooring of tankers.

Figure 52 illustrates the single-buoy mooring facility. This system utilizes a buoy which is anchored to the ocean floor. The tanker is connected
to the buoy with flexible hoses while the buoy has a pipeline extending
to the ocean bottom and then on to shore or a drilling platform. In
this manner the stresses on tankers from inclement weather conditions,
rough seas, strong currents and eddies can be minimized. Tankers are
able to rotate 360 degrees. This type of facility has been in use in many
areas outside the United States.[47]

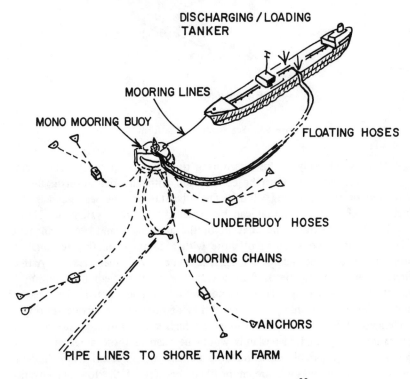

Figure 52. The single-buoy mooring facility.[39]

Figure 53 illustrates the sea island oil port. This facility is utilized in
the North Sea. The platform is set in place by pilings with tanker berths
on two sides. Oil pipelines are laid on the ocean floor from platform to
shore for tanker filling or withdrawals. The disadvantage of the sea island

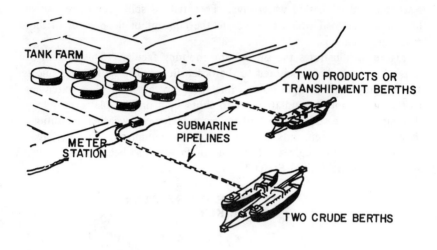

SEA ISLAND

Figure 53. Sea island oil platform.[39]

is that they normally are useless during times of rough oceans and inclement weather, and they must be constructed with an orientation in the direction of the prevailing winds. The advantage of the sea island is the high rate of oil transfer that can be achieved with this system.

Figure 54 illustrates the concept of the artificial island. A permanent island would be constructed offshore with a complete marine terminal, storage and docking facilities, barges, pipelines, and other vessels. Artificial islands can also be used in conjunction with single-buoy mooring facilities or sea islands. These islands would be constructed with weather and ocean dikes to inhibit the disruption of oil transfer services during inclement meteorological conditions and high seas. Although none are in operation today, artificial islands would be used in those areas of shallow ports where irreparable environmental damage would result from dredging or where tanker oil spills are more likely because of the low maneuverability of the tankers.

A significant technique of tanker oil spill prevention would be through improved tanker design. Aside from improving the tanker structurally, a modification of tanker compartment design could reduce oil pollution due to tanker casualty. Upon the rupturing of a tanker wall due to some sort of casualty, the entire compartment contents will spill into the ocean.

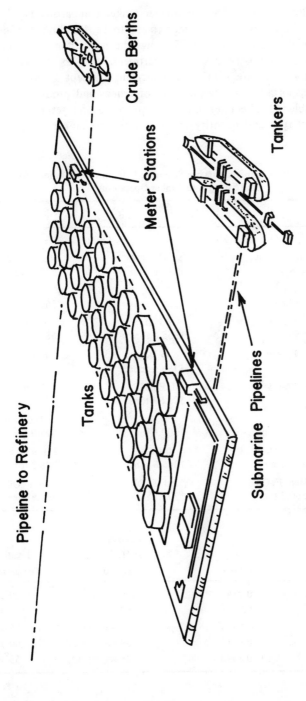

Figure 54.   Permanent artificial tanker port island.[39]

This has resulted in the calling for smaller tanker compartments. Regulations call for the largest tanker compartment to be 40,000 cubic meters or about 250,000 barrels. Figure 55 shows the manner in which tankers are divided into compartments. Compartments add to the safety and strength of the structure. (The compartment concept is also used to allow the transportation of several types of crudes and products without mixing.) Table 41 illustrates this concept by listing several oils being transported by type and quantity in their respective tanker compartment.

Table 41. Tanker Compartments can Segregate Different Oils Adding Efficiency and Strength to Tanker Operations[50]

Port	Center	Starboard
1.	Empty Empty	
2.	Med. F. Oil Med. F. Oil	Med. F. Oil
	436.4 tons 1314.6 tons	436.4 tons
3.	B.F.O. B.F.O.	B.F.O.
	564.8 tons 1355.9 tons	563.4 tons
4.	B.F.O. B.F.O.	B.F.O.
	586.9 tons 1355.8 tons	586.9 tons
5.	Empty G.O.	Empty
	1198.6 tons	
6.	Empty L.F.B.S.	Empty
	1212.7 tons	
7.	D.O. D.O.	D.O.
	529.0 tons 1200.6 tons	527.9 tons
8.	B.F.O. B.F.O.	B.F.O.
	271.8 tons 1345.8 tons	278.3 tons
9.	Med. F. Oil Med. F. Oil	Med. F. Oil
	500.2 tons 1295.1 tons	498.9 tons

Grade	S.G. at $60°$F	Viscosity at $100°$F Redwoods	Loading Temperature
B. F. O.	0.984	3400 sec	$120°$F
Med. F. Oil	0.9471	167 sec	$105°$F
D.O.	0.8473	37 sec	$55°$F
L.F.B.S.	0.8571	39 sec	$60°$F
G.O.	0.8343	35 sec	$53°$F

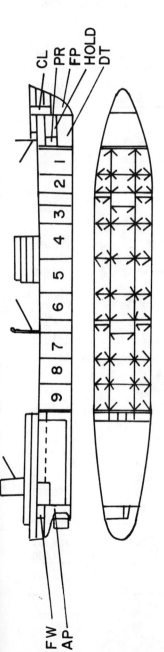

CL PR FP HOLD DT

FW AP

PARTICULARS

MARKING	DRAFT	DISPL	DW
TROPICAL FW	31'-3 1/2"	23013	16725
FRESH WATER	30'-8"	22478	16190
TROPICAL	30'-73/4	23013	16725
SUMMER	30'-01/4	22478	16190
WINTER	29'-43/4	21943	15655
WINTER NA	28'-11/2	21572	15284

DIMENSIONS

	MOULDED	EXTREME
LENGTH	515'-0"	547'-01/4"
BREADTH	69'-6"	69'-91/2"
DEPTH	37'-6"	

TONNAGE

	BRITISH	SUEZ	PANAMA
GROSS	11329·99		
UNDER OK	9973·66		
NET	6634·63		

Figure 55.   Top and side schematic of oil tanker.[40]

Tanker casualties caused by structural failure is a significant problem suggesting an inadequacy of tanker design. Table 42 describes the loss of structural integrity of several tankers in the 1969-1973 period. Reasons for the casualty and the amount of fuel allowed to discharge are given.

Table 42. Description of Tanker Structural Integrity Losses During 1969-1973 Period[51]

Description	Number	Oil Outflow (long tons)
A. Loss of structural integrity of hull primarily caused by external forces or where local material conditions deteriorated. No explosion or fire was associated with the accident. These may be broken down into:		
1. Structural failure of main hull girder from excess bending or shear loading	12	243,619
2. Local structural failure of hull envelope		
a. Failure of hull penetration	2	36,750
b. Local hull plating failure	2	39,169
c. Unknown local structure failure	1	34,000
3. Hull damage caused by collision or grounding		
a. Collision	2	4,138
b. Grounding	11	187,726
Subtotal	30	545,402
B. Loss of structural integrity from damage primarily caused by explosion or fire or where explosion or fire contributed to loss of structural integrity. These may be broken down into:		
1. Explosion or fire initiated in own ship cargo tanks	12	90,030
2. Explosion or fire set off by vessel collision or grounding		
a. Collision	4	136,163
b. Grounding	1	2,500
Subtotal	17	228,693

Explosions of tankers occur when the atmosphere within the tanker, a hydrocarbon-air mixture, is ignited. Most of the tanker explosions that have occurred are in empty tankers with large volumes of these explosive mixtures. A preventive measure would be to reduce the available oxygen within the tanker from the normal air content of 21% to less than 11%, thereby reducing the explosive risk. This can be accomplished through the introduction of inert gases to the tanker compartments. A system in effect in some tankers cycles the flue gas (4% O_2) from the tanker's engines into the oil storage compartments.[47]

Preventive measures against groundings and collisions both with other ships and structures include improved electronic navigation and operating practices. Complicated radar and harbor traffic control systems have become necessary as tankers grow larger and larger to the Very Large Crude Carrier (VLCC) size. Continuous communications may be required in congested areas or during berthing operations. Further, recent legislation requires ship-to-ship contact in the major harbors to make each captain aware of the position of his vessel relative to others.

OFFSHORE PLATFORMS

Most major offshore structures are designed to withstand the environmental conditions of the region of exploration. The 100-year storm forces are used as design criteria. Should these forces be exceeded, an oil spill is probable. The possibility of one storm occurring with forces greater than the 100-year storm is 26%.

Offshore structures can be designed to withstand the forces of specific natural events that will occur in the vicinity of the oil field. In the event of excessive storm forces or earthquakes, unprotected wells can blow out if the platform collapses shearing pipes and valves. Table 43 gives an estimate of platform collapse and well blowout possibilities in the Atlantic Ocean. Table 44 is a summary of the effects of natural phenomena on various elements of the oil production system. This summary is based on careful structural design and safety precautions with regard to the environment of the existing region and governmental criteria.

Safety valves and special casings are used to prevent oil blowouts. Cement casings are required to depths of at least 300 feet to withstand oil pressures in the event of a disruption. Properly designed, constructed and emplaced pipelines at the ocean floor are relatively insensitive to all natural phenomena. Most of the oil spills due to pipelines result from improper construction or corrosion-weakened pipewalls. Present pipe construction requires cathodic protection and/or protective coatings to minimize corrosion. In the event of leaks automatic shutdown valves can stop flow in faulty pipes. Further, scheduled inspections for weak spots in the pipe are necessary to prevent spill occurrences.

Table 43. An Estimate of Platform Collapse and Well Blowout Possibilities[11]

	Age of Field in Years			Remarks
	20	30	40	
Severe Storm Design Standard				
100-yr storm				
Margin of safety—1.5	0.09/0.0036/0.0009[a]	0.14/0.0056/0.0014	0.19/0.0076/0.0019	Average number of times severe storms will cause well blowout
Margin of safety—2.0	0.04/0.0016/0.0004	0.07/0.0028/0.0007	0.08/0.0032/0.0008	
200-yr storm				
Margin of safety—1.5	0.05/0.002/0.0005	0.07/0.0028/0.0007	0.09/0.0036/0.0009	
Margin of safety—2.0	0.02/0.0008/0.0002	0.03/0.0012/0.0003	0.04/0.0016/0.0004	
Earthquake Design				
M = 6.6				
Atlantic M_s—1.5	0.25/0.01/0.0025	0.38/0.015/0.0038	0.51/0.02/0.0051	Average number of times earthquakes will cause well blowout
Atlantic M_s—2.0	0.23/0.009/0.0023	0.35/0.014/0.0035	0.46/0.018/0.0046	
M = 7.2				
Atlantic M_s—1.5	0.10/0.004/0.0010	0.16/0.064/0.0016	0.21/0.008/0.0021	
Atlantic M_s—2.0	0.09/0.0036/0.0009	0.14/0.056/0.0014	0.18/0.007/0.0018	
Combined Severe Storm and Tsunami				
100-yr storm	0.001/0.00004/0.00001	0.0015/0.00006/0.000015	0.002/0.00008/0.00002	
200-yr storm	0.0005/0.00002/0.000005	0.0007/0.000028/0.000007	0.001/0.00004/0.00001	

aPlatform Collapse/Well Blowout R = 0.96/Well Blowout R = 0.99.

Table 44. Natural Phenomena Effects on an Offshore Oil Production System[11]

Element	Severe Storm	Earthquake Vibration	Earthquake Soil Stability	Tsunami	Volume of Oil at Risk per Event
Platform	Slight	Slight[a]	Slight[b]	None	500 to 1500 bbl/well/day
Pipeline	None	None	Variable[c]	None	10,000 bbl or more
Onshore Storage	Slight[d]	Slight[e]	Slight[b]	None	Up to 1,000,000[f] bbl or greater

[a]Provided earthquake-resistant design features are used.
[b]Provided careful soil analysis program is followed.
[c]The avoidance of slump, faulted, or poorly consolidated sediments can reduce potential impacts.
[d]Provided tanks are sited away from flood-prone areas.
[e]Provided free surface effect is reduced.
[f]Dikes give protection against damaging oil spill.

OIL SPILL CLEANUP

Once an oil spill has occurred, it must be cleaned up. There are several factors contributing to the fate of the oil after it enters a water body. Petroleum and petroleum products are composed of hydrocarbons, trace metals, sulfur, nitrogen and oxygen. Table 45 lists the percentages of the various petroleum and petroleum product components. An oil slick is subjected to several physical, chemical and biological processes. These can include spreading, dissolution, evaporation, emulsification, sedimentation photooxidation, biodegradation, and chemical and physical changes effected by the preceding processes.[11,46] Figure 56 identifies these natural forces relating to modification and dispersion of the oil spill.

Table 45. Percentage Components of the Various Oil Constituents[11]

	Paraffins	Cycloparaffins	Aromatics	NSO
Heavy Crude	10	45	25	20
Average Crude	30	50	15	5
#6 Fuel Oil (Bunker C)	15	45	25	15
#2 Fuel Oil	30	45	25	0
Gasoline	50	40	10	0
Kerosene	35	50	15	0

There are several approaches to oil spill cleanup depending on weather conditions, ocean conditions, and the environment of the spill region. An effective cleanup approach to an oil spill would:

1. lessen the initial adverse effects of the spill,
2. keep further oil damage to a minimum,
3. keep cleanup-resultant damage to a minimum,
4. provide for a regeneration of destroyed organisms,
5. restore the region to pre-spill conditions.[51]

Several materials are available which are utilized to control petroleum slicks on water.[11,47,48]

- Dispersants, or surface-active agents, promote the formation of fine oil droplets in water suspensions.
- Settling agents are materials added to the oil slick creating a mixture denser than water that will sink.
- Combustion agents are materials added to the oil slick creating a mixture that can be burned easily.

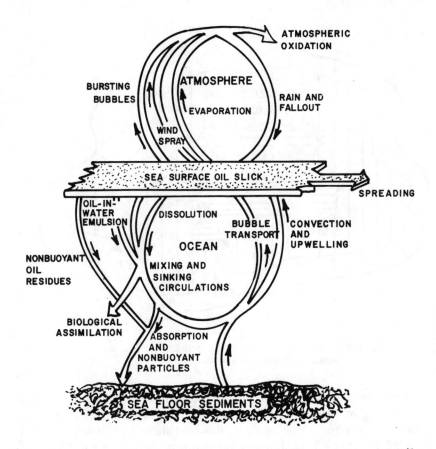

Figure 56. Modification and dispersion of an oil slick by natural phenomena.[41]

- Biodegradable agents are materials that enhance the oxidation of the oil slick by microorganisms.
- Gelling agents are materials which solidify the oil and promote ease of removal
- Sorping agents are materials which adsorb or absorb oil and are later removed from the water surface.
- Herding agents are materials added to the perimeter of the oil spill and force the spill into a small concentrated area (see Figure 57).

A major problem with some of the above oil slick control agents is that they can cause more environmental destruction than the oil itself. Therefore, before any chemical agent is used for oil spill cleanup, its ecological impact must be evaluated.

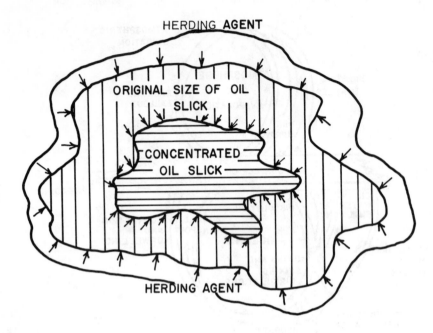

Figure 57. Herding agent forces the spreading oil slick into a smaller concentrated area for easier cleanup.

Barriers to control oil spills are diversified and in widespread use. Many refineries deploy barriers around water discharge points to inhibit small oil spills from spreading. Most of these, however, are in harbors where waters are relatively calm. Typical booms are shown in Figures 58 and 59. Rivers or open seas with swift moving currents make it more difficult to maintain the integrity of the barrier. Since the objective of the barrier is to hold back oil, it must remain vertical allowing the water to pass without the oil. This type of boom is strung around the spill site to contain the oil.

Another type of boom, as illustrated in Figure 60, is deployed not only to contain the oil spill, but can be used for recovery. A fabric is deployed between flotation devices. This fabric is impermeable to oil but allows water to pass through. The oil is then recovered with one of several available skimming devices as shown in Figure 61.

Deployment and positioning of the boom for an oil spill should be practiced and personnel should be trained in all facets of oil containment and recovery. Figure 62 shows how such a system might be put into operation in the event of a spill.

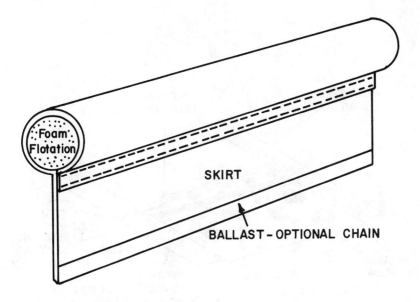

Figure 58. A typical oil boom to inhibit the spreading of oil.[37]

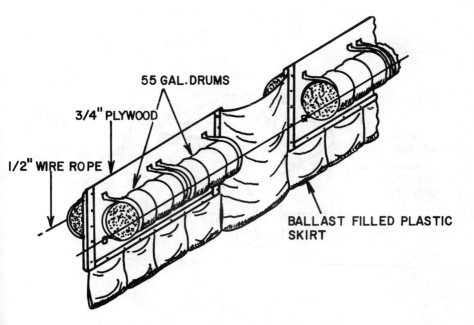

Figure 59. Another boom type, known as the Navy boom.[37]

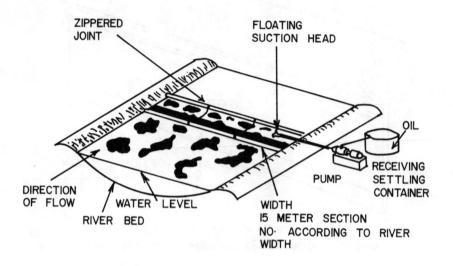

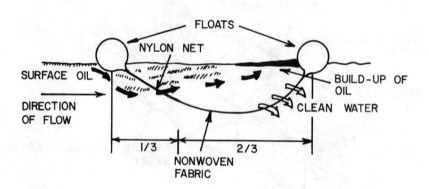

Figure 60. Oil containment and recovery boom[42]

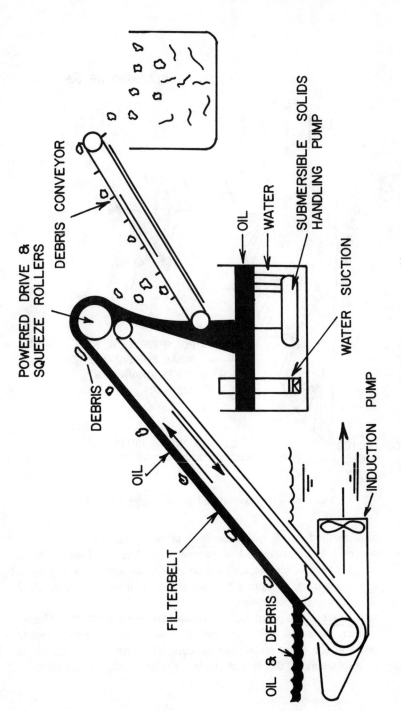

Figure 61. Oil skimming recovery device.[43]

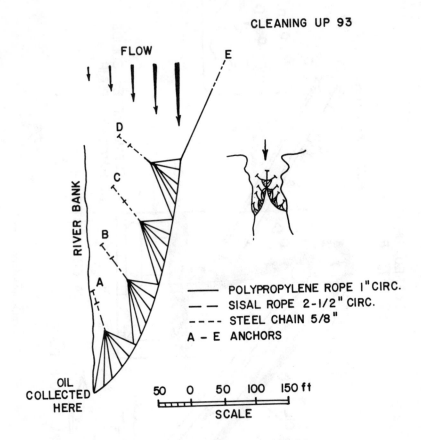

Figure 62.   Deployment and positioning of boom.

Another method of oil spill cleanup is primarily concerned with high seas recovery. A skimming and oil recovery vessel is deployed at the oil spill site. It sweeps over the spill collecting the oil and storing it for later disposal or processing. Figure 63 shows this system which can be either manually or remotely controlled.

Thus, the assessment must not only consider and anticipate environmental effects of an oil spill, but must also detail the preventive measures to be taken in the event an accident occurs. This description should include an assessment of the current technology to combat these effects.

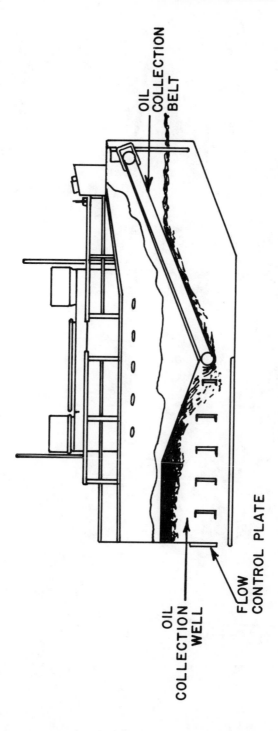

Figure 63. Vessel for oil recovery and storage.[44]

CHAPTER 9

NUCLEAR POWER

No environmental issue is as controversial or hotly contested as the nuclear power plant. Nuclear power, at one time the glamour energy source and cure-all, has come under attack as an unnecessary risk for disaster. The writing of an EIS is an imperative part of any proposal for nuclear energy use. Initially, we will discuss guidelines for preparing an EIS for a nuclear power installation.

THE PROPOSED FACILITY

This section of the EIS should deal with the purpose of the installation and aim to justify its need. Annual data with regard to the electrical requirements of the region to be serviced by the nuclear power plant should be supplied. Peak loading, total interruptible load, and expected load duration data including seasonal and daily variations are necessary. These data would be for the ten years prior to the EIS writing and for the projected regional demands on the installation for its expected life. To illustrate short-term system requirements, the expected annual load duration for at least two years following the start of the project's commercial operation should be supplied.[55]

The EIS is required to examine the proposed project's power system, the power pool or area of planning, and the region of plant location. Figure 64 is a map showing the region of concern for a New York State nuclear power plant. It identifies the service areas of the project. Table 46 gives past data and projects annual peak load demand through 1987. Forecasting methods should be described including assumptions of household formation, migration, personal income, industrial and commercial construction volume and location, and other factors. The system's capacity and its reserve capacity with respect to regional needs are described.

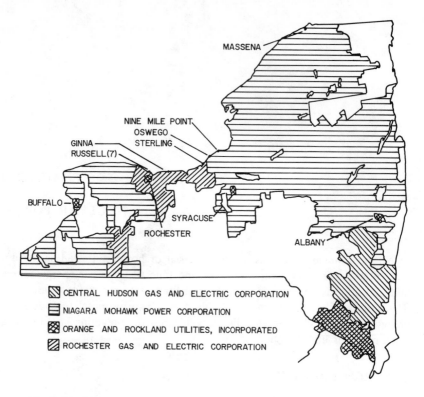

Figure 64. Region to be served by and the participating companies in the Sterling Power Plant Nuclear Unit of New York State.[56]

Details of generators and other system functions are necessary. The effects of construction delays on regional power requirements should be discussed also.

SITE LOCATION

The nuclear power plant's location is assessed with respect to the environmental parameters discussed in Chapter 3. All relevant information pertaining to land and water use requirements, historic, aesthetic and cultural landmarks, and the geology, topography, meteorology and ecology are presented. Included along with a map of the site area are explanations of the facility's property lines; identification of buildings and other structures; the location of industrial, recreational or residential structures; the minimum distance from each reactor to the site's boundary; and the location of highways, railways and waterways on, or adjacent to, the site.

Table 46. Annual Energy Requirements of the Region Projected Through 1987[56]

Year	RGE	Change (%)	OR.	Change (%)	CH	Change (%)	NM	Change (%)	Total	Change (%)
1964	2,534		1,219		1,745		20,570		26,068	
1965	2,797	10.4	1,351	10.8	1,950	11.7	22,068	7.3	28,166	8.0
1966	3,027	8.2	1,511	11.8	2,232	14.5	23,486	6.4	30,256	7.4
1967	3,329	10.0	1,652	9.3	2,325	4.2	24,028	2.3	31,334	3.6
1968	3,626	9.0	1,869	13.1	2,507	7.8	25,402	5.7	33,404	6.6
1969	3,966	9.4	2,083	11.5	2,750	9.7	26,712	5.2	35,511	6.3
1970	4,134	4.2	2,348	12.7	2,959	7.6	27,150	1.6	36,591	3.0
1971	4,382	6.0	2,553	8.8	3,127	5.7	27,543	1.4	37,605	2.8
1972	4,693	7.1	2,804	9.8	3,378	8.0	28,836	4.7	39,711	5.6
1973	4,928	5.0	2,971	6.0	3,530	4.5	30,457	5.6	41,886	5.5
1974	4,881	(1.0)	2,883	(3.0)	3,358	(4.9)	30,426	(0.1)	41,548	(0.1)
Forecast										
1975	5,058	3.6	3,094	7.3	3,685	9.7	31,674	4.1	43,511	4.7
1976	5,411	7.0	3,281	6.0	3,971	7.8	32,911	3.9	45,574	4.7
1977	5,789	7.0	3,592	9.5	4,281	7.8	34,175	3.8	47,839	5.0
1978	6,194	7.0	3,922	9.1	4,619	7.9	35,471	3.8	50,206	4.9
1979	6,626	7.0	4,258	8.6	4,982	7.8	36,796	3.7	52,662	4.9
1980	7,090	7.0	4,604	8.1	5,375	7.9	38,152	3.7	55,221	4.8
1981	7,498	5.8	4,952	7.6	5,806	8.0	39,528	3.6	57,784	4.6
1982	7,961	6.2	5,308	7.2	6,264	7.9	40,934	3.6	60,467	4.6
1983	8,435	6.0	5,675	6.9	6,752	7.8	42,370	3.5	63,232	4.7
1984	8,915	5.7	6,051	6.6	7,269	7.6	43,834	3.4	66,069	4.5
1985	9,413	5.6	6,435	6.3	7,823	7.6	45,329	3.4	69,000	4.4
1986	9,933	5.5	6,830	6.1	8,379	7.1	46,838	3.3	71,980	4.3
1987	10,480	5.5	7,233	5.9	8,943	6.7	48,375	3.3	75,031	4.2

The definition of site is the location of the nuclear facilities and surrounding lands which are controlled and restricted by licenses to limit the potential radiation doses of the normal operation of the facilities. For this reason the location of a nuclear power plant with respect to the population density of the surrounding areas is of major importance. Some ambiguities can be expected here since power shortages usually occur in the high-density metropolitan regions, but locating the installation in such a region can cause community concern. The effects on the ecology by radiation should be detailed also. The flora and fauna are catalogued. All the populations, habitats, communities, and ecosystems and their interrelationships are discussed.

THE NUCLEAR PLANT

The operating plant and transmission system are described in this section. A set of plant drawings and a detailed description are necessary. The architecture of the plant should be aesthetically pleasing, and the EIS should include an artist's rendition of the site's development plan.

The type of reactor and the electrical generation system are described. Since the radioactive environmental effects are of utmost importance in the EIS, particular concern is given to plant effluents and plant-related systems that interact with the environment. Plant water use, heat dissipation systems, radioactive waste systems, and other chemical and sanitary wastes are analyzed and all effluent release points are indicated on maps provided. Ranges of operational variations should be given as well as complete effluent inventories.

The EIS should contain sufficient data to allow a complete analysis of the effects of transmission lines and supporting facilities constructed between the nuclear installation and the existing electrical distribution systems. Included in this section is a discussion of parameters related to radiated electrical and acoustical noise, induced and conducted ground currents, and ozone production.

ENVIRONMENTAL EFFECTS OF SITE PREPARATION

The preparation of site and the construction of the nuclear plant and its support facilities will have both adverse and beneficial effects on the environment. An adverse effect would be one that causes a change or stress reducing the standard of living, lowering the quality of renewable resources, or impairing the recycling of depletable resources. Beneficial effects result in changes or stresses with opposite consequences than those previously mentioned.

The EIS should consider rights-of-way for construction of facilities and transmission lines, alternate rights-of-way, methodology of erecting structures, consideration of access roads and rail service, erosion attributable to construction activities, and loss of agricultural productivity. Further, measures designed to eradicate or reverse noise, dust, traffic, flooding, and surface and ground water modifications should be described. Any adverse environmental effects that are unavoidable and irreversible should be minimized and, if these losses occur, their effects should be evaluated with respect to short- and long-term impacts.

ENVIRONMENTAL EFFECTS OF PLANT OPERATION

The impacts of the operation of the proposed nuclear facility should be qualified and quantified and systematically presented. The discussion of each impact should be supported with data (included in the Appendix). An evaluation of this data should also be made with regard to its derivation, *i.e.,* whether it is based on theoretical, laboratory, or actual pilot plant and field analysis. Each impact is defined in terms of air, water and land effects.

The EIS should discuss the region affected by the nuclear installation with reference to its maintenance and enhancements for the effective life of the facility and beyond. Specifically, the EIS should consider:

- the effects of the operation of the heat dissipation system such as thermal pollution of surface waters, hydrology and meteorology, and evaporation of cooling tower waters including visible plume impacts;
- applicable state and federal water quality standards—including thermal discharges and their effects on water quality;
- the physical and biological effects of thermal pollution;
- the effects on man and the flora and fauna of the region due to normal operating radiation releases. Include in this section estimates of dosage rates for the above and note the various life activities of the region (within 50-mile radius);
- the effects of sanitary and chemical waste discharges;
- the effects of the uranium fuel cycle including mining and waste disposal;
- the effects of decommissioning and dismantling the plant at the end of its useful life.

EFFLUENT AND ENVIRONMENTAL MEASUREMENTS AND MONITORING PROGRAMS

As part of the EIS effort, a program should be established to determine the characteristics of the proposed site and the surrounding region. These measurements create the environmental baseline needed to assess the impact of the nuclear facility.

Parameters to be measured have been discussed in previous chapters. Radiological monitoring is necessary to determine the possible effects of this unique parameter. A monitoring program is to be established for the nuclear facility, in the EIS, for the time of the operation. In this manner, background radiological effects and the input of the nuclear facility can be assessed. Sampling frequency, design, methodology and instrumentation for the collection and analysis are described. Further, information as to instrument accuracy, sensitivity and reliability is to be included.

Should the site of the proposed nuclear facility lie within the confines of a region with an environmental monitoring program that is not directly supported by the proposed installation, a comparison and exchange of data would be necessary.

ENVIRONMENTAL EFFECTS OF ACCIDENTS

Accident prevention through proper design and controlled manufacture and operation is obviously required to protect the atmosphere around the installation. Even with strong preventive measures and maintenance, the possibility of a serious occurrence exists. All environmental risks associated with possible accidents should be considered. In the Appendix are listed some of the environmental considerations set forth by the Congress for the uranium fuel cycle and the transportation of radioactive materials. Possible accidents should be assessed as to their relative seriousness, based on the release of radioactivity and the environmental consequences.

Table 47 is an insert originally published in the *Federal Register*, December 1, 1971 (36FR22851) discussing the assumptions made in assessing possible accidents in nuclear facilities. Nine classes of environmental consequences are given along with specific parameters for EIS preparation.

ECONOMIC AND SOCIAL EFFECTS OF
PLANT CONSTRUCTION AND OPERATION

Based on a productive life of 30 years or more the economic and social effects of the nuclear facility should be assessed. Shown in Table 48 are the primary benefits to be considered in a cost/benefit analysis. Other benefits, economic and social, may include: creation of temporary and permanent jobs; recreational possibilities enhanced due to public use of parks, artificially created cooling lakes, etc.; special plant architectural design improving area aesthetics; creation and improvement of transportation facilities; and increased knowledge and awareness of the installation's environmental setting.

Table 47.   Discussion of Accidents in Applicants' Environmental Reports:
Assumptions

The complete text of the proposed Annex to Appendix D, 10 CFR Part 50, follows.
It was originally published in the *Federal Register* December 1, 1971 (36 FR 22851).

This Annex requires certain assumptions to be made in discussion of accidents in
Environmental Reports submitted pursuant to Appendix D by applicants[a] for construction permits or operating licenses for nuclear power reactors.[b]

In the consideration of the environmental risks associated with the postulated accident, the probabilities of their occurrence and their consequences must both be taken
into account. Since it is not practicable to consider all possible accidents, the spectrum of accidents, ranging in severity from trivial to very serious, is divided into
classes.

Each class can be characterized by an occurrence rate and a set of consequences.

Standardized examples of classes of accidents to be considered by applicants in
preparing the section of Environmental Reports dealing with accidents are set out in
tabular form below. The spectrum of accidents, from the most trivial to the most
severe, is divided into nine classes, some of which have subclasses. The accidents
stated in each of the eight classes in tabular form below are representative of the
types of accidents that must be analyzed by the applicant in Environmental Reports;
however, other accident assumptions may be more suitable for individual cases.
Where assumptions are not specified, or where those specified are deemed unsuitable,
assumptions as realistic as the state of knowledge permits shall be used, taking into
account the specific design and operational characteristics of the plant under consideration.

For each class, except Classes 1 and 9, the environmental consequences shall be
evaluated as indicated. Those classes of accidents, other than Classes 1 and 9, found
to have significant adverse environmental effects shall be evaluated as to probability,
or frequency of occurrence to permit estimates to be made of environmental risk or
cost arising from accidents of the given class.

Class 1 events need not be considered because of their trivial consequences.

Class 8 events are those considered in safety analysis reports and AEC staff safety evaluations. They are used, together with highly conservative assumptions, as the
design-basis events to establish the performance requirements of engineered safety features. The highly conservative assumptions and calculations used in AEC safety evaluations are not suitable for environmental risk evaluation, because their use would
result in a substantial overestimate of the environmental risk. For this reason, Class
8 events shall be evaluated realistically. Consequences predicted in this way will be
far less severe than those given for the same events in safety analysis reports where
more conservative evaluations are used.

The occurrences in Class 9 involve sequences of postulated successive failures more
severe than those postulated for establishing the design basis for protective systems
and engineered safety features. Their consequences could be severe. However, the
probability of their occurrence is so small that their environmental risk is extremely
low. Defense in depth (multiple physical barriers), quality assurance for design, manufacture, and operation, continued surveillance and testing, and conservative design
are all applied to provide and maintain the required high degree of assurance that potential accidents in this class are, and will remain, sufficiently remote in probability
that the environmental risk is extremely low. For these reasons, it is not necessary
to discuss such events in applicants' Environmental Reports.

Table 47, continued.

Furthermore, it is not necessary to take into account those Class 8 accidents for which the applicant can demonstrate that the probability has been reduced and therefore the calculated risk to the environment made equivalent to that which might be hypothesized for a Class 9 event.

Applicant may substitute other accident class breakdowns and alternative values of radioactive material releases and analytical assumptions, if such substitution is justified in the Environmental Report.

Accident Assumptions
Table of Contents

Accident
1.0 Trivial incidents
2.0 Small releases outside containment.
3.0 Radwaste system failure.
    3.1 Equipment leakage or malfunction.
    3.2 Release of waste gas storage tank contents.
    3.3 Release of liquid waste storage tank contents.
4.0 Fission products to primary system (BWR).
    4.1 Fuel cladding defects.
    4.2 Off-design transients that induce fuel failures above those expected.
5.0 Fission products to primary and secondary systems (PWR).
    5.1 Fuel cladding defects and steam generator leaks.
    5.2 Off-design transients that induce fuel failure above those expected and steam generator leak.
    5.3 Steam generator tube rupture.
6.0 Refueling accidents.
    6.1 Fuel bundle drop.
    6.2 Heavy object drop onto fuel in core.
7.0 Spent fuel handling accident.
    7.1 Fuel assembly drop in fuel storage pool.
    7.2 Heavy object drop onto fuel rack.
    7.3 Fuel cask drop.
8.0 Accident initiation events considered in design basis evaluation in the safety analysis report.
    8.1 Loss-of-coolant accidents.
    8.1(a) Break in instrument line from primary system that penetrates the containment.
    8.2(a) Rod ejection accident (PWR).
    8.2(b) Rod drop accident (BWR).
    8.3(a) Steamline breaks (PWRs outside containment).
    8.3(b) Steamline breaks (BWR).

Accident Assumptions
Accident—1.0 Trivial Incidents

These incidents shall be included and evaluated under routine releases in accordance with proposed Appendix 1.[c]

Table 47, continued.

Accident–2.0 Small Release Outside Containment

These releases shall include such things as releases through steamline relief valves and small spills and leaks of radioactive materials outside containment. These releases shall be included and evaluated under routine releases in accordance with proposed Appendix 1.

Accident–3.0 Radwaste System Failure

3.1 *Equipment leakage or malfunction* (includes operator error).
    (a) Radioactive gases and liquids: 25% of average inventory in the largest storage tank shall be assumed to be released.
    (b) Meteorology assumptions–χ/Q values are to be 1/10 of those given in AEC Safety Guide No. 3 or 4.^{d}
    (c) Consequences should be calculated by weighting the effects in different directions by the frequency the wind blows in each direction.

3.2 *Release of waste gas storage tank contents* (includes failure of release valve and rupture disks).
    (a) 100% of the average tank inventory shall be assumed to be released.
    (b) Meteorology assumptions: χ/Q values shall be 1/10 of those given in Safety Guide No. 3 or 4.
    (c) Consequences should be calculated by weighting the effects in different directions by the frequency the wind blows in each direction.

3.3 *Release of liquid waste storage tank contents*
    (a) Radioactive liquids: 100% of the average storage tank inventory shall be assumed to be spilled on the floor of the building.
    (b) Building structure shall be assumed to remain intact.
    (c) Meteorology assumptions: χ/Q values shall be 1/10 of those given in AEC Safety Guide No. 3 or 4.
    (d) Consequences should be calculated by weighting the effects in different directions by the frequency the wind blows in each direction.

Accident–4.0 Fission Products to Primary System (BWR)

4.1 *Fuel cladding defects.*
    Release from these events shall be included and evaluated under routine releases in accordance with proposed Appendix I.

4.2 *Off-design transients that induce fuel failures above those expected* (such as flow blockage and flux maldistributions).
    (a) 0.02% of the core inventory of noble gases and 0.02% of the core inventory of halogens shall be assumed to be released into the reactor coolant.
    (b) 1% of the halogens in the reactor coolant shall be assumed to be released into the steamline.
    (c) The mechanical vacuum pump shall be assumed to be automatically isolated by a high radiation signal on the steamline.
    (d) Radioactivity shall be assumed to carry over to the condenser where 10% of the halogens shall be assumed to be available for leakage

Table 47, continued.

from the condenser to the environment at 0.5%/day for the course of the accident (24 hours).

(e) Meteorology assumptions—χ/Q values shall be 1/10 of those given in AEC Safety Guide No. 3 dated November 2, 1970.

(f) Consequences should be calculated by weighting the effects in different directions by the frequency the wind blows in each direction.

Accident—5.0 Fission Products to Primary and Secondary Systems (Pressurized Water Reactor)

5.1 *Fuel cladding defects and steam generator leak.*

Release from these events shall be included and evaluated under routine releases in accordance with proposed Appendix I.

5.2 *Off-design transients that induce fuel failure above those expected and steam generator leak* (such as flow blockage and flux maldistributions).

(a) 0.02% of the core inventory of noble gases and 0.02% of the core inventory of halogens shall be assumed to be released into the reactor coolant.

(b) Average inventory in the primary system prior to the transient shall be based on operation with 0.5% failed fuel.

(c) Secondary system equilibrium radioactivity prior to the transient shall be based on a 20 gal/day steam generator leak and a 10 gpm blowdown rate.

(d) All noble gases and 0.1% of the halogens in the steam reaching the condenser shall be assumed to be released by the condenser air ejector.

(e) Meteorology assumptions: χ/Q values should be 1/10 of those given in AEC Safety Guide No. 4.

(f) Consequences should be calculated by weighting the effects in different directions by the frequency the wind blows in each direction.

5.3 *Steam generator tube rupture.*

(a) 15% of the average inventory of noble gases and halogens in the primary coolant shall be assumed to be released into the secondary coolant.

The average primary coolant activity shall be based on 0.5% failed fuel.

(b) Equilibrium radioactivity prior to rupture shall be based on a 20 gallon per day steam generator leak and a 10 gpm blowdown rate.

(c) All noble gases and 0.1% of the halogens in the steam reaching the condenser shall be assumed to be released by the condenser air ejector.

(d) Meteorology assumptions: χ/Q values shall be 1/10 of those given in AEC Safety Guide No. 4.

(e) Consequences should be calculated by weighting the effects in different directions by the frequency the wind blows in each direction.

Table 47, continued.

Accident—6.0 Refueling Accidents

6.1   *Fuel bundle drop.*

(a) The gap activity (noble gases and halogens) in one row of fuel pins shall be assumed to be released into the water. (Gap activity is 1% of total activity in a pin.)

(b) One week decay time before the accident occurs shall be assumed.

(c) Iodine decontamination factor in water shall be 500.

(d) Charcoal filter efficiency for iodines shall be 99%.

(e) A realistic fraction of the containment volume shall be assumed to leak to the atmosphere prior to isolating the containment.

(f) Meteorology assumptions: X/Q values shall be 1/10 of those given in AEC Safety Guide No. 3 or 4.

(g) Consequences should be calculated by weighting the effects in different directions by the frequency the wind blows in each direction.

6.2   *Heavy object drop onto fuel in core.*

(a) The gap activity (noble gases and halogens) in one average fuel assembly shall be assumed to be released into the water. (Gap activity shall be 1% of total activity in a pin.)

(b) 100 hours of decay time before object is dropped shall be assumed.

(c) Iodine decontamination factor in water shall be 500.

(d) Charcoal filter efficiency for iodines shall be 99%.

(e) A realistic fraction of the containment volume shall be assumed to leak to the atmosphere prior to isolating the containment.

(f) Meteorological assumptions: X/Q values shall be 1/10 of those given in AEC Safety Guide No. 3 or 4.

(g) Consequences should be calculated by weighting the effects in different directions by the frequency the wind blows in each direction.

Accident—7.0 Spent Fuel Handling Accident

7.1   *Fuel assembly drop in fuel storage pool.*

(a) The gap activity (noble gases and halogens) in one row of fuel pins shall be assumed to be released into the water. (Gap activity shall be 1% of total activity in a pin.)

(b) One week decay time before accident occurs shall be assumed.

(c) Iodine decontamination factor in water shall be 500.

(d) Charcoal filter efficiency for iodines shall be 99%.

(e) Meteorology assumptions: χ/Q values shall be 1/10 of those given in AEC Safety Guide No. 3 or 4.

(f) Consequences shall be calculated by weighting the effects in different directions by the frequency the wind blows in each direction.

7.2   *Heavy object drop onto fuel rack.*

(a) The gap activity (noble gases and halogens) in one average fuel assembly shall be assumed to be released into the water. (Gap activity is 1% of total activity in a pin.)

(b) 30 days decay time before the accident occurs shall be assumed.

Table 47, continued.

(c) Iodine decontamination factor in water shall be 500.

(d) Charcoal filter efficiency for iodines shall be 99%.

(e) Meteorology assumptions: χ/Q values shall be 1/10 of those given in AEC Safety Guide No. 3 or 4.

(f) Consequences should be calculated by weighting the effects in different directions by the frequency the wind blows in each direction.

7.3   *Fuel cask drop.*

(a) Noble gas gap activity from one fully loaded fuel cask (120-day cooling) shall be assumed to be released. (Gap activity shall be 1% of total activity in the pins.)

(b) Meteorology assumptions: χ/Q values shall be 1/10 of those given in AEC Safety Guide No. 3 or 4.

(c) Consequences should be calculated by weighting the effects in different directions by the frequency the wind blows in each direction.

Accident–8.0 Accident Initiation Events Considered in Design Basis Evaluation in the Safety Analysis Report

8.1   *Loss-of-coolant accidents*

Small Pipe Break (6 in. or less)

(a) Source term: the average radioactivity inventory in the primary coolant shall be assumed. (This inventory shall be based on operation with 0.5% failed fuel.)

(b) Filter efficiencies shall be 95% for internal filters and 99% for external filters.

(c) 50% building mixing for boiling water reactors shall be assumed.

(d) For the effects of Plateout, Sprays, Decontamination Factor in Pool, and Core Sprays, the following reduction factors shall be assumed:
For pressurized water reactors–0.05 with chemical additives in sprays, 0.2 for no chemical additives.
For boiling water reactors–0.2

(e) A realistic building leak rate as a function of time shall be assumed.

(f) Meteorology assumptions: χ/Q values shall be 1/10 of those given in AEC Safety Guide No. 3 or 4.

(g) Consequences should be calculated by weighting the effects in different directions by the frequency the wind blows in each direction.

Large Pipe Break

(a) Source term: The average radioactivity inventory in the primary coolant shall be assumed. (This inventory shall be based on operation with 0.5% failed fuel), plus release into the coolant of:
For pressurized water reactors–2% of the core inventory of halogens and noble gases.
For boiling water reactors–0.2% of the core inventory of halogens and noble gases.

(b) Filter efficiencies shall be 95% for internal filters and 99% for external filters.

Table 47, continued.

(c) 50% building mixing for boiling water reactors shall be assumed.

(d) For the effects of Plateout, Containment Sprays, Core Sprays (values based on 0.5% of halogens in organic form), the following reduction factors shall be assumed:

For pressurized water reactors—0.05 with chemical additives in sprays, 0.2 for no chemical additives.

For boiling water reactors—0.2.

(e) A realistic building leak rate as a function of time and including design leakage of steamline valves in BWRs shall be assumed.

(f) Meteorology assumptions: χ/Q values shall be 1/10 of those given in AEC Safety Guide No. 3 or 4.

(g) Consequences should be calculated by weighting the effects in different directions by the frequency the wind blows in each direction.

8.1(a) Break in instrument line from primary system that penetrates the containment (lines not provided with isolation capability inside containment).

(a) The primary coolant inventory of noble gases and halogens shall be based on operation with 0.5% failed fuel.

(b) Release rate through failed line shall be assumed constant for the four-hour duration of the accident.

(c) Charcoal filter efficiency shall be 99%.

(d) Reduction factor from combined plateout and building mixing shall be 0.1.

(e) Meteorology assumptions: χ/Q values shall be 1/10 of those given in AEC Safety Guide No. 3.

(f) Consequences shall be calculated by weighting the effects in different directions by the frequency the wind blows in each direction.

8.2(a) Rod ejection accident (pressurized water reactor).

(a) 0.2% of the core inventory of noble gases and halogens shall be assumed to be released into the primary coolant plus the average inventory in the primary coolant based on operation with 0.5% failed fuel.

(b) 1% of the halogens in the reactor coolant shall be assumed to be released into the condenser.

(c) The mechanical vacuum pump shall be assumed to be automatically isolated by high radiation signal on the steamline.

(d) Radioactivity shall be assumed to carry over to the condenser where 10% of the halogens shall be assumed to be available for leakage from the condenser to the environment at 0.5%/day for the course of the accident (24 hours).

(e) Meteorology assumptions: χ/Q values shall be 1/10 of those given in AEC Safety Guide No. 3.

(f) Consequences should be calculated by weighting the effects in different directions by the frequency the wind blows in each direction.

8.3(a) Steamline breaks (pressurized water reactors—outside containment). Break size equal to area of safety valve throat.

Table 47, continued.

Small Break

(a) Primary coolant activity shall be based on operation with 0.5% failed fuel. The primary system contribution during the course of the accident shall be based on a 20 gal/day tube leak.

(b) During the course of the accident, a halogen reduction factor of 0.1 shall be applied to the primary coolant source when the steam generator tubes are covered; a factor of 0.5 shall be used when the tubes are uncovered.

(c) Secondary coolant system radioactivity prior to the accident shall be based on:

(1)   20 gallons per day primary-to-secondary leak.

(2)   Blowdown of 10 gpm.

(d) Volume of one steam generator shall be released to the atmosphere with an iodine partition factor of 10.

(e) Meteorology assumptions: χ/Q values shall be 1/10 of those given in AEC Safety Guide No. 4.

(f) Consequences shall be calculated by weighting the effects in different directions by the frequency the wind blows in each direction.

Large Break

(a) Primary coolant activity shall be based on operation with 0.5% failed fuel. The primary system contribution during the course of the accident shall be based on a 20 gal/day tube leak.

(b) A halogen reduction factor of 0.5 shall be applied to the primary coolant source during the course of the accident.

(c) Secondary coolant system radioactivity prior to the accident shall be based on:

(1)   20 gallons per day primary-to-secondary leak.

(2)   Blowdown to 10 gpm.

(d) Volume of one steam generator shall be assumed to be released to the atmosphere with an iodine partition factor of 10.

(e) Meteorology assumptions: χ/Q values shall be 1/10 of those in AEC Safety Guide No. 3.

(f) Consequences shall be calculated by weighting the effects in different directions by the frequency the wind blows in each direction.

8.3(b) *Steamline breaks (boiling water reactor)*

Small Pipe Break (of 1/4 ft^2)

(a) Primary coolant activity shall be based on operation with 0.5% failed fuel.

(b) The main steamline shall be assumed to fail, releasing coolant until 5 seconds after isolation signal is received.

(c) Halogens in the fluid released to the atmosphere shall be at 1/10 the primary system liquid concentration.

(d) Meteorology assumptions: χ/Q values shall be 1/10 of those in AEC Safety Guide No. 3.

(e) Consequences shall be calculated by weighting the effects in different directions by the frequency the wind blows in each direction.

Table 47, continued.

Large Break

(a) Primary coolant activity shall be based on operation with 0.5% failed fuel.

(b) Main steamline shall be assumed to fail, releasing that amount of coolant corresponding to a 5 seconds isolation time.

(c) 50% of the halogens in the fluid exiting the break shall be assumed to be released to the atmosphere.

(d) Meteorology assumptions: χ/Q values shall be 1/10 of those in AEC Safety Guide No. 3.

(e) Consequences shall be calculated by weighting the effects in different directions by the frequency the wind blows in each direction.

---

[a] Although this Annex refers to applicants' Environmental Reports, the current assumptions and other provisions thereof are applicable, except as the content may otherwise require, to AEC draft and final Detailed Statements.

[b] Preliminary guidance as to the content of applicants' Environmental Reports was provided in the Draft AEC Guide to the Preparation of Environmental Reports for Nuclear Power Plants dated February 19, 1971, a document made available to the public as well as to the applicant. Guidance concerning the discussion of accidents in environmental reports was provided to applicants in a September 1, 1971, document entitled "Scope of Applicants' Environmental Reports with Respect to Transportation, Transmission Lines, and Accidents," also made available to the public.

[c] 36 FR 11113, June 8, 1971.

[d] Copies of such guide(s) dated November 2, 1970, are available at the Commission's Public Document Room, 1717 H Street N.W., Washington, D.C., and on request to the Director, Division of Reactor Standards, U.S. Nuclear Regulatory Commission, Washington, D.C. 20555. (These two guides have been revised and reissued as Revision 2, Regulatory Guide 1.3, and Revision 2, Regulatory Guide 1.4, both dated June 1974. Copies of these guides may be obtained by request from the U.S. Nuclear Regulatory Commission, Washington, D.C. 20555, Attention: Director of Office of Standards Development.)

Tables 49 through 51 show the type of information that must be supplied with regard to plant capital, operating and maintenance, and electrical generation costs.[55]

ALTERNATE ENERGY SOURCES AND SITES

In this section of the EIS the basis for the selection of the site and nuclear power as an energy source must be made from the available alternate sites and energy sources. The alternatives are considered in light of the previously discussed measures in Chapter 3. The EIS should present

Table 48. Primary Benefits to be Considered in a Nuclear Power Plant
Cost Benefit Analysis[55]

Direct Benefits	Indirect Benefits (As Appropriate)

Expected Average Annual Generation (kWh)
Capacity (kW)
Proportional Distribution of Electrical Energy
  (Expected Annual Delivery in kWh)
    Industrial
    Commercial
    Residential
    Other
Expected Average Annual Btu (Millions) of
  Steam Sold from the Facility
Expected Average Annual Delivery of Other
  Beneficial Products (Appropriate Physical
  Units)
Annual Revenues from Delivered Benefits
  Electrical Energy Generated
  Steam Sold
  Other Products
	Taxes (Local, State, Federal)
	Research
	Regional Product
	Environmental Enhancement:
	  Recreation
	  Navigation
	  Air Quality:
	    SO_2
	    NO_x
	    Particulates
	    Others
	Employment
	Education
	Others

its process for site and plant type selection as it is related to alternative costs and benefits. Table 52 is a list of environmental factors to be used in comparing alternative station systems.

Relative information should be included with respect to availability of alternatives and their relative merits. Two classes of alternatives are considered:

1. those that meet the new power requirements without demanding a new generating facility, and
2. those that demand a new facility.

The one alternative that is always considered is: no project.

PLANT DESIGN ALTERNATIVES

In this section of the EIS the proposed project is described in light of alternatives considered. Thus, an evaluation of designs and systems is made in accordance with the following: circulating water system, intake system for circulating water, discharge system for circulating water, other cooling systems, biocide systems, chemical waste treatment, sanitary waste treatment, liquid radioactive wastes, gaseous radioactive wastes, transmission facilities, and other systems. Table 53 considers the cost information for nuclear and alternative power generation methods.

Table 49.   Total Direct Cost Estimate Sheet of Radwaste Treatment System for Light-Water-Cooled Nuclear Reactors[55]

Description of Augment _____

	Direct Cost (1975 $1000) Reactor			
Item	Labor	Equipment/ Materials	Total	Basis for Cost Estimate
1. Process Equipment				
2. Building Assignment				
3. Associated Piping Systems				
4. Instrumentation and Controls				
5. Electrical Service				
6. Spare Parts				
Subtotal				
7. Contingency				
8. Total Direct Costs				

Table 50.   Annual Operating and Maintenance Cost Estimate Sheet for Radwaste Treatment System for Light-Water-Cooled Nuclear Reactors[55]

Description of Augment _____

	Cost (1975 $1000) Reactor			
Item	Labor	Other	Total	Basis for Cost Estimate
1. Operating Labor, Supervisory and Overhead				
2. Maintenance Material and Labor				
3. Consumables, Chemicals, and Supplies				
4. Utilities and Services Waste Disposal Water Steam Electricity Building Services Other				
5. Total Operating and Maintenance Annual Cost				

Table 51.   Estimated Costs of Electrical Energy Generation

	Mills/Kilowatt-Hour
Fixed Charges[a]	
Cost of money	_____
Depreciation	_____
Interim replacements	_____
Taxes	_____
Fuel Cycle Costs[b]	
For fossil-fueled plants, costs of high-sulfur coal, low-sulfur coal, or oil	_____
For nuclear stations:	
Cost of U_3O_8 (yellowcake)	_____
Cost of conversion and enrichment	_____
Cost of conversion and fabrication of fuel elements	_____
Cost of processing spent fuel	_____
Carrying charge on fuel inventory	_____
Cost of waste disposal[c]	_____
Credit for plutonium or U-233	_____
Costs of Operation and Maintenance[d]	
Fixed component	_____
Variable component	_____
Costs of Insurance	
Property insurance	_____
Liability insurance	_____

[a] Give the capacity factor assumed in computing these charges, and give the total fixed-charge rate as a percentage of station investment
[b] Include shipping charges as appropriate. Give the heat rate in Btu/kilowatt-hour.
[c] If no costs are available, the applicant may use the cost assumptions as listed in the most recent publication of *Nuclear Digest.*
[d] Give separately the fixed component that in dollars per year does not depend on capacity factor and the variable component that in dollars per year is proportional to capacity factor.

ENVIRONMENTAL APPROVALS AND CONSULTATION

The EIS lists all licenses, permits and other approvals of plant construction and operations required by the governmental regulatory agencies. The status of efforts to obtain a water quality certification under Section 401 and discharge permits under Section 402 of the Federal Water Pollution Control Act (FWPCA) are discussed. Information required for radioactive source term calculations for two reactor types is listed in Tables 54 and 55.

Table 52.   Environmental Factors to be Used in Comparing Alternative Station Systems

Primary Impact	Population or Resources Affected	Description	Unit of Measure[a]	Method of Computation
1. Natural Surface Water Body	(Specify natural water body affected)			
1.1 Impingement or entrapment by cooling water intake structure	1.1.1 Fish[b]	Juveniles and adults are subject to attrition.	Percent of harvestable or adult population destroyed per year for each important species	Identify all important species as defined in Section 2.2. Estimate the annual weight and number of each species that will be destroyed. (For juveniles destroyed, only the expected population that would have survived naturally need be considered.) Compare with the estimated weight and number of the species population in the water body.
1.2 Passage through or retention in cooling systems	1.2.1 Phytoplankton and zooplankton	Plankton population (excluding fish) may be changed due to mechanical, thermal and chemical effects.	Percent changes in production rates and species	Field studies are required to estimate (1) the diversity and production rates of readily recognizable groups (e.g., diatoms, green algae, zooplankton) and (2) the mortality of organisms passing through the condenser and pumps. Include indirect effects[c] which affect mortality.

[a]Applicant may substitute an alternative unit of measure where convenient. Such a measure should be related quantitatively to the unit of measure shown in this table.
[b]*Fish* as used in this table includes shellfish and other aquatic invertebrates harvested by man.
[c]Indirect effects could include increased disease incidence, increased predation, interference with spawning, changed metabolic rates, hatching of fish out of phase with food organisms.

Table 52, continued.

Primary Impact	Population or Resources Affected	Description	Unit of Measure[a]	Method of Computation
	1.2.2. Fish	All life stages (eggs, larvae, etc.) that reach the condenser are subject to attrition.	Percent of harvestable or adult population destroyed per year for each impor-	Identify all important species as defined in Section 2.2. Estimate the annual weight and number of each species that will be destroyed. (For larvae, eggs, and juveniles destroyed, only the expected population that would have survived naturally need be considered.) Compare with the estimated weight and number of the species population in the water body.
1.3 Discharge area and thermal plume	1.3.1 Water quality, excess heat	The rate of dissipation of the excess heat, primarily to the atmosphere, will depend on both the method of discharge and the state of the receiving water (i.e., ambient temperature and water currents).	Acres and acre-feet	Estimate the average heat in Btu's per hour dissipated to the receiving water at full power. Estimate the water volume and surface areas within differential temperature isotherms of 2, 3 and 5°F under conditions that would tend, with respect to annual variations, to maximize the extent of the areas and volumes.
	1.3.2 Water quality,	Dissolved oxygen concentration of receiving waters may be modified as a consequence of changes in the water	Acre-feet	Estimate volumes of affected waters with concentrations below 5, 3 and 1 ppm under conditions that would tend to maximize the impact.

temperature, the translocation of water of different quality, and aeration.

1.3.3 Fish (nonmigratory)	Fish[b] may be affected directly or indirectly because of adverse conditions in the plume.	Net effect in pounds per year (as harvestable or adult fish by species of interest)	Field measurements are required to establish the average number and weight (as harvestable or adults) of important species (as defined in Section 2.2). Estimate their mortality in the receiving water from direct and indirect effects.[c]
1.3.4 Wildlife (including birds and aquatic and amphibious mammals and reptiles)	Suitable habitats (wetland or water surface) may be affected	Acres of defined habitat or nesting area	Determine the areas impaired as habitats because of thermal discharges, including effects on food resources. Document estimates of affected population by species.
1.3.5 Fish (migratory)	A thermal barrier may inhibit migration, both hampering spawning and diminishing the survival of returning fish.	Pounds per year (as adult or harvestable fish by species of interest)	Estimate the fraction of the stock that is prevented from reaching spawning grounds because of station operation. Prorate this directly to a reduction in current and long-term fishing effort supported by that stock. Justify estimate on basis of local migration patterns, experience at other sites, and applicable state standards.

Table 52, continued.

Primary Impact	Population or Resources Affected	Description	Unit of Measure[a]	Method of Computation
1.4 Chemical effluents	1.4.1 Water quality, chemical	Water quality may be impaired.	Acre-feet, %	The volume of water required to dilute the average daily discharge of each chemical to meet applicable water quality standards should be calculated. Where suitable standards do not exist, use the volume required to dilute each chemical to a concentration equivalent to a selected lethal concentration for the most important species (as defined in Section 2.2) in the receiving waters. The ratio of this volume to the annual minimum value of the daily net flow, where applicable, of the receiving waters should be expressed as a percentage and the largest such percentage reported. Include the total solids if this is a limiting factor. Include in this calculation the blowdown from cooling towers and other closed-cycle cooling systems.

1.4.2 Fish	Aquatic populations may be affected by toxic levels of discharged chemicals or by reduced dissolved oxygen concentrations.	Pounds per year (by species of fish)	Total chemical effect on important species of aquatic biota should be estimated. Biota exposed within the facility, as well as biota in receiving waters, should be considered. Supporting documentation should include reference to applicable standards, chemicals discharged, and their toxicity to the aquatic populations affected.
1.4.3 Wildlife (including birds and aquatic and amphibious mammals and reptiles.	Suitable habitats for wildlife may be affected.	Acres	Estimate the area of wetland or water surface impaired as a wildlife habitat because of chemical contamination, including effects on food resources. Document the estimates of affected population by species.
1.4.4 People	Recreational water uses (boating, fishing, swimming) may be inhibited.	Lost annual user days and area (acres) or shoreline miles for dilution	The volume of the net flow to the receiving waters required for dilution to reach accepted water quality standards must be determined on the basis of daily discharge and converted to either surface area or miles of shore. Cross-sectional and annual minimum flow characteristics should be incorporated where applicable. The annual number of visitors to the affected area or shoreline

Table 52, continued.

Primary Impact	Population or Resources Affected	Description	Unit of Measure[a]	Method of Computation
				must be obtained. This permits estimation of lost user-days on an annual basis. Any possible eutrophication effects should be estimated and included as a degradation of quality.
1.5 Radionuclides discharged to water	1.5.1 Aquatic organisms	Radionuclide discharge may introduce a radiation level that adds to natural background radiation.	Rad per year	Sum dose contributions from radionuclides expected to be released.
	1.5.2 People, external	Radionuclide discharge may introduce a radiation level that adds to natural background radiation for water users.	Rem per year for individual; man-rem per year for estimated population at the midpoint of station operation	Sum annual dose contributions from nuclides expected to be released.
	1.5.3 People, ingestion	Radionuclide discharge may introduce a radiation level that adds to natural background radiation for ingested food and water.	Rem per year for individuals (whole body and organ); man-rem per year for population at the midpoint of station operation.	Estimate biological accumulation in foods and intake by individuals and population. Calculate doses by summing results for expected radionuclides.

1.6 Consumptive use	1.6.1 People	Drinking water supplies drawn from the water body may be diminished.	Gallons per year	Where users withdraw drinking water supplies from the affected water body, lost water to users should be estimated. Relevant delivered costs of replacement drinking water should be included.
	1.6.2 Agriculture	Water may be withdrawn from agricultural usage, and use of remaining water may be degraded	Acre-feet per year	Where users withdraw irrigation water from the affected water body, the loss should be evaluated as the sum of two volumes: the volume of the water lost to agricultural users and the volume of dilution water required to reduce concentrations of dissolved solids in station effluent water to an agriculturally acceptable level.
	1.6.3 Industry	Water may be withdrawn for industrial use.	Gallons per year	
1.7 Plant construction (including site preparation)	1.7.1 Water quality, physical	Turbidity, color or temperature of natural water body may be altered.	Acre-feet and acres	The volume of dilution water required to meet applicable water quality standards should be calculated. The areal extent of the effect should be estimated.
1.7	1.7.2 Water quality, chemical	Water quality may be impaired.	Acre-feet, %	To the extent possible, the applicant should treat problems of spills and drainage during construction in the same manner as in Item 1.4.1.

Table 52, continued.

Primary Impact	Population or Resources Affected	Description	Unit of Measure[a]	Method of Computation
1.8 Other impacts				The applicant should describe and quantify any other environmental effects of the proposed station that are significant.
1.9 Combined or interactive effects				Where evidence indicates that the combined effect of a number of impacts on a particular population or resource is not adequately indicated by measures of the separate impacts, the total combined effect should be described.
1.10 Net effects				See discussion in Section 5.7.
2. Ground Water				
2.1 Raising/lowering of ground water levels	2.1.1 People	Availability or quality of drinking water may be decreased, and the functioning of existing wells may be impaired.	Gallons per year	Volume of replacement water for local wells actually affected should be estimated.
	2.1.2 Vegetation	Trees and other deep-rooted vegetation may be affected.	Acres	Estimate the area in which ground water level change may have an adverse effect on local vegetation. Report this acreage on a separate schedule by land

use. Specify such uses as recreational, agricultural and residential.

No.	Item	Sub-item	Description	Units	Instructions
2.2	Chemical contamination of ground water (excluding salt)	2.2.1 People	Drinking water of nearby communities may be affected.	Gallons per year	Compute annual loss of potable water.
		2.2.2 Vegetation	Trees and other deep-rooted vegetation may experience toxic effects.	Acres	Estimate area affected and report separately by land use. Specify such uses as recreational, agricultural and residential.
2.3	Radionuclide contamination of ground water	2.3.1 People	Radionuclides that enter ground water may add to natural background radiation level for water and food supplies.	Rem per year for individuals (whole) body and organ); man-rem per year for population at the midpoint of station operation.	Estimate intakes by individuals and populations. Sum dose contributions for nuclides expected to be released.
		2.3.2 Vegetation and animals	Radionuclides that enter ground water may add to natural background radiation level for local plant forms and animal population.	Rad per year	Estimate uptake in plants and transfer to animals. Sum dose contributions for nuclides expected to be released.
2.4	Other impacts on ground water				The applicant should describe and quantify any other environmental effects of the proposed station that are significant.

Table 52, continued.

Primary Impact	Population or Resources Affected	Description	Unit of Measure[a]	Method of Computation
3. Air				
3.1 Fogging and icing (caused by evaporation and drift)	3.1.1 Ground transportation	Safety hazards may be created in the nearby regions in all seasons.	Vehicle-hours per year	Compute the number of hours per year that driving hazards will be increased on paved highways by fog and ice due to cooling towers and ponds. Documentation should include the visibility criteria used for defining hazardous conditions on the highways actually affected.
	3.1.2 Air transportation	Safety hazards may be created in the nearby regions in all seasons.	Hours per year, flights delayed per year	Compute the number of hours per year that commercial airports will be closed to visual (VFR) and instrumental (IFR) air traffic because of fog and ice from cooling towers. Estimate number of flights delayed per year.
	3.1.3 Water transportation	Safety hazards may be created in the nearby regions in all seasons.	Hours per year, number of ships affected per year.	Compute the number of hours per year ships will need to reduce speed because of fog from cooling towers or ponds or because of warm water added to the surface of the river, lake or sea.

3.1.4 Vegetation		Damage to timber and crops may occur through introduction of adverse conditions.	Acres by crop	Estimate the acreage of potential plant damage by crop.
3.2 Chemical discharge to ambient air	3.2.1 Air quality chemical	Pollutant emissions may diminish the quality of the local ambient air.	% and pounds or tons	The actual concentration of each pollutant in ppm for maximum daily emission rate should be expressed as a percentage of the applicable emission standard. Report weight for expected annual emissions.
	3.2.2 Air quality, odor	Odor in gaseous discharge or from effects on water body may be objectionable.	Statement	A statement must be made as to whether odor originating in station is perceptible at any point offsite.
3.3 Radionuclides discharged to ambient air and direct radiation from radioactive materials (in plant or being transported)	3.3.1 People, external	Radionuclide discharge or direct radiation may add to natural background radiation level.	Rem per year for individuals (whole body and organ); man-rem per year for population at the midpoint of station operation	Sum dose contributions from nuclides expected to be released.
	3.3.2 People, ingestion	Radionuclide discharge may add to the natural radioactivity in vegetation and in soil.	Rem per year for individuals (whole body and organ); man-rem per year for population at the midpoint of station operation	For radionuclides expected to be released, estimate deposit and accumulation in foods. Estimate intakes by individuals and populations and sum results for all expected radionuclides.

Table 52, continued.

Primary Impact	Population or Resources Affected	Description	Unit of Measure[a]	Method of Computation
	3.3.3 Vegetation and animals	Radionuclide discharge may add to natural background radioactivity of local plant and animal life.	Rad per year	Estimate deposit of radionuclides on and uptake in plants and animals. Sum dose contributions for radionuclides expected to be released.
3.4 Other impacts on air				The applicant should describe and quantify any other environmental effects of the proposed plant that are significant.
4. Land				
4.1 Site selection	4.1.1 Land, amount	Land will be preempted for construction of nuclear power station, station facilities and exclusion zone.	Acres	State the number of acres preempted for station, exclusion zone, and accessory facilities such as cooling towers and ponds. By separate schedule, state the type and class of land preempted (e.g., scenic shoreline, wet land, forest land, etc.).
4.2 Construction activities (including site preparation)	4.2.1 People (amenities)	There will be a loss of desirable qualities in the environment due to the noise and movement of men, material and machines.	Total population affected, years	The disruption of community life (or alternatively the degree of community isolation from such irritations) should be estimated. Estimate the number of residences, school, hospitals, etc.,

			within area of visual and audio impacts. Estimate the duration of impacts and total population affected.
4.2.2 People (accessibility of historical sites)	Historical sites may be affected by construction	Visitors per year	Determine historical sites that might be displaced by generation facilities. Estimate effect on any other sites in plant environs. Express net impact in terms of annual number of visitors.
4.2.3 People (accessibility of archeological sites)	Construction activity may impinge upon sites of archeological value.	Qualified opinion	Summarize evaluation of impact on archeological resources in terms of remaining potential value of the site. Referenced documentation should include statements from responsible county, state or federal agencies, if available.
4.2.4 Wildlife	Wildlife may be affected.	Qualified opinion	Summarize qualified opinion including views of cognizant local and state wildlife agencies when available, taking into account both beneficial and adverse effects.
4.2.5 Land (erosion)	Site preparation and station construction will involve cut and fill operations with accompanying erosion potential.	Cubic yards and acres	Estimate soil displaced by construction activity and erosion. Beneficial and detrimental effects should be reported separately.

Table 52, continued.

Primary Impact	Population or Resources Affected	Description	Unit of Measure[a]	Method of Computation
4.3 Station operation	4.3.1 People (amenities)	Noise may induce stress.	Number of residents, school populations, hospital beds	Use applicable state and local codes for offsite noise levels for assessing impact. If there is no code, consider nearby land use, current zoning, and ambient sound levels in assessing impact. The predicted sound level may be compared with the published guidelines of the EPA, AIHA and HUD.
	4.3.2 People (esthetics)	The local landscape as viewed from adjacent residential areas and neighboring historical, scenic and recreational sites may be rendered esthetically objectionable by station structures.	Qualified opinion	Summarize qualified opinion, including views of cognizant local and regional authorities when available.
	4.3.3 Wildlife	Wildlife may be affected.	Qualified opinion	Summarize qualified opinion, including views of cognizant local and state wildlife agencies when available, taking into account both beneficial and adverse effects.

4.3.4 Land, flood control	Salts discharged from cooling towers	Health and safety near the water body may be affected by flood control.	Reference to Flood Control District approval	Reference should be made to regulations of cognizant Flood Control Agency by use of one of the following terms: Has *No Implications* for flood control, *Complies* with flood control regulation.
4.4				
4.4.1 People		Intrusion of salts into ground water may affect water supply.	Pounds per square foot per year	Estimate the amount of salts discharged as drift and particulates. Report maximum deposition. Supporting documentation should include patterns of deposition and projection of possible effect on water supplies.
4.4.2 Vegetation and animals		Deposition of entrained salts may be detrimental in some nearby regions.	Acres	Salt tolerance of vegetation in affected area must be determined. That area, if any, receiving salt deposition in excess of tolerance (after allowance for dilution) must be estimated. Report separately an appropriate tabulation of acreage by land use. Specify such uses as recreational, agricultural and residential. Where wildlife habitat is affected, identify populations.
4.4.3 Property resources		Structures and movable property may suffer degradation from corrosive effects.	Dollars per year	If salt spray impinges upon a local community, property damage may be estimated by applying to the local value of

Table 52, continued.

Primary Impact	Population or Resources Affected	Description	Unit of Measure[a]	Method of Computation
4.5 Transmission route selection				buildings, machinery and vehicles a differential in average depreciation rates between this and a comparable seacoast community.
	4.5.1 Land, amount	Land will be pre-empted for construction of transmission line systems	Miles, acres	State total length and area of new rights-of-way. Estimate current market value of land involved.
	4.5.2 Land use and land value	Lines may pass through visually sensitive (that is, sensitive to presence of transmission lines and towers) areas, thus impinging on the present and potential use and value of neighboring property.	Miles, acres, dollars	Total length of new transmission lines and area of rights-of-way through various categories of visually sensitive land. Estimate minimum loss in current property values of adjacent areas.
	4.5.3 People (esthetics)	Lines may present visually undesirable features.	Number of such features	Estimate total number of visually undesirable features, such as number of major road crossings in vicinity of intersection of interchanges; number of major waterway crossings; number of crest, ridge, or other high point crossings; and number of "long views"

of transmission lines perpendicular to highways and waterways.

4.6	Transmission facilities construction			
4.6.1	Land adjacent to rights-of-way	Constructing new roads for access to rights-of-way may have environmental impact.	Miles	Estimate length of new access and service roads required for alternative routes.
4.6.2	Land, erosion	Soil erosion may result from construction activities.	Tons per year	Estimate area with increased erosion potential traceable to construction activities.
4.6.3	Wildlife	Wildlife habitat and access to habitat may be affected.	Number of important species affected	Identify important species that may be disturbed (Section 2.2).
4.6.4	Vegetation	Vegetation may be affected.		
4.7	Transmission line operation			
4.7.1	Land use	Land preempted by rights-of-way may be used for additional beneficial purposes such as orchards, picnic areas, nurseries, and hiking and riding trails.	%, dollars	Estimate percent of rights-of-way for which no multiple-use activities are planned. Annual value of multiple-use activities less cost of improvements.
4.7.2	Wildlife	Modified wildlife habitat may result in changes.	Qualified opinion	Summarize qualified opinion including views of cognizant local and state wildlife agencies when available.
4.8	Other land impacts			The applicant should describe and quantify any other environmental effects of the proposed station that are significant.

Table 52, continued.

Primary Impact	Population or Resources Affected	Description	Unit of Measure[a]	Method of Computation
4.9 Combined or inter-active effects				Where evidence indicates that the combined effects of a number of impacts on a particular population or resource are not adequately indicated by measures of the separate impacts, the total combined effect should be described. Both beneficial and adverse interactions should be indicated.
4.10 Net effects				See discussion in Section 5.7.

[a]Applicant may substitute an alternative unit of measure where convenient. Such a measure should be related quantitatively to the unit of measure shown in this table.

[b]Fish as used in this table includes shellfish and other aquatic invertebrates harvested by man.

[c]Indirect effects could include increased disease incidence, increased predation, interference with spawning, changes metabolic rates, hatching of fish out of phase with food organisms.

Table 53. Cost Information for Nuclear and Alternative Power Generation Methods

1. Interest during construction	_____ %/year,
	_____ compound rate
2. Length of construction workweek	_____ hours/week
3. Estimated site labor requirement	_____ man-hours/kWe
4. Average site labor pay rate (including fringe benefits) effective at month and year of NSSS order	_____ $/hour
5. Escalation rates	
Site labor	_____ %/year
Materials	_____ %/year
Composite escalation rate	_____ %/year

6. Power Station Cost[a]

Direct Costs	Unit 1	Unit 2	Indirect Costs	Unit 1	Unit 2
a. Land and land rights	___	___	a. Construction facilities, equipment and services	___	___
b. Structure and site facilities	___	___	b. Engineering and construction management services	___	___
c. Reactor (boiler) plant equipment	___	___	c. Other costs	___	___
d. Turbine plant equipment not including heat rejection systems	___	___	d. Interest during construction (@___%/ year)	___	___
e. Heat rejection system	___	___	**Escalation**		
f. Electric plant equipment	___	___	Escalation during construction @ ___%/year	___	___
g. Miscellaneous equipment	___	___			
h. Spare parts allowance	___	___	**Total Cost**		
i. Contingency allowance	___	___	Total Station Cost, @ Start of Commercial Operation	___	___
Subtotal	___	___			

[a]Cost components of nuclear stations to be included in each cost category listed under direct and indirect costs in Part 6 above are described in "Guide for Economic Evaluation of Nuclear Reactor Plant Designs," U.S. Atomic Energy Commission, NUS-531, Appendix B, available from National Technical Information Service, Springfield, Virginia 22161

Table 54. Data Needed for Radioactive Source Term Calculations
for Pressurized Water Reactors[55]

The applicant should provide the information listed. The information should be consistent with the contents of the safety analysis report (SAR) and the environmental report (ER) of the proposed pressurized water reactor (PWR). Appropriate sections of the SAR and ER containing more detailed discussions or backup data for the required information should be referenced following each response. Each response, however, should be independent of the ER and SAR.[a] This information constitutes the basic data required to calculate the releases of radioactive material in liquid and gaseous effluents (the source terms). All responses should be on a per-reactor basis. Indicate systems shared between reactors. The following data should be provided:

I. **General**

1. The maximum core thermal power (MWt) evaluated for safety considerations in the SAR. (Note: All of the following responses should be adjusted to this power level.)

2. Core properties:
  a. The total mass (lb) of uranium and plutonium in an equilibrium core (metal weight)
  b. The percent enrichment of uranium in reload fuel, and
  c. The percent of fissile plutonium in reload fuel.

3. If methods and parameters used in estimating the source terms in the primary coolant are different from those given in Regulatory Guide 1.112, "Calculation of Releases of Radioactive Materials in Gaseous and Liquid Effluents from Light-Water-Cooled Power Reactors," describe in detail the methods and parameters used. Include the following information:
  a. Station capacity factor,
  b. Fraction of fuel releasing radioactivity in the primary coolant (indicate the type of fuel cladding),
  c. Concentration of fission, activation and corrosion products in the primary and secondary coolant (μCi/g). Provide the bases for the values used.

4. The quantity of tritium released in liquid and gaseous effluents (Ci/yr per reactor.

II. **Primary System**

1. The total mass (lb) of coolant in the primary system, excluding the pressurizer and primary coolant purification system at full power.

2. The average primary system letdown rate (gpm) to the primary coolant purification system.

3. The average flow rate (gpm) through the primary coolant purification system cation demineralizers. (Note: The letdown rate should include the fraction of time the cation demineralizers are in service.)

4. The average shim bleed flow (gpm).

III.   Secondary System

1. The number and type of steam generators and the carryover factor used in the applicant's evaluation for iodine and nonvolatiles.

2. The total steam flow (lb/hr) in the secondary system.

3. The mass of steam in each steam generator (lb) at full power.

4. The mass of liquid in each steam generator (lb) at full power.

5. The total mass of coolant in the secondary system (lb) at full power. For recirculating U-tube steam generators, do not include the coolant in the condenser hotwell.

6. The primary to secondary system leakage rate (lb/day) used in the evaluation.

7. Description of the steam generator blowdown and blowdown purification systems. The average steam generator blowdown rate (lb/hr) used in the applicant's evaluation. The parameters used for steam generator blowdown rate (lb/hr).

8. The fraction of the steam generator feedwater processed through the condensate demineralizers and the decontamination factors (DF) used in the evaluation for the condensate demineralizer system.

9. Condensate demineralizers:
   a. Average flow rate (lb/hr),
   b. Demineralizer type (deep bed or powdered resin),
   c. Number and size (ft^3) of demineralizers,
   d. Regeneration frequency,
   e. Indicate whether ultrasonic resin cleaning is used and the waste liquid volume associated with its use, and
   f. Regenerant volume (gal/event) and activity.

IV.   Liquid Waste Processing Systems

1. For each liquid waste processing system (including the shim bleed, steam generator blowdown, and detergent waste processing systems), provide in tabular form the following information:
   a. Sources, flow rates (gpd) and expected activities (fraction of primary coolant activity, PCA) for all inputs to each system,
   b. Holdup times associated with collection, processing and discharge of all liquid streams,
   c. Capacities of all tanks (gal) and processing equipment (gpd) considered in calculating holdup times,
   d. Decontamination factors for each processing step,
   e. Fraction of each processing stream expected to be discharged over the life of the station,
   f. For demineralizer regeneration provide: time between regenerations, regenerant volumes and activities, treatment of regenerants, and fraction of regenerant discharged (include parameters used in making these determinations), and
   g. Liquid source term by radionuclide in Ci/yr for normal operation, including anticipated operational occurrences.

2. Piping and instrumentation diagrams (P&IDs) and process flow diagrams for the liquid radwaste systems along with all other systems influencing the source term calculations.

V. Gaseous Waste Processing System

1. The volumes (ft^3/yr) of gases stripped from the primary coolant.

2. Description of the process used to hold up gases stripped from the primary system during normal operations and reactor shutdown. If pressurized storage tanks are used, include a process flow diagram of the system indicating the capacities (ft^3), number, and design and operating storage pressures for the storage tanks.

3. Description of the normal operation of the system, *e.g.,* number of tanks held in reserve for back-to-back shutdown, fill time for tanks. Indicate the minimum holdup time used in the applicant's evaluation and the basis for this number.

4. If HEPA filters are used downstream of the pressurized storage tanks, provide the decontamination factor used in the evaluation.

5. If a charcoal delay system is used, describe this system and indicate the minimum holdup times for each radionuclide considered in the evaluation. List all parameters, including mass of charcoal (lb), flow rate (cfm), operating and dew point temperaturew, and dynamic adsorption coefficients for Xe and Kr used in calculating holdup times.

6. Piping and instrumentation diagrams (P&IDs) and process flow diagrams for the gaseous radwaste systems, along with other systems influencing the source term calculations.

VI. Ventilation and Exhaust Systems

For each building housing systems that contain radioactive materials, the steam generator blowdown system vent exhaust, and the main condenser air removal system, provide the following:

1. Provisions incorporated to reduce radioactivity releases through the ventilation or exhaust systems.

2. Decontamination factors assumed and the bases (include charcoal adsorbers, HEPA filters, mechanical devices).

3. Release rates for radioiodine, noble gases, and radioactive particulates (Ci/yr) and the bases.

4. Release points to the environment, including height, effluent temperature and exit velocity.

5. For the containment building, provide the building free volume (ft^3) and a thorough description of the internal recirculation system (if provided), including the recirculation rate, charcoal bed depth, operating time assumed, and mixing efficiency. Indicate the expected purge and venting frequencies and duration and continuous purge rate (if used).

VII. Solid Waste Processing Systems

1. In tabular form, provide the following information concerning all inputs to the solid waste processing system: source, volume (ft^3/yr per reactor), and activity (Ci/yr per reactor) of principal radionuclides, along with bases for values used.

2. Provide information on onsite storage provisions (location and capacity) and expected onsite storage times for all solid wastes prior to shipment.

3. Provide piping and instrumentation diagrams (P&IDs) for the solid radwaste system.

---

[a]The ER or SAR may be referenced as to the bases for the parameters used; however, the parameters should be given with the responses in this appendix.

Table 55. Data Needed for Radioactive Source Term Calculations for Boiling Water Reactors[55]

The applicant should provide the information listed. The information should be consistent with the contents of the safety analysis report (SAR) and the environmental report (ER) of the proposed boiling water reactor (BWR). Appropriate sections of the SAR and ER containing more detailed discussions of the required information should be referenced following each response. Each response, however, should be independent of the ER and SAR.[a] This information constitutes the basic data required to calculate the releases of radioactive material in liquid and gaseous effluents (the source terms). All responses should be on a per-reactor basis. Indicate systems shared between reactors. The following data should be provided:

I. General

1. The maximum core thermal power (MWt) evaluated for safety considerations in the SAR. (Note: All of the following responses should be adjusted to this power level.)

2. Core properties:
   a. The total mass (lb) of uranium and plutonium in an equilibrium core (metal weight),
   b. The percent enrichment of uranium in reload fuel, and
   c. The percent of fissile plutonium in reload fuel.

3. If methods and parameters used in estimating the source terms in the primary coolant are different from those given in Regulatory Guide 1.112, "Calculation of Releases of Radioactive Materials in Gaseous and Liquid Effluents from Light-Water-Cooled Power Reactors," describe in detail the methods and parameters used. Include the following information:
   a. Plant capacity factor,
   b. Isotopic release rates of noble gases to the reactor coolant at 30-minute decay (μCi/sec), and
   c. Concentration of fission, corrosion and activation products in the reactor coolant (μCi/sec). Provide the bases for the values used.

4. The quantity of tritium released in liquid and gaseous effluents (ci/yr per reactor).

II. Nuclear Steam Supply System

1. Total steam flow rate (lb/hr).

2. Mass of reactor coolant (lb) and steam (lb) in the reactor vessel at full power.

III. Reactor Coolant Cleanup System

1. Average flow rate (lb/hr).

2. Demineralizer type (deep bed or powdered resin).

3. Regeneration frequency.

4. Regenerant volume (gal/event) and activity.

IV. Condensate Demineralizers

1. Average flow rate (lb/hr).

2. Demineralizer type (deep bed or powdered resin).

3. Number and size (ft^3) of demineralizers.

4. Regeneration frequency.

5. Indicate whether ultrasonic resin cleaning is used and the waste liquid volume associated with its use.

6. Regenerant volume (gal/event) and activity.

V. Liquid Waste Processing Systems

1. For each liquid waste processing system, provide in tabular form the following information:

a. Sources, flow rates (gpd) and expected activities (fraction of primary coolant activity, PCA) for all inputs to each system,

b. Holdup times associated with collection, processing and discharge of all liquid streams,

c. Capacities of all tanks (gal) and processing equipment (gpd) considered in calculating holdup times,

d. Decontamination factors for each processing step,

e. Fraction of each processing stream expected to be discharged over the life of the station,

f. For waste demineralizer regeneration, time between regenerations, regenerant volumes and activities, treatment of regenerants, and fractions of regenerants discharged (include parameters used in making these determinations), and

g. Liquid source term by radionuclide in Ci/yr for normal operation, including anticipated operational occurrences.

2. Piping and instrumentation diagrams (P&IDs) and process flow diagrams for the liquid radwaste systems along with all other systems influencing the source term calculations.

VI. Main Condenser and Turbine Gland Seal Air Removal Systems

1. The holdup time (hr) for offgases from the main condenser air ejector prior to processing by the offgas treatment system.

2. Description and expected performance of the gaseous waste treatment systems for the offgases from the condenser air ejector and mechanical vacuum pump. The expected air inleakage per condenser shell, the number of condenser shells, and the iodine source term from the condenser.

3. The mass of charcoal (tons) in the charcoal delay system used to treat the offgases from the main condenser air ejector, the operating and dew point temperatures of the delay system, and the dynamic adsorption coefficients for Xe and Kr.

4. Description of cryogenic distillation system, fraction of gases partitioned during distillation, holdup in system, storage following distillation, and expected system leakage rate.

5. The steam flow (lb/hr) to the turbine gland seal and the source of the steam (primary or auxiliary).

6. The design holdup time (hr) for gas vented from the gland seal condenser, the iodine partition factor for the condenser, and the fraction of radioiodine released through the system vent. Description of the treatment system used to reduce radioiodine and particulate releases from the gland seal system.

7. Pipings and instrumentation diagrams (P&IDs) and process flow diagrams for the gaseous waste treatment system along with all other systems influencing the source term calculations.

VII. Ventilation and Exhaust Systems

For each station building housing system that contains radioactive materials, provide the following:

1. Provisions incorporated to reduce radioactivity releases through the ventilation or exhaust systems.

2. Decontamination factors assumed and the bases (include charcoal adsorbers, HEPA filters, mechanical devices).

3. Release rates for radioiodines, noble gases, and radioactive particulates (Ci/yr) and the bases.

4. Release point to the environment including height, effluent temperature and exit velocity.

5. For the containment building, indicate the expected purge and venting frequencies and duration, and continuous purge rate (if used).

VIII. Solid Waste Processing Systems

1. In tabular form, provide the following information concerning all inputs to the solid waste processing system: source, volume (ft^3/yr per reactor), and activity (Ci/yr per reactor) of principal radionuclides along with bases for values.

2. Onsite storage provisions (location and capacity) and expected onsite storage times for all solid wastes prior to shipment.

3. Piping and instrumentation diagrams (P&IDs) and process flow diagrams for the solid radwaste system.

---

[a]The ER or SAR may be referenced as to the bases for the parameters used; however, the parameters should be given with the responses in this appendix.

SPECIAL CONSIDERATIONS AND
RISKS OF NUCLEAR POWER

The possibility of a sabotage attempt on a nuclear facility is becoming a real concern for the nuclear industry and the Nuclear Regulatory Commission (NRC). The social, economic and political climate of world affairs makes possible the eventuality of an attack on a nuclear installation. Political terrorists could infiltrate a plant's work force, and even the most insignificant incident could generate a wave of adverse public opinion and fear such that nuclear energy would be effectively useless as a resource. Methods that have been considered to safeguard nuclear facilities include:

- redundant plant safety and communication systems,
- a psychiatrist in residence,
- a continuing program of psychiatric examination for all nuclear employees,
- a safety program to make employees aware of threats and safeguards,
- increased plant security,
- improved reporting of unusual plant activities, and
- more extensive security clearance measures for personnel.

Another unique problem of safety to the nuclear installation is the transportation of plutonium and other radioactive materials. Many times these materials have been transported through metropolitan areas with large populations, stimulating adverse public reaction. Many cities have instituted or considered laws prohibiting the movement of these materials within their limits. Further, the NRC has banned the air transportation of nuclear materials across the United States, as have officials in Canada and Mexico. Generally, it is believed that the effects of an accident involving air-transported radioactive cargo would be devastating although the possibility of an occurrence is very slight.

The construction of a nuclear power plant on the San Andreas fault would be ridiculous. However, the possibility of earthquake damage to installations in other regions of the country may not have been examined as closely as necessary. Many experts feel that one area of the country is just as apt to have an earthquake as another. Although an area may not have experienced earthquake activity in recent history, its geology may be such that one could be inevitable. Further, earthquake phenomena are not clearly understood and no area of the country is safe from earthquakes.

Thus, a major revision in plant design may be warranted, and improved safeguards are necessary. The costs of safety may prove to be so great that the desirability and feasibility of nuclear power will be diminished.

ENVIRONMENTAL EFFECTS OF OFFSHORE NUCLEAR PLANTS

The problems associated with land-based nuclear power plants are multiplied when they are moved offshore. Much of the concern over these plants is not attributed to the detrimental effects on marine life, but toward the potential dangers to humans in nearby areas. An offshore facility must be planned in light of the previously described EIS parameters. Special considerations for safety must be evaluated. Inclement weather conditions (hurricanes, high winds) and high seas must be anticipated and appropriate safeguards incorporated into the system's design.

Unique handling and transportation procedures for the radioactive fuel and wastes may be necessary and costly. Aside from the safety considerations of the construction and operation of a nuclear facility (*i.e.*, regarding radioactivity), the thermal pollution effects are of utmost concern.

THE EFFECTS OF THERMAL POLLUTION

The Federal Water Pollution Control Act of 1972 and the National Pollutant Discharge Elimination System (NPDES) established a national program to control the discharge of water pollutants. Thus, in the filing of permits the following information is usually required: normal water temperatures, water flow rates, seasonal variations in temperatures, heat input sources, and dissipative capacity of waters. Figure 65 is a simplified flow diagram of the cooling system of a nuclear power plant with thermal discharges to surface waters.

The introduction of thermal waters to water bodies creates adverse conditions in water quality. Some physiological effects on aquatic life include:

* increased sensitivity to some toxic substances,
* alterations of organism metabolic rates,
* enzyme and hormone structure modifications, and
* diffusion problems for inter- and intracellular activity.

The thermal pollution of waters also results in several undesirable biological effects:

* Temperature increases and variations can destroy aquatic life and render these natural resources useless for recreational and commercial purposes.
* Temperature increases lower the dissolved oxygen content of the water, effectively destroying many aquatic life forms.
* Temperature increases along with nutrients promote the growth of aquatic plants to extreme proportions.

Organisms can be affected seriously by variations in their environmental conditions. Many life forms are more sensitive to these variations in their

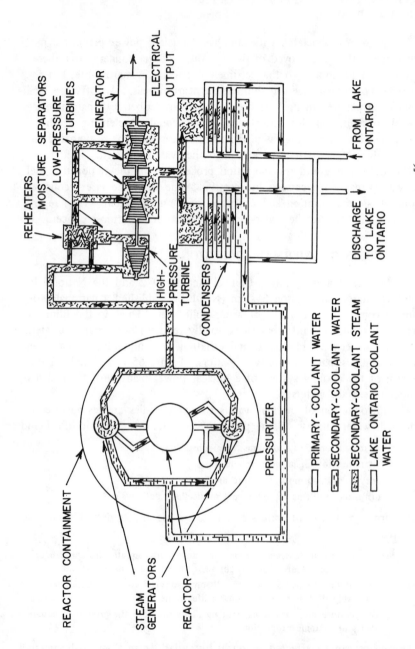

Figure 65.   Flow diagram of nuclear power plant cooling system.[56]

egg or larvae phases, usually during the spring spawning season. The adaptation of species to latitudinal temperature differences are common. For example, an oyster in Long Island Sound spawns at 61°F, but in southern waters at 68°F. The horseshoe crab has a lethal temperature of 115.3°F in Florida waters, but a corresponding temperature of 105.8°F in Massachusetts.

Oyster larvae make up part of the plankton of estuarine waters. All free-floating, microscopic organisms transported by waters constitute plankton. Plankton consists of both animal and plant life, representing an essential component of biological structure. Energy fixation in the marine life cycle is primarily carried out by the plant plankton. Plankton is also the major food source of other species. In any particular bay area there are probably several hundred plankton species, some subject to seasonal variation.

Off the New Jersey coast plankton counts would not be reduced unless water temperatures reached 100°F for extended periods. Higher temperatures can spur high growths for some algal species, but only with nutrient and sunlight availability. Many fish species in this area have upper avoidance breakdown temperatures. Upper avoidance temperatures represent the highest summer water temperatures in which particular organisms will be found. Waters with higher temperatures will be avoided by the organisms. The upper breakdown temperatures are water temperatures which occur for short time intervals and affect the locomotory control of an organism. An organism cannot escape these conditions and will perish eventually.

Thus, behavioral effects are generated by thermal discharges. Many cold-water species have no adaptability and must migrate to survive. Others have the ability to make physiological adjustments or find a protective habitat within the environment. Alterations to behavioral patterns might involve vertical movement, migratory behavior, or spawning and development patterns.

Another important impact of thermal discharges is the degradation of water quality reducing its availability to domestic, recreational and industrial users. Impacts will vary somewhat if the receiving waters are fresh, estuarine or marine. However, most effects are common in principle to all water bodies. For example, higher temperatures in estuarine and marine waters can change water salinity. Further, many problems that have been associated with thermal discharges involve the sudden variation of temperatures. After slow adaptation to rising temperatures at the effluent point of a power plant, fish kills result when the plant is shut down for maintenance and temperatures drop drastically.

BIOLOGICAL EFFECTS OF RADIATION

The amount of radiation energy absorbed by part or all of the body is the radiation exposure. It can be classified as either chronic or acute. Chronic exposure means daily absorption for long periods of time of relatively small doses of radiation. An acute exposure would be a single large dose of radiation occurring within a 24-hr period.

The unit of measurement for radiation in humans is the rem. It measures the ionization produced in air by the passage of X- and gamma radiation. A relatively small quantity of radiation would be 5 rems or less, whereas 25 rems is a large dose although perhaps no observable effects will result. Changes in blood characteristics occur with a dosage of about 50 rems; nausea and vomiting result from a dosage of 100 rems; 300-500 rems give the exposed individual an even chance for survival without medical attention; and an exposure to 650 rems is lethal. Table 56 gives the expected effects of various acute whole-body radiation exposures. The effects of continued exposure to radiation on individual life spans are given in Table 57. It is noted that generally there is a reduction in life expectancy with exposure to radiation, and increased exposures result in a further reduction.

Radiation from a nuclear power installation can eneter the ecosystem in a variety of ways. Figure 66 generalizes the radiation exposure pathways for organisms other than man. As a result of gaseous and liquid effluents, radiation can be ingested or can act directly upon life forms. Figure 67 generalizes these pathways for man. Radiation enters the food chain and, depending upon several factors, the internal exposure can produce varying effects. These factors include:

1. the nature of the radioactive material,
2. the quantities of radioactive material ingested,
3. its manner of entry into the body,
4. locations in the body of deposition and accumulation, and
5. its biological half-life.

The time period required for the body to rid itself of half the absorbed radiation is the biological half-life.

CELLULAR EFFECTS OF RADIATION

Radiation can be introduced to the body through inhalation, ingestion and absorption. Once in the body, radiation can have serious effects on normal cellular activities. It can change the structure or electrical charges of an atom, thereby altering the bonding mechanism of molecules and causing the molecule to break down. New and different molecules may be

Table 56. Expected Effects Due to Various Acute Whole-Body
Radiation Exposures[57]

Dose in Rems	Effect
0 to 50	No obvious effect except, possibly, minor blood changes.
80 to 120	Vomiting and nausea for about 1 day in 5 to 10% of exposed personnel. Fatigue, but no serious disability.
130 to 170	Vomiting and nausea for about 1 day, followed by other symptoms of radiation sickness in about 25% of personnel. No deaths anticipated.
180 to 220	Vomiting and nausea for about 1 day, followed by other symptoms of radiation sickness in about 50% of personnel. No deaths anticipated.
270 to 330	Vomiting and nausea in nearly all personnel on first day, followed by other symptoms of radiation sickness. About 20% deaths within 2 to 6 weeks after exposure; survivors convalescent for about 6 months.
400 to 500	Vomiting and nausea in all personnel on first day, followed by other symptoms of radiation sickness. About 50% deaths within 1 month; survivors convalescent for about 6 months.
550 to 750	Vomiting and nausea in all personnel within 4 hours from exposure, followed by other symptoms of radiation sickness. Up to 100% deaths; few survivors convalescent for about 6 months.
1000	Vomiting and nausea in all personnel within 1 to 2 hours. Probably no survivors from radiation sickness.
5000	Incapacitation almost immediately. All fatalities within 1 week.

Table 57. Effects on Life Span of Continued Radiation Exposure[57]

Physicians having no known professional contact with radiation	65.7 years
Specialists having some exposure to radiation	63.3 years
Radiologists[a]	60.5 years
U.S. population over 25 years of age	65.6 years

[a]These results reflect the inadequate protection used in the past and do not reflect present-day life expectancies of people occupationally exposed to radiation.

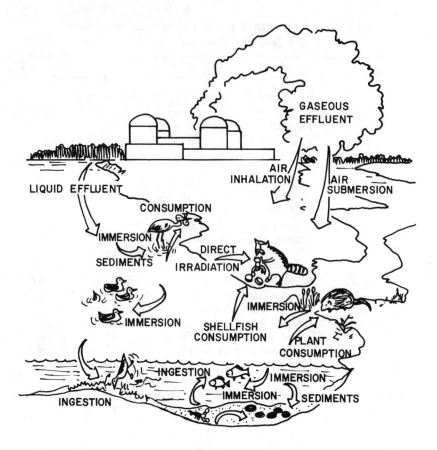

Figure 66. Generalized exposure pathways for organisms other than man.[57]

formed from these parts (see Figure 68). In this manner, radiation passing through a cell attacks the major component of the cell—water. As illustrated in Figure 69, the hydrogen-oxygen bonds are disrupted resulting in H+ and OH⁻ radicals and the formation of hydrogen dioxide (HO_2) and hydrogen peroxide (H_2O_2). These compounds can chemically break down cellular protein matter and alter enzyme activity. Figure 70 shows the breakdown of cell molecules by radiation-affected water ions.

Since enzymes control cellular activity, such as growth and reproduction, the entire cell life cycle is adversely disrupted. Normal cell division is shown in Figure 71. Radiation can cause a cell to divide prematurely, too late, or not at all. Further, the new cells might be altered such that they have an abnormal growth rate, are unable to reproduce, or may reproduce at an abnormal rate. Figures 72 and 73 illustrate these effects.

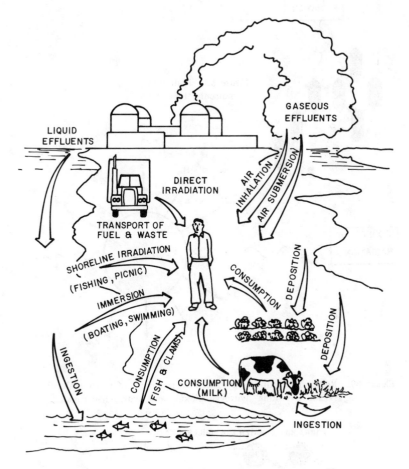

Figure 67. Generalized exposure pathways for man.[57]

Different cells are susceptible to the effects of radiation in varying degrees. The time of cell division, its nourishment and its metabolic rate
can greatly influence the ability of the cell to withstand exposures. The
following list of cells of the human body is given in relative order of sensitivity to radiation. The first cells listed[57] are more susceptible than the
next, and so on.

1. white blood cells formed in the spleen and lymph nodes (lymphocytes),
2. white blood cells formed in the bone marrow (granulocytes),
3. basal cells found in the gonads, skin, bone marrow, etc.,
4. oxygen-absorbing cells of the lungs (alveolar),

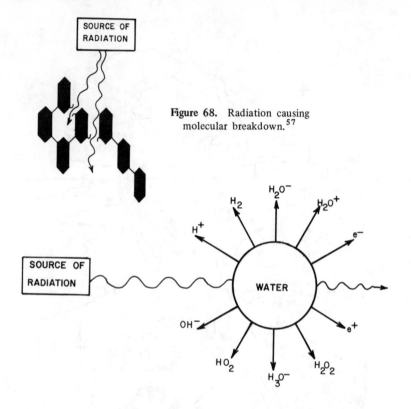

Figure 68. Radiation causing molecular breakdown.[57]

Figure 69. Radiation disrupting hydrogen-oxygen bond of water molecules, forming H+ and OH⁻ ions, hydrogen dioxide and hydrogen peroxide.[57]

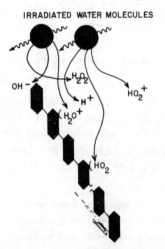

IRRADIATED WATER MOLECULES

Figure 70. Irradiated water molecules breaking up cell molecules.[57]

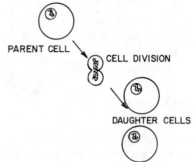

Figure 71. Normal cell division.[57]

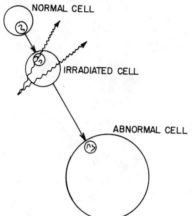

Figure 72. Radiation effected abnormal cell growth.[57]

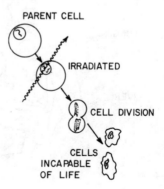

Figure 73. Radiation-effected abnormal cell reproduction.[57]

5. bile duct cells,
6. kidney tubules cells,
7. endothelial cells,
8. structural tissue cells,
9. muscle cells,
10. bone cells, and
11. nerve cells.

Generally, from this list, those tissues and organs which are most sensitive to radiation can be determined. Figure 74 locates these areas on the human body and they include:[57]

1. blood and bone marrow,
2. lymphatic system (spleen and other tissues),
3. skin and hair follicles,
4. gastrointestinal tract,
5. adrenal glands,
6. thyroid gland,
7. lungs,
8. urinary tract,
9. liver and gall bladder,
10. bones,
11. eyes, and
12. reproductive organs.

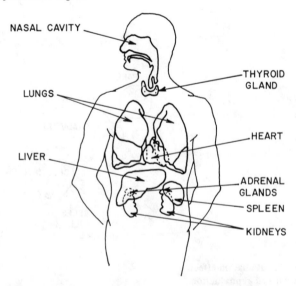

Figure 74. Radiation-sensitive areas of the human body.[57]

Table 58 lists radioisotopes and the organs and tissues in which they may be concentrated.

**Table 58. Radioisotopes and the Organs and Tissues
In Which They May Be Concentrated[57]**

Lungs	Bone
Nickel 63	Beryllium 7
Radon 222	Carbon 14
Polonium 210	Fluorine 18
Uranium 238	Phosphorus 32
Plutonium 239	Calcium 45
	Vanadium 48
	Zinc 65
Kidneys	Gallium 72
Chromium 51	Strontium 89
Manganese 56	Strontium 90 and
Germanium 71	Yttrium 90
Arsenic 76	Yttrium 91
Rhodium 105	Niobium 95
Ruthenium 106 and	Molybdenum 99
Rhodium 106	Tin 113
Technetium 127	Barium 140 and
Tellurium 129	Lanthanum 140
Iridium 190, 192	Praseodymium 143
Gold 198	Cerium 144 and
Uranium 238	Praseodymium 144
	Promethium 147
Liver	Samarium 151
Manganese 56	Europium 154
Nickel 59	Holmium 166
Cobalt 60	Thulium 170
Copper 64	Lutecium 177
Silver 105	Tungsten 185
Cadmium 109 and	Lead 203
Silver 109	Radium 226
Silver 111	Uranium 233
	Thorium 234
	Plutonium 239
	Americium 241
	Curium 242

GENETIC EFFECTS

Chromosomes are threadlike structures that are contained in the nuclei of most human cells. Genes are components of the nuclei which align themselves along the chromosomes. These components determine hereditary characteristics of cells and the whole individual. Radiation can cause changes in genes, mutations in cells, or it can increase the normal mutation rate. Breaks in chromosomes can also occur. Mutations can result in life-span reductions, sterilization, and even death.

CANCER

There are three cancer-causing forms of radiation:

1. carcinogenic—a direct effect,
2. cocarcinogenic—a boost effect, and
3. remote carcinogenic—a remote effect.

A direct effect would be cancer caused by exposure to a large dose of radiation. A boost effect would be cancer caused as a result of a small exposure to radiation, spurring cancer agents. A remote effect would be cancer caused in one part of the body by radiation-induced cancer in another part of the body.

Radiologists and others who work with radioactive materials have relatively large incidences of cancer. The biological effects of radiation can be devastating, and a thorough EIS examining all safety precautions and accident prevention measures in detail is imperative.

CHAPTER 10

RADIOACTIVE WASTE TREATMENT PRACTICES*

This chapter describes the current state-of-the-art for the design of liquid and solid radioactive waste treatment systems. Changes in equipment and system design resulting from operating experiences are presented. Stringent federal regulations which restrict radioactive effluent releases have resulted in more emphasis being placed on coordinated water management programs in nuclear power plants. Actual releases of liquid radioactive wastes are quantified, and the conclusion reached is that releases are only a small fraction of permissible limits.

INTRODUCTION

This chapter deals with the treatment for disposal of liquid and solid radioactive wastes in commercial nuclear power plants. System designs, equipment descriptions and relevant experience from operating nuclear power plants will be discussed. The regulatory framework within which these systems are designed and operated will also be discussed.

With only a few exceptions, nuclear power plants operating in the United States today are either Pressurized Water Reactors (PWR) or Boiling Water Reactors (BWR). Both PWRs and BWRs use light water as the coolant which circulates through the reactor core removing heat. The coolant either produces steam directly by boiling or transfers heat to secondary water to produce steam for turning the turbines that generate electricity. In the BWR, the coolant itself is the source of steam, which after going through the turbine is condensed and returned to the reactor.

---

*Contributed by Richard A. Edelman, Supervisor, Licensing and Environmental Liquid Metal Fast Breeder Reactor Project, Burns and Roe, Inc., Oradell, N.J.

In a PWR, the primary coolant is pressurized and the temperature is maintained below saturation, thus restricting bulk boiling. The heat is transferred from the primary coolant to a steam generator to produce steam in a secondary coolant loop external to the reactor system. Figures 75 and 76 depict schematically the coolant flow paths for a BWR and PWR.

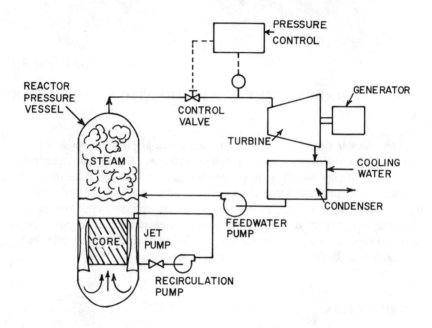

Figure 75. Schematic diagram of a BWR system.

Nuclear reactors generate radioactive materials in the fuel as a consequence of fission of uranium and plutonium. The coolant may receive fission product activity from traces of uranium on the outer surface of the fuel cladding, or through the release of fission products through defects in cladding integrity; this activity may contain radioactive nuclides (or isotopes) of krypton, xenon, cesium, rubidium, barium, strontium, iodine and bromine.

In addition, radioactive materials are generated in the coolant and in structural materials by absorption of neutrons leading to induced radioactivity. Activity directly induced in the coolant by activation of water and air generally consists of gases (Ar-41, F018, N-13, N-16 and O-19) which have short half-lives. Activity of considerably longer half-life may enter the coolant from corrosion or erosion of the structural materials

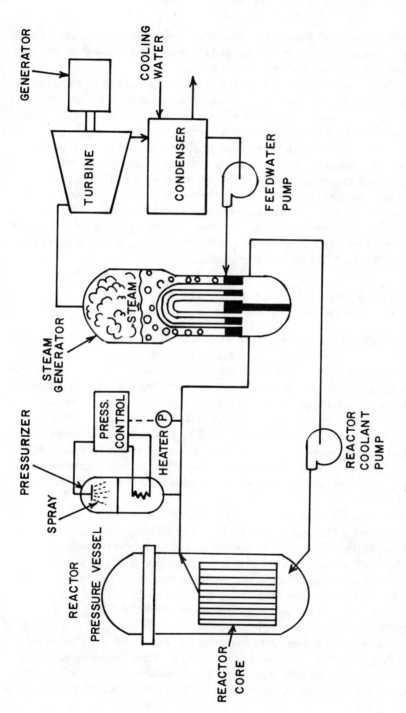

Figure 76. Schematic diagram of a PWR system.

used for the fuel elements, reactor vessel, piping and equipment. Radio-active isotopes of zinc, manganese, nickel, iron, cobalt, chromium and copper are added to the coolant by this process.

A third source of radioactivity is through the activation of chemicals added to the reactor water. Corrosion inhibitors, such as lithium, are added to the coolant. In most PWRs the primary coolant contains boric acid as a chemical shim for reactivity control.

The radioactive materials generated in the coolant from these various sources are pumped around the piping system and appear in other parts of the plant through leaks and vents, as well as in the effluent streams from processes used to purify the primary coolant. The radioactive materials, or wastes, are collected and processed within the plant under controlled and monitored conditions to assure that the quantity of radioactive nuclides ultimately released to the environment is reduced to well below the limits established in federal regulations.

INPUTS TO THE LIQUID RADIOACTIVE
WASTE TREATMENT SYSTEM

As discussed previously, radioactive wastes originate in the reactor either by fission product leakage or activation of corrosion products and are carried by the primary coolant to various plant systems. The wastes are collected throughout the plant and routed to a radioactive waste treatment building for processing. Collection points consist of various floor drain sumps, equipment sumps, filter sludge tanks and drain lines. In order to meet revised regulatory requirements and to minimize any environmental impact, essentially all radioactive liquid wastes are processed to concentrate the radioactive contamination for solidification, and controlled disposal, and to maximize the amount of purified liquid available for reuse in the plant.

The desired output from the liquids treatment process is water suitable for recycling. This water must be low in conductivity and low in activity (radioactivity). The inputs to the liquid waste treatment system are generally either low-conductivity high-activity (reactor grade) water from equipment sumps and floor drains from the reactor containment, or high-conductivity low-activity (nonreactor grade) water from floor drains throughout the plant, laboratory and sample drains, and laundry drains. Segregation of liquid wastes by chemical content, conductivity, and activity is the essential requirement to minimize the amount of waste processed and keep operating costs down. For example, floor drain wastes are normally kept separate from equipment drain wastes as the former are expected to have high solids content while the latter are expected to be of good quality. Likewise, both chemical and detergent wastes are normally treated separately. Chemical wastes may require neutralization and solids removal as preliminary steps. Detergent wastes are normally treated separately, as they cause foaming in evaporators.

In a PWR leakage of primary coolant into the secondary coolant in the steam generator is the only source of radioactivity in the secondary coolant system. Water or steam leakage from the secondary system provides significant inputs to the liquid radwaste treatment system.

Tables 59 and 60 list the sources of wastes and the approximate volume for a large size (1100-MWe) BWR and PWR nuclear power plant. The specific radioactive nuclides potentially found in the liquid wastes are shown in Table 61.

Table 59. BWR Liquid Radwaste System Inputs

Source	Average[a] Volume (gpd)
Reactor Building Equipment Drain	2,000
Drywell Equipment Drain Sump	5,800
Radwaste Building Equipment Drain Sump	1,000
Turbine Building Equipment Drains	5,700
Reactor Building Floor Sump	2,000
Drywell Floor Sump	2,900
Radwaste Building Floor Drain	1,000
Turbine Building Floor Drains	2,000
Laboratory Drains	500
Condensate Demineralizer Regeneration	1,800
Ultrasonic Resin Cleaning	15,000
Demineralizer Backwash Resin Transfer	4,200
Detergent Waste (laundry, decontamination, showers)	450

[a]From U.S. Atomic Energy Commission, "Principal Parameters used in Source Term Calculations," U.S. AEC Directorate of Licensing.

LIQUID RADIOACTIVE WASTE TREATMENT—SYSTEM DESIGN

Boiling Water Reactors

Boiling water reactors are manufactured in the U.S. by the General Electric Company. The current generation of BWRs is known as the BWR/6 design and is marketed for nuclear power plants in the 1100- to 1300-MWe capacity range. The system description which follows is designed to handle the wastes generated by the BWR/6 reactor.

Table 60.   PWR Liquid Radwaste System Inputs

Source	Average[a] Volume (gpd)
Containment Building Sump	40
Auxiliary Building Floor Drains	200
Laboratory Drains and Waste Water	400
Sample Drains[b]	35
Turbine Building Floor Drains[c]	7,200
Miscellaneous Sources	700
Steam Generator Blowdown	0.06% of main steam flow
Detergent Waste (laundry, decontamination, showers)	450

[a] From U.S. Atomic Energy Commission, "Principal Parameters used in Source Term Calculations," U.S. AEC Directorate of Licensing.
[b] Fifteen gallons per day for continuous purge recycle.
[c] For once-through steam generator systems, equals 3,200 gallons per day.

The liquid radwaste system is made up of three subsystems designed to treat clean radwaste (low conductivity), dirty radwaste (high conductivity and solids content), and detergent-bearing wastes. A process flow diagram for these subsystems is shown in Figure 77.

Waste Collector Subsystem

The subsystem which collects and processes radwaste of relatively low conductivity is the waste collector subsystem. Equipment drains and condensate demineralizer washwater are typical of wastes found in this subsystem. These wastes are collected in the low-conductivity tanks and then filtered on a precoated traveling-bed or candle-type filter for removal of insolubles. The second treatment is for the removal of soluble contaminants by the use of two mixed-resin deep-bed demineralizers in series. The resulting liquid is routed to the condensate storage tank for reuse unless instrumentation indicates the conductivity is too high, in which case it is recycled for further treatment. Radiation instrumentation is also provided to prevent routing to the condensate storage tank if limits on radioactivity concentration are exceeded. A second filter train is provided for the dewatering of system resin sludges, with the filtrate routed to a collector tank for additional processing.

Floor Drain-Neutralizer Subsystem

The subsystem which collects and processes radwaste of high conductivity and solids content is the floor drain-neutralizer subsystem. Floor drains, decontamination solutions, condensate demineralizer regeneration

Table 61. Expected Form of Various Activities Potentially Found in Radwaste

Nuclide	Form	Nuclide	Form
Activation Products		**Fission Products**	
Na-24	Soluble	Sr-89	Soluble
P-32	Soluble	Sr-90	Soluble
Cr-51	Insoluble	Sr-91	Soluble
Mn-54	Insoluble	Sr-92	Soluble
Mn-56	Insoluble	Zr-95	Insoluble
Co-58	Insoluble	Zr-97	Insoluble
Co-60	Insoluble	Nb-95	Insoluble
Fe-59	Insoluble	Mo-99	Insoluble
Ni-65	Insoluble	Tc-99m	Soluble
Zn-65	Soluble	Tc-101	Soluble
Zn-69m	Soluble	Ru-103	Insoluble
Ag-110m	Insoluble	Ru-106	Insoluble
W-187	Insoluble	Te-129m	Soluble
		Te-132	Soluble
		Cs-134	Soluble
Halogens		Cs-136	Soluble
Br-83	Soluble	Cs-137	Soluble
Br-84	Soluble	Cs-138	Soluble
Br-85	Soluble	Ba-139	Soluble
I-131	Soluble	Ba-140	Soluble
I-132	Soluble	Ba-141	Soluble
I-133	Soluble	Ba-142	Soluble
I-134	Soluble	Ce-141	Insoluble
I-135	Soluble	Ce-143	Insoluble
		Ce-144	Insoluble
		Pr-143	Insoluble
		Nd-147	Insoluble
		Np-239	Soluble

solutions, and laboratory drains are typical of wastes found in this subsystem. The wastes are collected in high-conductivity tanks and chemically adjusted to a basic pH as required. The initial treatment is concentration by a forced-circulation waste evaporator with submerged steam-heated elements to reduce the volume containing contaminants and to decontaminate the distillate. The distillate is then demineralized to remove residual soluble contaminants and then combined with the waste collector subsystem just prior to the second waste demineralizer as shown in Figure 77. If conductivity instrumentation indicates the activity is too high, it is recycled to the distillate tank.

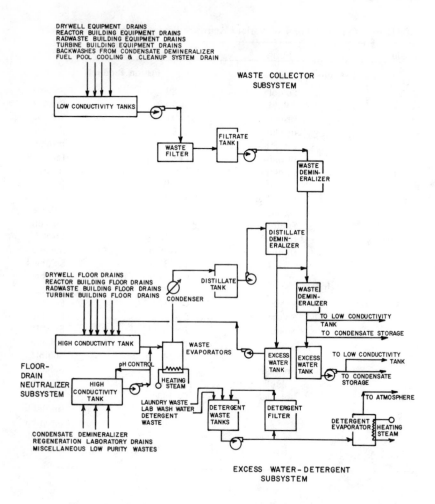

Figure 77. Basic liquid radwaste treatment system for a BWR.

Excess Water-Detergent Subsystem

The excess water-detergent subsystem collects and processes detergent-bearing water from personnel or clothing-laundering operations with detergent-type decontamination solutions. In addition, it processes water in excess of plant inventory requirements which usually results from a special, infrequent plant occurrence.

Detergent wastes are collected in one of two collector tanks, filtered, and then treated by the detergent evaporator. Distillate from the detergent evaporator is discharged as a vapor to the building ventilation system.

The evaporator bottoms are routed to the concentrated waste tank for treatment by the solid radwaste system.

Excess water collected from the effluent of the second demineralizer in the waste collector system is stored in excess water storage tanks for use as makeup for the condensate storage tanks. When excess water, above the storage tank capacity is present, it can be evaporated in the detergent waste evaporator.

Pressurized Water Reactors

Pressurized water reactors in the U.S. are manufactured by three companies—Westinghouse, Combustion Engineering, and Babcock and Wilcox. Although different names may be given to the subsystems composing the liquid radwaste system, the designs are basically the same. As is the case for BWRs, modern PWR plants attempt to segregate liquid wastes prior to processing into two categories: (1) low-conductivity liquid wastes and (2) high-conductivity liquid wastes. These categories are sometimes referred to as deaerated and aerated wastes, respectively, since deaerated drains tend to be cleaner and have a lower conductivity.

The bulk of the radioactive liquids discharged from the reactor coolant system are processed and recycled by the boron recycle system. Other liquid wastes processed will be designated as reactor-grade water (recycle) sources, nonreactor-grade water sources, and steam generator blowdown (different manufacturers use different names for the treatment systems for these sources). A typical flow diagram for a PWR liquid radwaste system is shown in Figure 78.

Boron Recycle System

The primary coolant is continuously purified by passing a side stream through filters and demineralizers in the reactor coolant treatment systems. This is necessary to maintain the purity of the primary coolant to prevent fouling of heat transfer surfaces in the steam generators. Chemicals are added to the primary coolant to inhibit corrosion and to improve fuel economy. Lithium hydroxide is added for pH control to reduce corrosion. Boron, in the form of boric acid, is added as a neutron absorber for reactivity shim control. As the fuel cycle progresses, boron is removed from the primary coolant through the reactor coolant treatment system loop (shim bleed). The shim bleed is processed through an evaporator and the boron in the evaporator bottoms is either reused or packaged as solid waste. Most of evaporator distillate is recycled to the reactor coolant system as makeup water. Some distillate is released to the environment,

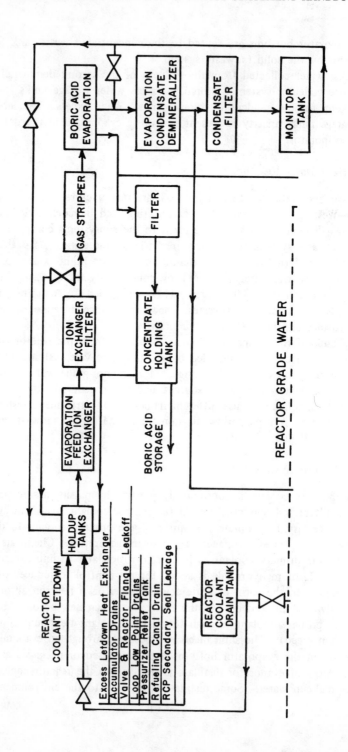

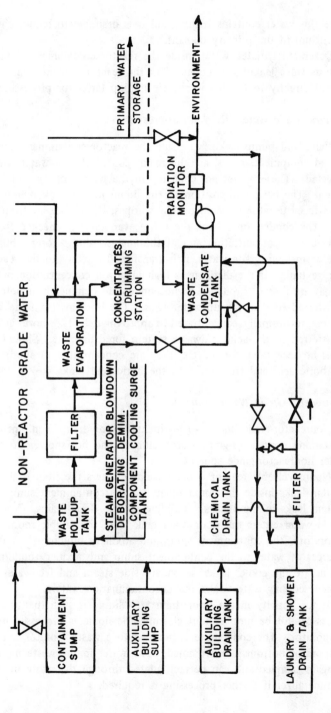

Figure 78. Basic liquid radwaste treatment for a PWR'

since this water contains tritium and it is desirable to reduce the build-up of tritium in the primary coolant.

Deaerated tritiated water inside the reactor containment, from sources such as valve leakoffs, is collected in the reactor coolant drain tank and routed directly to the boron recycle holdup tanks for processing.

Reactor-Grade Water (Recycle) Sources

Valve and pump leakoffs outside the reactor containment, and controlled equipment drains, are the major contributor of water which can be recycled. These wastes are collected in the waste holdup tank for processing. The basic composition of the liquid in this tank is evaporated in a waste evaporator to remove radioisotopes, boron and air from the water. The condensate leaving the waste evaporator is checked for boron and activity concentration and, if the analysis shows compatibility with reactor makeup grade water, it is transferred directly to the reactor water storage tank. If the distillate is high in boron concentration or activity, it may be passed through a waste condensate demineralizer before being transferred to the makeup water tank. The bottoms from the waste evaporator are normally concentrated to approximately 12% boric acid and are transferred to the solid radwaste system for drumming. Should the bottoms be acceptable for recycle, they are concentrated to approximately 4% boric acid and transferred to the boric acid tanks.

Nonreactor-Grade Water Sources

Nonreactor-grade liquid wastes include floor drains, equipment drains containing nonreactor-grade water, laundry and hot shower drains, and other nonreactor-grade sources.

Nonrecycleable reactor coolant leakage enters the floor drain tank from system leaks inside the containment via the containment sump and from system leaks in the auxiliary building via the floor drains. Nonreactor-grade water enters the floor drain tank from the sample room, chemical laboratory, the containment sump and auxiliary building floor drains. Sources of water to the containment sump and auxiliary building floor drains are fan cooler leaks, secondary side steam and feedwater leaks, component cooling water leaks and decontamination water.

If the activity in the floor drain tank liquid is such that the discharge limits cannot be met without cleanup, the liquid is processed in the waste evaporator. The condensate is pumped to a waste monitor tank and the evaporator bottoms are drummed. The water in the waste monitor tank is again sampled and can be recirculated through the waste monitor tank demineralizer if further processing is required.

Laundry and hot (contaminated) shower drains are the largest volume source of liquid wastes and normally need no treatment for removal of radioactivity. This water is transferred to a second waste monitor tank via the laundry and hot shower filter.

The laboratory drain subsystem and sampling room each contain two sinks. One is used for radioactive liquid wastes which is directed to the chemical drain tank (for laboratory drains) or waste holdup tank (sampling room). The second sink is used for nonradioactive liquid wastes and is directed to the floor drain tank. Liquid wastes released from the waste monitory tanks to the discharge canal are monitored for radiation. A discharge valve will close automatically when the radioactivity concentration in the liquid discharge exceeds a preset limit.

Steam Generator Blowdown

In some radwaste designs, steam generator blowdown is directed to the waste holdup tank and is treated with other nonreactor-grade liquid wastes. Other systems, however, will use a separate system to collect steam generator blowdown and potentially radioactive liquids from the secondary plant (including turbine building floor drains). In the unusual event of high radioactive content from operation with leaks in the steam generator tubes and defective fuel in the core, a separate system provides for processing of steam generator blowdown liquid wastes through the waste evaporator. This provides additional operating flexibility in the plant design. Normally, these wastes are filtered, demineralized, and sampled prior to discharge.

LIQUID RADIOACTIVE WASTE TREATMENT– EQUIPMENT DESIGN

The removal of radioactivity and other contamination from the liquid streams in a nuclear power plants is accomplished by a combination of treatment methods. These methods include holdup for decay, evaporation, demineralization, filtration and reverse osmosis. The design of pumps, valves and tanks necessary to accomplish treatment will require use of noncorroding materials (*i.e.,* stainless steel) and must have very low leakage capabilities. In other respects the pumps, valves and tanks used in liquid radwaste service are not unique and a detailed discussion is not warranted. The basic treatment processes used and appropriate equipment descriptions will be described individually below.

Holdup for Decay

Storage tanks provide a method for reducing the radioactivity of waste solutions by providing time for radionuclides to decay. For example, the half-life of cesium-137 is 30 years while that for iodine-131 is approximately 8 days. By retaining the radioactive liquid wastes in a tank for 30 days, large fractions of a relatively short-lived isotope such as iodine-131 decay, whereas this decay would have little effect on cesium-137 or other long-lived isotopes.

Evaporation

Evaporation is used to concentrate solutions containing radioactive contaminants from waste liquid by boiling off the water, leaving behind most dissolved solids. Materials having vapor pressures similar to water are difficult to separate from water by evaporation. Because of these factors, iodine, ruthenium and tritium are among radioactive materials that are poorly separated from waste liquid by evaporation.

An overall decontamination factor of more than 10^4 between condensate (distillate) and thick liquor (concentrate) is generally expected from single-effect evaporators for nonvolatile radioactive contaminants. Higher factors can be obtained by the use of deentrainment devices to reduce liquid carryover.[58] Liquid carryover is normally caused by (1) too vigorous boiling, (2) foaming, (3) inadequate deentrainment height, or (4) liquid flashing due to unexpected pressure fluctuations. Depending on the amount of dissolved solids in the waste fed to an evaporator, a volume reduction of 10 to 50 can usually be achieved in the concentration.

Evaporation is most suitable for processing liquids which have a high total solids concentration and require a high decontamination factor. Evaporation is relatively costly and therefore is not suitable for the processing of large volumes of low-activity wastes.

The factors that have to be considered in the selection of the type of evaporator to be used are:

1. characteristics of the feed, such as composition and amount of dissolved materials and their salting, scaling, corroding and foaming characteristics;
2. amount of liquid to be treated per hour, its fluctuations and number of working hours per day;
3. volume reduction and decontamination factor required;
4. ease of maintenance;
5. space limitations; and
6. economic limitations.

There are several different types of evaporators used in operating liquid radwaste systems. Evaporators are usually provided in capacities ranging from 15 to 50 gpm. Manufacturers of evaporators for liquid radwaste service include the following:

Artisan Industries	Riley-Beaird, Inc.
HPD, Inc.	Swenson Div., Whiting Corp.
Luwa Corp.	Systems Engineering Research Facilities
Joseph Oat Corp.	Unitech Div., Ecodyne Corp.
Pfaudler Div., Sybron Corp.	Westinghouse Electric Corp.

In the past, natural circulation evaporators were used in many BWRs for processing low-solids-content solutions. As operating practices changed to the handling of waste streams with higher total solids content, problems have occurred with tube plugging, corrosion and instrumentation line plugging. Based on experience, natural circulation evaporators are not recommended for new radwaste systems. The different types of evaporators used are described below:

Forced-Circulation Vertical or Horizontal-Tube Evaporators

In these evaporators, liquid is recirculated by a pump through the heating tubes at reasonably high velocities (4 to 10 ft/sec) with little boiling per pass. The heating element is normally installed as a separate unit which has the advantage of allowing greater ease of cleaning or replacement of tubes. The horizontal tube heating element is furnished where low head room requirements apply. The horizontal tube bundle is usually divided into two or more liquid passes, giving lower circulation rates but higher pump heads than the single-pass vertical units.

The advantages of forced-circulation evaporators are high heat-transfer coefficients, positive circulation, and relative freedom from salting, scaling and fouling. The disadvantages are higher cost and power required for the circulation pump. A diagram of horizontal and vertical tube forced circulation evaporators is included as Figure 79.

Vapor-Compression Evaporators

A higher total heat efficiency in evaporators can be achieved by reuse of the energy of vaporization to provide a heat source for further evaporation. In vapor-compression evaporators the energy potential of low-pressure vapor rising from the evaporator is increased by compressing the vapor, thereby making the latent heat of condensation available at a higher temperature. Two types of vapor-compression evaporators are currently used, mechanical compression by a blower and compression by a

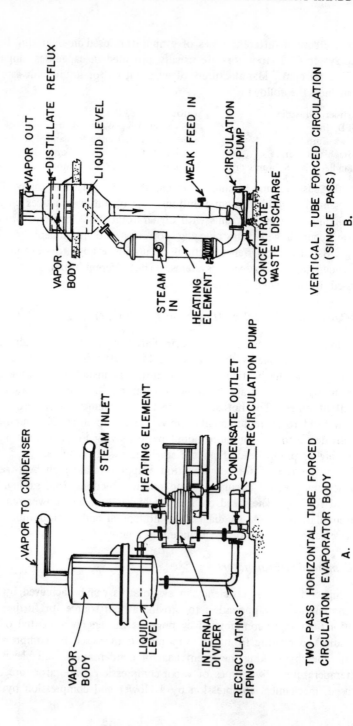

Figure 79. A—horizontal tube forced-circulation evaporator; B—vertical tube forced-circulation evaporator.
Courtesy: Ecodyne Corp.

steam injector. A vapor-compression evaporator, as shown in Figure 80, can be expensive and its use requires a detailed economic balance to determine its feasibility.

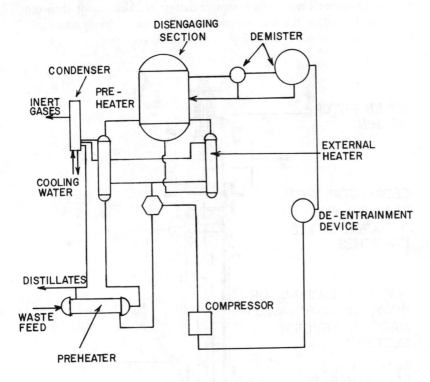

Figure 80. Vapor-compression evaporator.[58]

Wiped-Film Evaporator

This type of evaporator uses mechanical energy to improve heat transfer. The heating surface consists of a single (vertical-type) or tapered (horizontal-type) cylinder of large diameter in which an agitating blade or a series of wipers is rotated, either maintaining or riding on the film of liquid on the wall. A schematic of a vertical-type wiped-film evaporator is shown in Figure 81.

The liquid waste fed into the cylinder is agitated vigorously by the rotating blades, becoming filmy by centrifugal force, and flows along the heating surface. The rotating blades also serve to break down foam and throw entrainment out of the central vapor passage. The heat is transferred from the outside of the cylinder and the vapor formed is exhausted from the vapor outlet, passing through the spaces between the blades. An advantage of this evaporator is that the continuous formation of the film permits much higher concentration of thick liquor than can be handled in other types of evaporators. The major disadvantage of this type of evaporator is the high cost of construction.

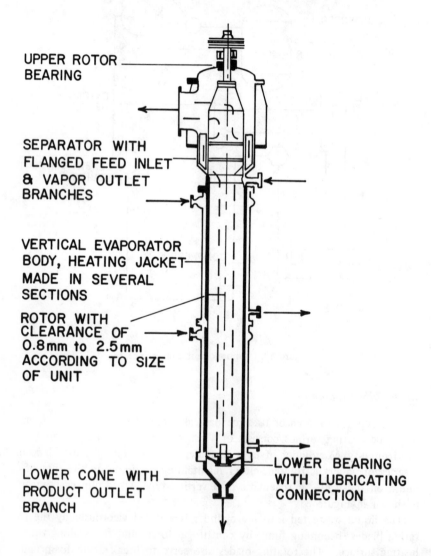

Figure 81. Vertical wiped-film evaporator.[58]

Vertical Falling-Film Evaporator

In this type of evaporator (Figure 82), the feed flows down the vertical tube walls as a film. Heat-transfer coefficients in the falling-film evaporator are high, and operation on low-temperature differentials can be achieved. High-entrainment separation efficiency can be expected as entrainment occurs mainly in the tubes themselves, and the spray from the film on one side of the wall impinges on the film on the other side.

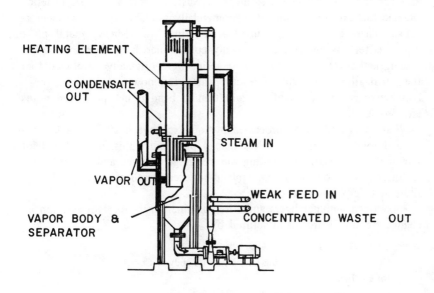

FALLING FILM EVAPORATOR

Figure 82. Vertical falling-film evaporator.

Short residence time and low cost are advantages of this type of evaporator. The chief problem is the achievement of uniform feed distribution to the top of all the tubes to avoid dryout of a portion of tube. Means of distribution include sprays and perforated plates above the top tube sheets or orifices inserted at the inlet to each tube.

Demineralization

Demineralization is carried out in separate demineralizers or in combination filter/demineralizers. Demineralizers consist of tanks filled with a bed of cation and/or anion resin beads, and are used to remove both radioactive and nonradioactive dissolved ions by ion exchange. Filter/demineralizers utilize a powdered resin which is precoated on filter media to achieve ion exchange for removal of dissolved ions and filtration for removal of suspended solids.

Ion exchange is the exchange of ions that occurs across the boundary between a solid particle and a liquid. Dissolved electrolytes, when in contact with ion exchanger solids, will exchange stoichiometrically equivalent amounts of ions of the same sign. A wide range of materials is available for the ion exchange treatment of wastes. Nuclear-grade synthetic organic ion exchange resins are commercially available from a number of firms. These are very high purity bead resins, low in heavy metal and organic matter content and are specially conditioned by a combination of cycling and solvent washing. These resins were originally developed for the purification treatment of reactor water, but now are also used for radioactive waste treatment. In addition to bead resins, powdered resins are also used.

Waste solutions to be processed by ion exchange must have a low suspended solids concentration, less than 4 ppm, since this material will interfere with the process by coating the exchanger surface and by sorbing radionuclides. For this reason, mixed-bed demineralizers usually have a filter installed upstream.

Mixed-bed demineralizers and filter demineralizers for radwaste ion exchange treatment are manufactured by:

> Cochrane Division, Crane Co.
> Graver Water Conditioning Co., Div. of Ecodyne Corp.
> Ionics, Inc.
> L. A. Water Conditioning Co.
> DeLaval Turbine, Inc., Filtration Systems Div.
> Hydromation Filter Company

Mixed-Bed (Bead-Resin) Demineralizers

Mixed-bed demineralizers contain both the cation and anion exchangers in the same vessel. Synthetic organic resins of the strong acid and strong base type are normally used. During operation of mixed bed demineralizers, cation and anion exchangers are intimately mixed to ensure that the acid solution formed after contact with the H^+ form cation resin is immediately neutralized by the OH^- form anion resin. The waste solution is therefore maintained at a neutral pH and special materials of construction are not required.[59]

Demineralizer resins are generally replaced or regenerated when their removal efficiency decreases below preset levels. Deep-bed condensate demineralizers ($\sim$310 ft^3 of resin) are commonly regenerated. Resins in smaller demineralizers used for other purposes are completely replaced, thereby eliminating additional chemical radwastes resulting from regenerations. Spent resins from nonregenerable systems or from attrition in regenerable systems must be disposed of as solid wastes.

When resins are regenerated, they are either treated in place or are drained to a separate facility. Mixed beds must be backwashed to remove particulate matter and to separate the cation and anion resins. Classification by this means is possible because of the difference in specific gravity of the resins. Regenerants used are dilute mineral acids for the cation exchangers and dilute caustic solutions for the anion exchangers. Sulfuric acid and sodium hydroxide are most commonly used.

Resins which are regenerated lead to streams of acidic and caustic, highly radioactive chemical wastes which must be neutralized and processed for radionuclide removal. At the same time, large volumes of rinse, backwash and resin transfer water must be processed before discharge or reuse. Ultrasonic cleaning systems for recycling resins have been used to reduce the volume of resins disposed of as solid waste, particularly during plant start-up operations where high solids loadings exist.

Since regeneration produces wastes of greater volume than the exhausted resins, it has often been found more economical to dispose of the resin rather than regenerate it. Exhausted resins can be drained from specially designed columns or, if originally packaged in cartridge-type containers, can be removed and replaced as a unit. Final disposal is achieved by binding the resins into a solid matrix in the solidification system.

Mixed-bed demineralizers for radwaste service will typically have a resin volume of 30 ft^3 and operate at temperatures of 120 to 140°F. The normal flow rate will be 20 gpm, with design flow rates up to 50 gpm. The resins will be in a stainless steel tank constructed to Sections VIII or III of the ASME Boiler and Pressure Vessel Code.

Powdered Ion Exchange Filter/Demineralizers

Powdered ion exchange resins, as employed in the Powdex* process, are precoated on cylindrical filter elements. Usually 0.2 to 0.3 lb of dry resin per ft^2 of area are used. The cation-to-anion ratio may be varied; however, for nuclear service, a cation-to-anion ratio of 3:1 is usually used. The Powdex process utilizes the rapid ion exchange rates characteristic of high surface area/fine particle size resins, while eliminating pressure drop problems by using thin layers of ion exchange material.

---

*Registered trademark of Graver Water Conditioning Co., Ecodyne Corp.

When used as filter/demineralizers in radwaste service, the powdered ion exchange units will remove both dissolved solids (by demineralization) and suspended solids. The powdered resins are not regenerated, but instead are disposed of as solid wastes. In the Powdex process, a backwash procedure employs air and water for cleaning the elements. Complete cleaning is important to obtain even precoating, and thus full utilization of the resin's ion exchange capacity.

Filtration

The purpose of all filtration is to achieve the highest possible decontamination factor by removing the maximum amount of solids in the waste stream. Filters are capable of removing suspended solids from liquid waste streams, depending on the size of the solid particles and the type of filter. Filters are especially useful for limiting the amount of solids entering demineralizers or evaporators. Suspended solids tend to collect on resins causing increased pressure drop across demineralizers and thus decreased flow. Solids also tend to plate out on heat transfer surfaces in evaporators.

Three general types of filters are used in radwaste service. These are disposable filters, precoat filters, and backflushable or metallic recleanable filters. Each type of filter has advantages and disadvantages which are discussed individually below.

The manufacturers of radwaste filters include the following:

Chemapec, Inc. (Funda Filters)

Croll-Reynolds Engineering Co.

Cuno Division, AMF Corp.

DeLaval Turbine, Inc., Filtration Systems Div.

GAF Corporation

Hydromation Filter Co.

NEWCO, Inc. (Schenk Filters)

Pall Trinity Micro Corp.

Ronningen-Petter Division, Dover Corp.

United States Filter Corp.

Vacco Industries

Disposable Cartridge Filters

Disposable filters are designed with replaceable cartridge elements which are disposed of as solid radwaste when contaminated. These elements are usually constructed of pressed paper, matted fibers, or porcelain materials. Wound or spun cartridge elements generally have a particle retention of

25μ absolute and operate at flow rates from 20 to 200 gpm depending on the size of the filter. The design pressure and temperature are typically 150 psig and $200°F$, respectively, and are constructed to Section III, Class 3, or Section VIII of the ASME Boiler and Pressure Vessel Code.

The primary advantage of disposable cartridge filters is the low installed cost. The total installed cost of a complete cartridge filter system is approximately one-fifth the initial cost of a precoat filter system, and one-tenth that of a metallic recleanable filter system.[60] The disadvantage of this system, however, is the higher operating costs. Each time the cartridge must be replaced, expenses are incurred for downtime, element replacement, disposal costs, and costs associated with radioactive exposure of plant personnel. The costs for transportation and burial of solid radwaste has been increasing rapidly over the last few years. This will have a significant impact on the economic attractiveness of disposable filters.

Replacement of the spent radioactive filter cartridge involves remote disassembly of the filter vessel from behind a shield wall. This operation includes removal of a shield plug on the floor above the filter, removal of the filter cover using long-handled tools, and lifting the filter cartridge into a shield cask or overhead lead shield for transport to the drumming station.

The actual filter replacement process varies from plant to plant. Recent emphasis on reduced personnel radiation exposure has increased the need to standardize removal techniques. Filter manufacturers are being encouraged to develop remote handling equipment.

Precoat Filters

Precoat filters employ elements designed as retainers to prevent a precoat material from being flushed downstream. The precoat material, or filter aid, is generally solka-floc or diatomaceous earth. The precoat is applied by means of the recirculation of a suspension containing the filter aid. The actual filtering of system contamination is accomplished by the precoat material and, when the filter becomes contaminated, the precoat material is replaced.

Precoat material can be applied to vertical filter elements constructed of wire mesh or woven wire. The precoat is held in place by the liquid flow, and hence recirculation pumps must be used during "off" periods.

Two effective types of precoat filters which do not rely on flow to maintain the precoat are centrifugal filters and flat-bed or travelling-bed filters. Centrifugal filters are comprised of filter plates which are stacked on an axially located hollow shaft. The filter plates are fitted with woven wire screens or with cotton or synthetic fiber cloth. The plates are coated with the filter aid material.

The precoat filters are cleaned by a combination of centrifugal force and simultaneous backwash acting upon the cake. The ejected cake leaves the base of the vessel through a large port as a slurry (water backwash) or dry cake (air or gas backwash). The centrifugal force is provided by spinning the filter plate with the motive force to the hollow shaft being supplied mechanically or hydraulically. A process flow schematic is shown in Figure 83.

An advantage of the centrifugal-type precoat filters, manufactured by Schenk and Funda, is a self-cleaning process which eliminates manual cleaning. The centrifugal discharge process takes only a few seconds, and therefore downtime is significantly less than with a disposable cartridge filter. Another advantage is that due to the horizontal plate design, the filter precoat cake is not affected by pressure fluctuations and the cake does not tend to crack or channel. The chief disadvantage is the higher cost than disposable filters and the generation of additional solid waste (*i.e.*, precoat material) as compared to the metallic recleanable filters discussed in the next section.

The flat-bed or travelling-bed filter uses as a filtering element a fine dacron mesh made into an endless belt which is attached to winding drums. The filter element is precoated by using solka floc or diatomaceous earth as a filter aid. During filtration the cake thickness on the belt increases until a preset pressure drop value is reached. The liquid flow is then stopped and air is introduced to evacuate the liquid in the filter and dry the cake. The filter belt is then moved by an indexing system until the dry cake drops off the belt into a discharge hopper or drum. An air stream jet can be used to remove the cake when it does not drop freely. The advantages and disadvantages of the flat-bed filter are similar to those of the centrifugal-type precoat filters.

Metallic Recleanable Filters

Metallic recleanable or backflushable filter elements are designed to be cleaned to their original level of cleanliness, regardless of the number of loadings to which the element has been subjected. The filter elements consist of a stack of chemically etched discs with micron-size openings between the discs. As manufactured by Vacco Industries, each disc is either 0.002 or 0.003 in. in thickness and is chemically etched on one surface to produce a pore size identical to the required micron rating. Hundreds of identical discs are stacked together, etched surface against nonetched surface, and compressed to ensure flow through etched pores only.

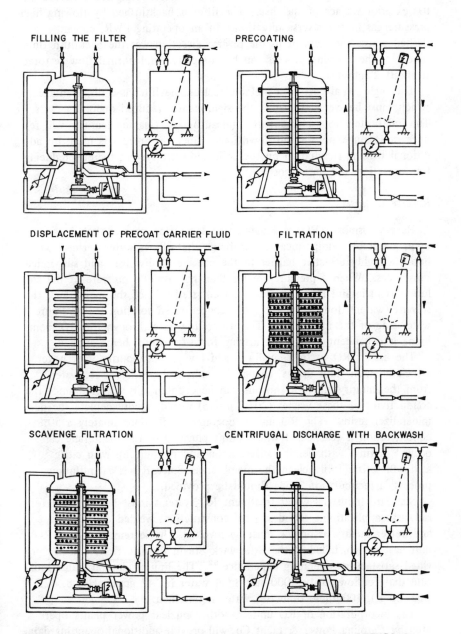

Figure 83. Process flow for a centrifugal precoat filter. Courtesy: NUWCO, Inc.

As the dirty liquid passes through the filter stacks, flowing from outside to inside, all particles greater than the micron rating are collected on the exterior surface of the disc. The filter is backflushed by flowing high-pressure gas in the reverse direction. In an operating BWR, coolant flashed across the element in the reverse direction was the backflush fluid used.[61] The cleaning process can be completed and normal flow restored if about 2 minutes.

The high initial cost of a metallic recleanable filter is a disadvantage which must be weighed against the significantly reduced operating costs. The indefinitely long element life and automatic cleaning features will reduce radiation exposure and downtime. Since no precoat is used, no additional loading is imposed on the solid radwaste system thereby reducing transportation and burial costs.

Reverse Osmosis

Reverse osmosis (RO) is a membrane separation process that depends upon a suitable semipermeable membrane that allows water permeation but is a highly effective barrier to the passage of dissolved and suspended substances. When a pressure greater than the osmotic pressure of the solution is maintained on the more concentrated side of the membrane, the solvent (water) flows from the more concentrated solution to the "pure" water side of the membrane. The water flow rate through the RO membrane is dependent on the net driving force (pressure) being applied.

The use of RO to treat medium- and low-chemical-purity wastewater associated with floor drains, chemical drains, and possibly wastewater from the laundry and hot showers can be economically justified since the conductivity of these wastes (>100 μmho) would rapidly deplete the demineralizer resins. The RO unit is operated until approximately a 50% concentration is reached at which point plugging of the membrane can occur. A rinse with demineralized water and periodically with citric acid will satisfactorily clean the membrane. Since RO is a continuous process, a recirculation mode should be provided.

Standard industrial water treatment RO systems must be upgraded for a radwaste application. Stainless steel construction, welded connections, and better quality assurance must be invoked. A Degremont-Infilco RO unit has been installed at the Brunswick Steam Electric Plant and has provided satisfactory operating experience.[62] The limited data available indicate that for low-purity 6000-mho inlet water (at 47 gpm, pH 6.8, 425 psig, 80°F), 100-μmho or better water is obtained.

The incorporation of RO units at other nuclear power plants operated by Carolina Power & Light Co. will provide additional operating data.

It is anticipated that the use of RO units at new nuclear plants will increase as the value of these units in a coordinated water management system is confirmed by operating experience.

SOLID RADIOACTIVE WASTE TREATMENT SYSTEM DESIGN

In both BWRs and PWRs, the major portion of the solid radioactive wastes results from measures taken to purify the reactor coolant and from treatment of the resulting liquid wastes. Dry compressible radioactive wastes such as rags, paper, clothing, etc. make up the remaining portion. Specifically, the primary sources of solid radwaste are:

1. evaporator bottoms,
2. spent bead resins (from mixed-bed demineralizers),
3. spent powdered resins,
4. filter sludges,
5. spent filter cartridges, and
6. miscellaneous paper, rags, tools, clothing, etc.

Table 62 identifies the expected annual volumes of solid radwastes to be handled by a typical system. Figure 84 shows a flow diagram of a solid radwaste treatment system.[63] Spent resins, evaporator bottoms and filter sludges are stored in tanks for decay of radioactivity, and then transferred to the solid radwaste packaging system which operates in a batch process mode. Solidification of wastes takes place in either 55-gal drums or casks with 50 to 200-ft^3 liners.

Figure 84 shows that spent resins may be dewatered in the shipping container, bypassing the solidification system. This practice has been widespread, but the trend is toward incorporation of spent resins into a solid matrix. The large burial facilities operated by Nuclear Engineering Co., at Morehead, Kentucky, and Nuclear Fuel Services at West Valley, New York, no longer accept dewatered resins.

Spent filter cartridges are placed into a drum or cask with liners. The filter may be shipped in this form with proper shielding added, or the remaining free volume of the shipping container can be filled with concrete.

Compressible solids are placed in 55-gal drums and compressed by a hydraulic baler or mechanical compactor which is completely enclosed. The baler or compactor is vented to the plant ventilation system through a particulate filter to prevent the spread of airborne radioactivity. Noncompressible solids such as contaminated tools and repair parts are placed in suitable containers for offsite shipment.

Shipping containers used for solid radwastes must meet all requirements of Title 49 (Department of Transportation) of the Code of Federal Regulations. Package shielding must be such that radiation levels from packages

Table 62. Quantities of Solid Wastes Per Year

Source	Waste Input to Solid Radwaste System	Solid Waste Volume Shipped from Station	Comments
Spent Bead Resins	$600 ft^3$ gross displacement volume (includes 35% void space)	$600 ft^3$	The $600-ft^3$ shipped volume includes $60 ft^3$ of evaporator bottoms and $150 ft^3$ of solidification agent.
Powdex Resins	$600 ft^3$	$800 ft^3$	Shipped volume based on a 3:1 volume ratio of waste to solidification agent.
Evaporator Bottoms	$3305 ft^3$	$4327 ft^3$	Shipped volume based on a 3:1 volume ratio of waste to solidification agent. $60 ft^3$ of bottoms used to solidify resins was not taken into account.
Filter Cartridges	29 cartridges	29 drums ($214 ft^3$)	One cartridge per drum
Miscellaneous Paper, Cloth, etc.	$4960 ft^3$	$992 ft^3$	A volume compaction ratio of 5:1 in baler.

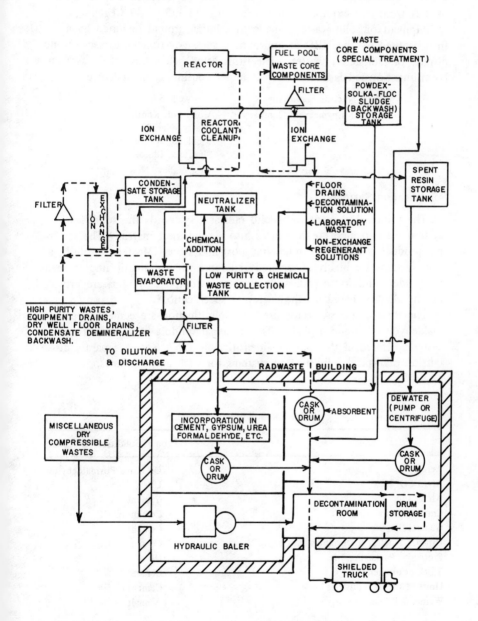

Figure 84. Typical system for treatment of solid radioactive wastes.[63]

will not exceed 200 mrem/hr on contact nor 10 mrem/hr at a distance of 3 ft from any external surface (Section 173.393 of 49 CFR).

Shipment of solid waste is made to a burial ground operated by a firm licensed by the NRC or appropriate state government to receive and dispose of radioactive wastes. The firms listed below bury the waste on government-owned land and keep it under continuous surveillance:

Chem-Nuclear Services q	Barnwell, S.C.
Nuclear Engineering Company	Morehead, Kentucky
	Beatty, Nevada
	Richland, Washington
	Sheffield, Illinois
Nuclear Fuel Services, Inc.	West Valley, New York

Solidification Systems

The principal element of a solid radwaste system is the solidification system used to incorporate wastes into a solid inert matrix. The resulting product is safer to handle, transport and store. Most new solidification systems are highly automated and represent a dramatic improvement over older solid waste processing systems. Complete systems are provided by the vendors based on the customer specifications.

The three principal agents used for solidification are cement, urea formaldehyde (UF) and asphalt. The principal vendors of solidification systems offered in the U.S. are identified in Table 63. A separate discussion of typical solidification systems for each agent follows.

Table 63. Solidification Systems

Vendor	Type of Solidification Agent
Aerojet	UF (Urea Formaldehyde)
ANEFCO, Inc.	UF
Atcor, Inc.	Cement
Chem-Nuclear	Cement or UF
General Electric	Cement
Hittman Nuclear & Development Corp.	Cement or UF
Protective Packaging, Inc.	UF
Stock Equipment	Cement
United Nuclear Industries	Cement or UF
Werner & Pfleiderer Corp.	Asphalt

Cement Systems

Incorporation of radwaste in cement has been widely and routinely used at nuclear power plants for over a decade. Processes for incorporating radwaste in cement are typically operated at ambient temperatures, using equipment ranging from conventional batch mixed to semiautomatic continuous units. Since transport of a fresh cement paste from a mixer into final storage drums can present problems with premature hardening, processes in which mixing takes place in the final storage drum or cask liner have been developed. Incorporation in cement increases the volume of material to be disposed of by a factor of about 2. However, the cement provides shielding and reduces radiation levels on the outside of the shipping container.

Radiation stability of radwaste cement products has not been found to be a problem. Leach rates are relatively low for most radionuclides; however, the more soluble nuclides such as cesium generally require a day additive to ensure a low leach rate. Cement is noncombustible, and the presence of oxidizing agents such as nitrates is not detrimental.

Operation. The initial conditioning of the liquid radwaste is to adjust the moisture content to ensure a proper cement mixture. This is accomplished by the addition or removal of water. Mixing of the waste with cement can take place in the drum or cask, in an in-line screw conveyor mixer, or a conventional concrete mixer. The cement storage bin, of about 500 ft^3 capacity, is filled by pneumatic means to receive truckloads of cement. The bin is equipped with a vibrating device to assure continuous flow of cement.

With an in-line mixer, the waste product and dry cement are introduced to the mixer/feeder simultaneously. As the materials are conveyed to the discharge pipes, the screw flight and paddle arrangement within the mixer/feed performs a thorough mixing action. The cement-waste mix is discharged to the drum at a rate of about 2 ft^3/min until a level detector, sensing the correct fill height, terminates operation. A vibrator is attached to the drum during filling to assure settling of the contents. Capping of the drum is done automatically, with the clamp bolt tightened remotely with a pneumatic wrench. After sealing, a decontamination spray of hot water or steam is used to remove any surface radioactive contamination.

When a large-volume cask liner is used, filling is accomplished through a port in the cask top. Depending on building design, the cask will be mounted on the trailer during filling, or must be transferred from the trailer to the filling station by a cart. A schematic diagram of a 50-ft^3 capacity cask is shown in Figure 85.

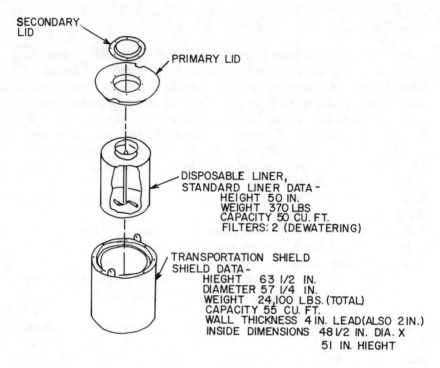

Figure 85. Shipping cask with 50 ft³ disposable liner.

Cement solidification systems are designed with flushing provisions in the waste feed or metering tank, and in other portions of the system which could contain wet cement. Flush water is directed to a miscellaneous waste tank for processing as a liquid waste. These systems also will include provisions for manual (hand-crank) operation of the mixer/feeder, and for manually moving drums in the event of a power failure.

Potential Problems. The operating problems summarized below have occurred with cement solidification systems. Vendor designs should be evaluated to insure that the following potential problems have been solved.

1. Solidification in the cement feed line can occur when liquid waste vapor backs up into the cement feed line.
2. Solidification of waste in the mixer has occurred requiring installation of a temporary system.
3. The ratio of cement-to-waste varies with the waste stream and is a critical parameter in determining the nature of the solid product.
4. Cement dusting problems have occurred at many plants.

Urea Formaldehyde Systems

Incorporation of radwaste in urea formaldehyde polymers is a relatively new process, and only limited results of operating experience and data on physiochemical properties such as leach rate, radiation stability, and effect of oxidizing agents are available. Urea formaldehyde has a shelf-life of about 10 months, with its viscosity building up with time.

Urea formaldehyde (UF) in the presence of a catalyst, such as sodium bisulfate, acts as a solidifying agent. Since the presence of water isn't required, about 50% more waste can be shipped per unit volume than with cement. The reduced shipping costs tend to offset the higher cost of UF as compared to cement.

Operation. The layout and operation of a UF sodlidification system is similar to that of a cement system. The UF from a storage tank is mixed with concentrated waste as it is pumped into a disposable liner or drum. Concurrently, the sodium bisulfate catalyst is sprayed into the liner or drum at the appropriate ratio. In some designs, the catalyst is injected into the waste-UF mixture just prior to its discharge into the shipping container. Solidification into a relatively lightweight cream-colored mass takes place in approximately 1 to 2 hr.

Potential Problems. The major problem with UF systems is the difficulty encountered in solidifying filter sludges consisting of solka-floc and/or diatomaceous earth. Vendor designs should be evaluated to insure that resolution of this potential problem has been adequately demonstrated. Also, the sodium bisulfate catalyst is a mildly corrosive agent and must be handled with care. Corrosion of the storage tank, lines and fittings have occurred at one plant where 304 stainless steel rather than Type 316 SS was selected.

Asphalt Systems

Incorporation of radwaste in asphalt has been used at nuclear research and power plant sites in Europe and over a decade. Although no asphalt process has been used by U.S. nuclear power plants, it is a viable process. The radiation stability of radwaste-asphalt products are lower than those of cement products, in some instances by several orders of magnitude.[64]

Since asphalt is combustible, the risk of fire during processing, storage and transportation must be considered. Research and development work on flammability was performed by Eurochemic (research center in Belgium), with (572°F). Since the highest temperature reached in the process system is 175°C (347°F), a safety margin of about 125°C (225°F) exists.[65]

Although the capital costs of an asphalt system are comparable to those for cement and UF systems, estimates for BWRs indicate the volume of material to be shipped can be reduced by a factor of 10 over cement,[66] thereby reducing operating costs. Since the wastes shipped are more concentrated, additional shielding will probably be required.

Operation. The heart of an asphalt solidification system is the extruder-evaporator. The extruder-evaporator simultaneously provides homogeneous mixing, liquid evaporation and solidification. A schematic diagram for the asphalt solidification system is shown on Figure 86.

The feed streams consist of asphalt and radioactive wastes, each contained in a separate holding tank. The two important factors in the feed system are the pH of the radioactive liquid concentrates and the ratio to be slightly alkaline, in the range of 8 to 10. The ratio of asphalt to radwaste is normally 50:50, although the waste content can go up to 60%.

All of the processing takes place in the extruder-evaporator at near ambient pressure. The feed materials enter the extruder in the first and/or second barrel section where they are enmeshed in the mixing action of the self-cleaning screws. Simultaneous with the mixing, the excess water is evaporated through steam domes located downstream from the feed barrel. The evaporation rate neeced to achieve a 99.5% moisture removal is the prime factor in determining the size extruder required.

The extruder-evaporator discharges the asphalt/radwaste mix into standard 55-gal drums or other size containers. The filled drum is allowed to cool and settle for up to 24 hr before it is capped for storage and transport to a disposal site.

Potential Problems. Although the operating experience in Europe has been good, no U.S. plants employ this process. Licensing of an asphalt solidification system by NRC is a necessary first step before widespread use of this system will occur.

Compaction of Dry Wastes

As previously mentioned, dry waste materials such as paper, clothing, rags, etc. are compacted in 55-gal drums for offsite disposal. The compaction is achieved utilizing either an electromechanical or hydraulic drive mechanism. A force of up to 9 tons is applied. The waste material is compacted to one-third to one-tenth of its original volume depending on the material present. A high-efficiency particulate air filter and blower assembly

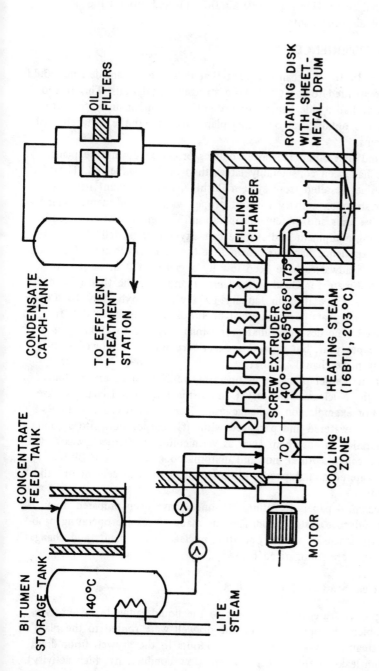

Figure 86. Asphalt solidification system. Courtesy: Werner & Pfleiderer Corp.

are incorporated to provide a positive air flow up and around the drum for removal of airborne contaminants.

OPERATING EXPERIENCE

In the early 1970s, it became evident that radwaste system designs would have to be augmented to meet the more stringent regulations. This forced redesign activity had a side benefit of presenting an opportunity to factor in the experience obtained at operating plants to prevent the recurrence of similar problems.

It was found that liquid radwaste volumes produced during the plant startup phase had been considerably larger than initially expected. These increased volumes resulted from system flushing, frequent plant tripouts, continued construction activity without proper water management, leaking pump and valve seals and system misoperation. The incorporation of additional processing equipment and storage capacity was indicated.

Operating experience has also shown that the design estimates of the volume of solid radwastes have been low by a factor of 2 or more. For example, the estimate for Quad Cities Units 1 and 2 at the time of licensing for 20% power operation in January 1972 was approximately 16,200 ft/yr, but the first years of commercial operation saw nearly 38,000 ft actually shipped. The design estimate for Ginna was 660 to 1000 ft of solid radwaste per year, but operating experience for 1972 showed shipments totaling more than 12,000 ft.

Several of the earlier practices at operating reactor plants are no longer preferred by the Nuclear Regulatory Commission (NRC) and changes have been made. For example, ion exchange resins from both BWRs and PWRs were normally dewatered and packaged without incorporation into a solid matrix. Sometimes an adsorbent such as vermiculite or Microcell was added. Filter sludges, cellulose filter aids and diatomaceous earths from BWRs were also dewatered and shipped without incorporation in cement or other binding agents.

Several technical papers by utility personnel have been prepared documenting the radwaste system modifications made to resolve operating problems. The experiences at several operating nuclear power plants and changes in system design are summarized below:

Dresden Nuclear Station, Units 2 and 3,[6][7]

1. Improper segregation of inputs to the liquid radwaste system was an early problem. Floor drains from the drywell were routed to the plant's floor drain treatment system, though the inputs to the drywell floor drain emanate from leaks in the primary system (low conductivity, high activity).

The high activity hampered operation of the floor drain system. The problem was resolved by rerouting the drywell floor drains to an equipment drain sump.

2. Sump pumps in the radwaste basement were a problem because the overflow of the sludge and resin tanks was to the radwaste floor drain sump. With the potential for personnel radiation exposure being so great if the installed pump malfunctions or is unable to transport the collected material, it is essential to design for the worst-case condition. Sewage lift-type pumps were installed to replace the regular sump pumps initially used.

3. The condensate demineralizers are a deep-bed resin type. An ultrasonic resin cleaner was added to clean the resins and produce greater throughput between regenerations. The original backwash routing was to the equipment drain collector tank. While the backwash was equipment grade water (low conductivity) it was very high in suspended solids. This condition caused filter throughput problems due to the high dirt load introduced into the system. The backwash water was rerouted to a tank to allow for sludge settling. After settling, the clear water is decanted back to the equipment drain collector. These changes imply that where high dirt load is a possibility, the radwaste system design should include provisions for a phase separator tank.

4. Radwaste filters were the focal point of many problems. Short filter runs would often create a condition where the volume of water used for backwashing the filters was greater than the water processed. An evaluation of the problems encountered led to several changes.

a. The location of the precoat filters about 20 ft above the collector tanks and process pumps necessitated the replacement of the isolation valves in the pump discharge line with a valve having more reliable leak-tight shutoff characteristics. This was necessary because draining of the filter vessel due to valve leakage would cause loss of the precoat material.

b. The piping configuration must preclude the introduction of air into the system, since air bubbles collect on the surface of the precoat material and are eventually forced through the precoat leaving a path for dirty water to penetrate the filter. At Dresden, a 6-in. filter inlet line had a pipe segment which drained during backwash and trapped the air during subsequent filling of the vessel. Venting of potential air traps was used to correct this problem.

c. The filter precoat material was backwashed to sludge tanks, with the sludge dewatered with a centrifuge, and the liquid effluent from the centrifuge going back to a collector tank and then to a filter. This arrangement resulted in some suspended solids being recirculated, which led to short filter runs. The filter medium was changed from solka floc to diatomaceous earth to improve filter throughput. It was found that fresh powered resins (Powdex) could be added to the sludge tanks to act as a flocculating agent. This caused the fine suspended solids to settle. Because the Powdex enhanced the settling characteristics of the sludge tanks, the use of Powdex as a bodyfeed material was initiated. This change added ion exchange capabilities to the filter, thus extending the lifeof the deep-bed demineralizer.

d. The existing 180-ft^2 filters were replaced by 380-ft^2 filters. This was done to increase throughput per precoat, to reduce the amount of time and volume of water used in backwashing, and to take advantage of improvements in filter design. The volume for one backwash of the larger filter is less than the volume for two backwashes of the smaller filter.

Monticello Nuclear Plant, BWR[68]

The original treatment method for floor drains (high conductivity, low activity) was to filter, sample, dilute and release to the Mississippi River. Laundry and chemical wastes were handled in the same manner. In an effort to minimize releases, modifications were made to provide for recycling of the floor drains. Design changes were made to combine the treatment of floor drains and equipment drains in one processing drain. Laundry tanks are pumped to the chemical waste tank where the contents are either used for making concrete in the solid radwaste system or, if an excess of water exists, the contents of the chemical waste tank are processed a small amount at a time to the floor drain collector tank for recycling.

Several practices used at Monticello to achieve the goal of minimizing the generation of wastes include the following.

1. Laundry soap which is high in dissolved solids is used sparingly.
2. Floor areas served by the floor drain collection tanks are painted with epoxy to prevent the dusting and high silica problems of bare concrete.
3. Administrative restrictions are placed in many areas on smoking to reduce the loading of solid waste.
4. Rather than regeneration of resins, which requires large quantities of water, the resins are disposed of as solid waste.
5. Wet wiping of spills on floors is done rather than hosing them down.

As was done at Dresden, the original solka-floc precoat for filters at Monticello was changed to powdered resins to add ion exchange capabilities and increase the length of the filter runs. The use of powdered resin as a flocculant aid in the condensate-phase separators reduced the level of suspended solids in the liquid radwaste system.

Problems encountered in the solid radwaste system were associated with solidification of cemented waste within the mixer, and areas around the drum where the waste would splatter. Improvements made included a new cement mixer with improved flush arrangement, new level monitoring instrumentation and drum-fill splatter shields.

Nine-Mile Point, BWR[69]

Prior to upgrading of the liquid radwaste system to meet 10 CFR, Part 50, Appendix I, a waste centrifuge was used to remove solids from the filter sludge tank. It was found that a high concentration of "fines" was carried over in the centrifugate discharged to the floor drain collector tank. Crud that was supposed to be removed ended up being recycled. A traveling-belt filter precoated with diatomaceous earth was installed and has efficiently removed sludge and fines from the filter sludge tank. As a result the waste centrifuge is now used very little.

Problems with candle-type filters were experienced when solid precoat material built up between the septa of the filter, and created a bridge such that little filtering surface was left. The bridging was attributed to poor backwashing of the filters. After changing the stainless steel septa to polypropylene, better backwashing and filter performance was obtained. The better backwashing of the plastic septa is probably due to the fact that the poly "sock" of the septa flexes during backwashing.

Waste concentrator tube plugging became a problem when Nine-Mile Point began to evaporate floor drain water to achieve as low as practicable release of liquids. The plugging of the heat exchanger tubes was caused by calcium silicate—the solidifying agent used on the solid radwaste area and ending up in the concentrator. Plugging problems ceased after the solidifying agent was changed to urea-formaldehyde.

R. E. Ginna Nuclear Plant, PWR[70]

Waste evaporator performance at this plant and H. B. Robinson Steam Electric Plant, Unit 2, did not provide a sufficient decontamination factor (DF). The unit was very sensitive to the presence of soap and detergents, with foaming causing a carryover of activity into the distillate tank. To correct this, piping from laundry and hot shower tanks to the waste condensate tanks was separated from the chemical drain tank. A demineralizer was added to the distillate discharge from the evaporator to improve system DF. An additional modification was to replace the concentrator's mechanical vacuum pump with an air ejector system to improve vacuum and concentrator throughput.

H. B. Robinson Steam Electric Plant, Unit 2, PWR[71]

Problems were encountered with both the waste evaporator (WE) and boric acid evaporator (BAE) systems. Specific problems were:

1. insufficient supply of steam to hot water converter (WE),
2. considerable foaming and carryover into distillate tank (WE),
3. difficulty in maintaining a vacuum in concentrator (WE and BAE), and
4. inability to achieve decontamination factors on the design order of 10^6 or design flow rate of 20 gpm (WE).

An effort to correct the foaming problem by the use of an antifoaming agent, DOW H-10, was implemented. The vacuum problems on the WE and BAE were corrected by replacing the mechanical vacuum pumps with steam ejectors. Additional modifications performed on the WE systems, based on experience at other plants (see Ginna) included:

1. Modification of the concentrator internals to relocate the reflux line, add a new moisture separator section, provide a new sieve moisture separator, and install a heater bundle shrod (these modifications reduced moisture carryover and improved DF).
2. Replacement of the Teflon* disc on the concentrator level control valve with a 316SS disc (this modification eliminated level control problems due to damage to the Teflon disc).
3. Installation of a higher capacity steam valve and increasing the steam supply piping (this increased unit output).
4. Provision of additional connections for sampling and chemical addition (improves control of foaming and plugging).

Additional tanks for waste condenstate were installed to increase holdup capacity and enhance operating flexibility.

Zion Nuclear Station, Unit 1,[72]

Difficulties were experienced with the operation of the evaporator and demineralizers. Borated liquid waste flow to the radwaste system exceeded predicted amounts. Modifications included the addition of a second, larger waste evaporator. Also, four 30-gpm reverse osmosis units, and two mixed-bed polishing demineralizers were added to the steam generator blow-down demineralizer system. These additions increased the boron removal ratio and provided sufficient capacity to handle the volume of liquid wastes being generated.

---

*Registered trademark, of E. I. duPont de Nemours & Company, Wilimington, Delaware.

Summary of Operating Experience Lessons

The experiences described above, and other examples reported in the literature,[73] are being used at newer plants to improve the performance of liquid and solid radwaste systems. A summary of lessons learned includes:

1. Frequent hosing of floors should be avoided, because it adds to the liquid radwaste/volume. Spills and leaks should be removed by dry mopping or with minimal water use.

2. Maximize internal water reuse. Wash fuel casks with recycle water, except for final rinse with clean condensate. Use recycle water as the seal water for radwaste pumps.

3. Equipment containing nonradioactive ciruclating or service water must not be drained into a floor drain system that becomes part of the liquid radwaste system.

4. Drycake discharge filters (traveling-bed filters) save a lot of processing water and can discharge directly into a solid waste shipping container.

5. Waste concentrators should be designed to handle all types of liquid waste. They should be equipped with forced circulation and a submersed heat exchanger.

6. The use of powdered resin (Powdex) on precoat filters reduces liquid radwaste discharges.

7. Use demineralized makeup water sparingly. Every gallon produced by the makeup system must either be lost by evaporation through the plant ventilation system or it must be discharged as liquid radwaste.

8. Build as much as possible of the solid radwaste system on the ground floor to avoid moving solid waste upstairs and downstairs.

9. Design the solid waste handling system around the loading dock. This is where all solid waste must eventually end up; therefore, it is a good place to develop the design.

RULES AND REGULATIONS GOVERNING THE RELEASE OF RADIOACTIVITY

Releases of radioactivity in nuclear power plant effluents have always been controlled by the Nuclear Regulatory Commission (formerly the Atomic Energy Commission) to protect the health of the general public. Since 1970, the NRC has proposed and then implemented more stringent requirements on the amount of radioactivity that could be released as either gaseous or liquid effluents. The reduction in radioactivity released has meant additional processing and purification in-plant. In addition, changes in environmental regulations related to thermal pollution have lead to the elimination of once-through cooling in new plants, and the use instead of closed-condenser cooling water systems. Allowing only for periodic blowdown of the heat sink, this has effectively reduced the dilution flow by 90 to 95%. Since discharge limits have applied to the concentration of radioactivity, the total quantity of radioactive effluent has been reduced. As a result the volume of water handled by the liquid radwaste system and the generation of solid radwastes has increased.

Background

Title 10 of the Code of Federal Regulations (10 CFR) contains regulations which establish standards for protection against radiation hazards arising from nuclear power plant activities. Part 20 of 10 CFR specifies external and internal radiation exposure limits for individuals who work within the plant's controlled boundaries. Limits are placed on the radiation doses that can be received, and also on the concentrations of airborne radioactive materials. Part 20 also specifies limits on radiation dose and concentration in unrestricted areas which are not controlled by the nuclear plant owner. Part 20 requires that the applicant for a license to operate a nuclear power plant demonstrate that any member of the general public will not receive a radiation dose in excess of 0.5 rem (500 millirem) in one year.

In December 1970, the AEC published amendments to the licensing requirements in 10 CFR, Part 50, which specify design and operating requirements for nuclear power plants, to keep levels of radioactivity in effluents to unrestricted areas as low as practicable. These amendments provided qualitative guidance, but not numerical criteria, for determining whether the requirements were met. In June 1971 the AEC published a proposed new Appendix I to 10 CFR, Part 50, to establish numerical guides for design objectives for new nuclear power plants. The proposed amendments specified that existing power plants would have to establish limiting conditions for operation which would conform to the numerical guides, within 36 months after formal issue of the amendments.

Extensive rulemaking hearings were instituted by the AEC to evaluate the practicality and environmental impact of releasing radioactive material in effluents from nuclear power plants within the levels defined in proposed Appendix I. An Environmental Impact Statement was prepared and issued by the AEC in July 1973, documenting the information used to substantiate slightly revised design objectives.[74] In the meantime, operating plants and plants in the design and construction state initiated actions to augment the plant's radioactive waste treatment systems to meet the anticipated stringent requirements.

Current Regulations—10 CFR, Part 50, Appendix I

Specific numerical guidelines for radioactive releases, which establish design objectives for radioactive waste treatment systems, and limiting conditions for plant operation were published May 5, 1975, and became effective on June 4, 1975. With respect to liquid radwaste systems, the regulations state:

> "The calculated annual total quantity of all radioactive material above background to be released from each light-water-cooled nuclear power reactor to unrestricted areas will not result in an estimated annual dose commitment from liquid effluents for any individual in an unrestricted area from all pathways of exposure in excess of 3 millirems to the total body or 10 millirems to any organ."

The adopted design objective guides do not contain specific limits upon quantities of radioactive material to be released, since the rulemaking hearing record showed this has little, if any, independent significance.

In addition to the effluent release objectives, the adopted regulation includes a requirement for further augmenting the radwaste system based on a cost benefit analysis. The plant must "...include in the radwaste system all items of reasonably demonstrated technology that, when added to the system sequentially and in order of diminishing cost benefit return, can for a favorable cost benefit ration effect reductions in dose to the population reasonably expected to be within 50 miles of the reactor." The NRC will use as an interim measure, a conservative value of $1000 per total-body man-rem and $1000 per man-thyroid rem, unless in individual cases a lower value can be demonstrated.

Cost/benefit analyses performed by the NRC staff have shown that when current plants meet the design objectives, additional radwaste equipment cannot be added for less than $1000/man-rem. As a result, NRC modified Appendix I on September 4, 1975 to exempt from the cost/benefit analysis provisions all nuclear plants which have a construction permit application docketed prior to June 4, 1976.

A further modification to Appendix I became effective on January 19, 1976 when the term "as low as is reasonably achievable" was substituted for "as low as practicable." The new term is more understandable and is consistent with terminology used by the International Commission on Radiological Protection.

Design Guidance Provided by NRC

In order to achieve effective implementation of its regulations, NRC issues design guidance in the form of regulatory guides, and technical position papers issued by individual regulatory staff review branches. Pertinent examples of such guidance and a summary of design criteria in each document follow:

Branch Technical Position, ETSB 11-1—
Design Guidance for Radioactive Waste Management Systems

The majority of liquid radwaste system piping and components should be designed to "Quality Group D (Augmented)" requirements. This classification calls for conventional fabrication codes except that welded joints for pressure-retaining components should be used to the maximum practicable extent, welding should be in accordance with ASME Pressure and Vessel Code Section IX, and piping systems should be hydrostatically tested. Portions of radwaste systems with low radioactivity inventories, such that complete release to the environment would not exceed allowable limits, and equipment used to collect detergent wastes do not require any augmented requirements. Also portions of the steam generator blowdown system on PWRs should be designed to ASME Code Section III, Class 2 requirements and be capable of withstanding the design basis (safe shutdown) earthquake for the plant. Retention capabilities to prevent uncontrolled release due to spillage should include dikes or ponds around outdoor tanks, and curbs around indoor equipment and tanks.

For solid radwaste systems, "Quality Group D (Augmented)" classification should be applied to all equipment which handles radioactive liquids (for transfer back to liquid radwaste system) and solids (spent resins and evaporator bottoms). Equipment used to solidify aqueous and nonaqueous liquid wastes can be built to conventional standards.

Branch Technical Position, ETSB 11-3—
Design Guidance for Solid Radioactive Waste Management Systems

Criteria in this branch position establish that all wastes must be in a solid immobile form prior to shipment offsite. Spent resins and filter sludge should be combined with a suitable binding agent (*e.g.,* cement, urea-formaldehyde) and formed into a solid matrix. Absorbants, such as vermiculite, are not acceptable substitutes for solidification. In order to assure complete solidification, a process control program must be included which will detect the presence of free liquid within the shipping container.

With regard to the compaction of dry wastes, special provisions, such as a ventilated shroud around the waste container, are needed to assure contaminated airborne dusts are not released to the process area. Storage areas for solidified waste should accommodate at least 30 days waste generation at normal generation rates. Storage tanks for spent resins should accommodate at least 60 days waste generation.

Regulatory Guide 1.21– Measuring, Evaluating and
Reporting Radioactivity in Solid Wastes and Releases
of Radioactive Materials in Liquid and Gaseous Effluents

In this guide the regulatory staff describes programs for measuring, reporting, and evaluating releases of radioactive materials in liquid effluents, and guidelines for reporting the categories and curie content of solid wastes. All major and potentially significant paths for liquid radioactive releases should be monitored. Measurements of effluent volume, rates of release, and specific radionuclides are required. The sensitivities of analyses of radioactive materials in liquid effluents are identified. This guide establishes that during the release of radioactive wastes, the effluent control monitor should be set to alarm and to initiate automatic closure of the waste discharge valve prior to exceeding specified limits.

Regulatory Guide 1.110– Cost/Benefit Analysis for Radwaste
Systems for Light-Water-Cooled Nuclear Power Reactors

In this guide the NRC staff outlines a method for performing the cost/benefit analysis required by Appendix I to 10 CFR, Part 50. The methodology and cost parameters are presented in Appendix A, and the basis for the costs calculated are given in Appendix B to this guide. The cost considered includes direct equipment costs, cost of building space, supportive services, maintenance, interest, operating and other costs generally considered in analyzing capital and operating costs in power plant estimating.

Additional regulatory guides have been issued to define acceptable methods for calculating doses to man from routine radioactive effluent releases to evaluate compliance with Appendix I. Since these guides bear only indirectly on radwaste system design, they are not discussed in this paper.

RADIOACTIVITY RELEASES IN LIQUID EFFLUENTS
FROM NUCLEAR POWER PLANTS

Releases of radioactivity in airborne and liquid effluents, as well as the number of shipments and activity in solid wastes, are routinely reported to the NRC by all nuclear power plants. This information is compiled and evaluated in a report by NRC that is issued yearly. The information for calendar year 1973 is presented here as the information for calendar year 1974, reported to NRC in early 1975, has not yet been published.[75]

The release data given are for the 37 nuclear power plants which were in commercial operation in 1973. Table 64 is a summary of liquid releases and approximate contributions of fixed fission and activation products,

Table 64. Summary of Radioactivity in Liquids Released to the Environment in
Calender Year 1973

	Liquid Waste Volume (10^5 liters)	Activity (Ci)	Percent of Activity Contributed by		
			Mixed Fission and Activation Products	Tritium	Dissolved Gases
Boiling-Water Reactors					
Oyster Creek 1	124.0	41.0	5.4	86.8	7.2
Nine Mile Point 1	140.0	87.0	46.7	53.2	0.1
Millstone 1	98.0	40.4	90.8	9.1	0.1
Dresden 1	31.0	27.8	33.3	66.7	0
Dresden 2, 3	256.0	50.9	49.4	50.6	0
Genoa	24.6	139.0	25.7	74.2	0.1
Monticello	$-^a$				
Big Rock Point	8.0	21.0	6.5	93.5	0
Humboldt Bay 3	19.2	54.0	4.4	95.6	0
Pilgrim 1	8.8	1.3	67.9	32.1	0
Quad Cities 1, 2	330.0	46.0	46.5	53.0	0
Vermont Yankee	1.2	0.2	0.01	99.9	0
Peach Bottom 2	162.0	0.0001	10.3	89.7	0
Pressurized Water Reactors					
Maine Yankee	970.0^b	154.0	0.1	99.7	0.01
Palisades	190.0	212.0	11.7	87.1	1.2
Yankee (Rowe)	170.0	694.0	<0.001	99.9	0.01
Indian Point 1	88.0^c	139.0	0.9	99.0	0.1
Indian Point 2	79.0	31.0	9.5	88.3	2.3
R. E. Ginna 1	17.0	286.0	0.01	99.9	0.0001
Haddam Neck	270.0	3904.0	<0.01	99.9	<0.01
H. B. Robinson 2	34.0	432.0	0.01	99.8	0.01
San Onofre 1	90.0	4138.0	0.01	98.4	1.1
Point Beach 1,2	58.3	577.0	<0.01	99.8	0.001
Surry 1, 2	160.0	449.0	0.001	99.6	0.1
Ft. Calhoun 1	33.0	16.0	2.0	97.9	0.1
Prairie Island 1	244.0	$-^d$			
Oconee 1,2	140.0	79.0	10.7	89.3	0
Turkey Point 3,4	82.0	329.0	0.001	99.9	0.001
Zion 1, 2	0	$-^d$			
Gas-Cooled Reactor					
Peach Bottom 1	7.4	3.8	0.1	99.4	0

[a]No liquid releases.
[b]From test tank secondary plant blowdown—secondary plant leakage.
[c]This plant was shut down the entire year.
[d]No significant activity.

Table 65. Liquid Effluent Trend: Mixed Fission and Activation Products (Ci)

	1970	1971	1972	1973	Average[a]
Boiling-Water Reactors					
Oyster Creek 1	18.5	12.0	10.0	4.2	11.2
Nine Mile Point	28.0	32.2	34.6	40.8	33.9
Millstone 1		19.7	51.5	33.4	34.9
Dresden 1	8.2	6.2	6.8	9.2	7.6
Dresden 2, 3		23.0	22.0	25.9	23.6
Genoa	6.4	17.1	48.5	35.9	27.0
Monticello		0.01	$<10^{-6}$	$-^b$	0.005
Big Rock Point	4.7	3.5	1.1	2.7	3.0
Humboldt Bay 3	2.4	1.8	1.4	2.4	2.0
Pilgrim 1			1.5^c	0.9	1.2
Quad Cities 1,2			2.4	21.4	11.9
Vermont Yankee			0^c	$<10^{-4}$	$<10^{-5}$
Peach Bottom 2				$<10^{-4c}$	$<10^{-4}$
Pressurized-Water Reactors					
Maine Yankee			0.02	$<10^{-7}$	0.01
Palisades			6.8	27.8	17.3
Yankee (Rowe)	0.03	0.01	0.02	0.004	0.02
Indian Point 1^d	7.8	81.1	25.4	0.6	28.7
Indian Point 2				2.2^c	2.2
R. E. Ginna 1	10.0	0.9	0.3	0.07	2.8
Haddam Neck	6.7	5.9	4.8	3.0	5.1
H. B. Robinson 2		0.7	0.8	0.6	0.7
San Onofre 1	7.6	1.5	30.3	16.0	13.9
Point Beach 1,2		0.1	1.5	0.8	0.8
Surry 1,2			0.2^c	0.1	0.1
Ft. Calhoun 1				0.01^c	0.01
Prairie Island				$<10^{-8\ c}$	$<10^{-8}$
Oconee 1,2				2.8^c	2.8
Turkey Point 3,4				0.03^c	0.03
Zion 1,2				$<10^{-4c}$	$<10^{-4}$
Gas-Cooled Reactor					
Peach Bottom 1	0.006	0.007	0.02	0.0001	

[a]Average calculated over the period of unit operations.
[b]No liquid releases.
[c]Did not operate a full year.
[d]This unit was shut down during all of 1973.

tritium, and dissolved noble gases. Trend data on a plant-by-plant basis are shown in Table 65 for 1970 through 1973. Wide variations in the amounts of radioactivity released are due to (1) the differences in fuel performance, (2) the amount of power produced, (3) the extent to which effluent treatment systems were used, (4) the improvements made in installed effluent treatment systems and (5) the improved methods used for measuring radioactive effluents.

The data reported show, and the results of NRC's inspection program have confirmed, that the established limits on amounts of radioactivity released were not exceeded and were low in comparison with the 10 CFR, Part 20, limits.

SUMMARY

The design of current radioactive waste treatment systems for liquid and solid wastes in operating plants and those plants under construction has changed considerably over the last several years. With the adoption of new and more stringent federal regulations (10 CFR, Part 50, Appendix I), increasing emphasis has been placed on the minimization of liquid discharges and the maximization of water reuse within the plant. Operating experiences have identified the need to better educate and train plant personnel in operating procedures which implement the philosophy of minimum water use and minimum radiation exposure.

Resolution of equipment problems at operating plants have been factored into new plant designs. As a result the volume of wastes processed has been reduced. Also, new and more sophisticated equipment (*i.e.,* urea-formaldehydr and asphalt solidification systems, reverse-osmosis units) have been adopted for radwaste service with a resulting reduction in waste volumes.

Although difficulties have been encountered in radwaste systems on operating nuclear power plants, the overall performance record has been excellent. No radioactive effluent releases to the environment have been maintained at only a small fraction of the legal limits.

CHAPTER 11

PETROLEUM REFINERY ENVIRONMENTAL ASSESSMENT

The guidelines presented in Chapter 3 for discussion of the various environmental parameters are general for all cases of writing an assessment. The following discussion is specific to the petroleum refinery industry to aid in the environmental assessment of many of its unique problems.

IMPACT IDENTIFICATION

The identification of the potential impacts is the first step in an environmental assessment of a proposed project. A process flow diagram of the petroleum refinery's waste streams is included in the EIS. Characteristic wastes with regard to the products developed are listed in Table 66. Data such as these are necessary inputs to the EIS.

Effluents should be defined for each process in the refining operation. Thus, we will examine the sources of the major air, water and solid waste effluents. Table 67 identifies and estimates the percentage usage of various manufacturing processes by petroleum refineries. In Table 68 is a list of the common pollutant emissions from typical hydrocarbon unit operations.

WASTEWATER

In the handling, transportation and storage of crude oil and products, residues can affect wastewaters through spills and leaks, tank-cleaning operations, and ballast waters from tankers. Oil, finished product, water and other residues on storage tank bottoms (*i.e.*, product, intermediate and crude storage tanks) are a major source of watewaters. Filters and filter media also contribute to effluents.

In the crude desalting operation, the desalter wastewaters are produced, and each subsequent refining process generates its characteristic

Table 66. Petroleum Refinery Products and the Wastes Generated
by Their Processing[76]

Products	Wastes
Aldehydes and alcohols by the Oxo method	Rectification residues with dissolved hydrocarbons and aldehydes
Hydrogen cyanide from natural gas and ammonia	Distillation residues (from the cyanide stripping process) with a small HCN and unreacted hydrocarbon content
Chlorinated derivatives of methane and ethylene	Column effluents containing chlorinated hydrocarbons, salinated to a certain extent with lime sludge
Aceylene from hydrocarbon cracking	Carbon black, hydrocarbons and hydrogen cyanide
Ethylene and propylene by thermal cracking	Wash liquors, oily wastes
Polymerization and alkylation	Alkaline wastes (NaOH), hydrocarbons, benzene derivatives, catalysts (phosphoric acid, aluminium chloride)
Alcohols from olefins by sulfonation and hydrolysis	Large quantities of wastes containing sodium sulphate, polymerized hydrocarbons and butyl or isobutyl alcohol
Aldehydes and alcohols by oxidation of hydrocarbons	Organic acids, formaldehyde, acetaldehyde, acetone, methanol and higher alcohols
Butylene from butane and butadiene from butylene	Small quantities of wastes with a high hydrocarbon content
Aromatic hydrocarbons by reforming processes	Condensates polluted with catalysts, hydrogen sulfide and ammonia

wastewaters. Crude oil fractionation generates wastewaters from overhead accumulators, oil sampling lines, and barometric condensers. Thermal cracking of crude oil produces wastewaters from the overhead accumulator. The waste stream from catalytic cracking must also be considered due to steam stripping.

Hydrocarbon rebuilding produces polymerization wastewaters, sulfuric acid alkylation and neutralization wastewaters, overhead accumulator wastewaters, and hydrofluoric acid alkylation rerun-unit wastewaters. Wastewaters are generated by various processes including hydrocarbon rearrangement, solvent refining, hydrotreating, grease manufacturing, asphalt production, and product finishing. Product-finishing operations include drying and sweetening (H_2S removal) processes, acid treatment,

Table 67. Estimated Percentage of Petroleum Refineries Using Various Manufacturing Processes[77]

	1950	1963	1967	1972	1977	Technology[a]
Crude Oil Desalting		100%	100%	100%	100%	
Chemical desalting		5	2	0	0	O
Electrostatic desalting		95	97	100	100	T,N
Crude Distillation	100%	100	100	100	100	
Atmospheric fractionator	100	100	100	100	100	O,T,N
Vacuum fractionator		60	64	70	75	O,T,N
Vacuum flasher						O,T,N
Thermal Cracking	59	48	45	40	35	
Thermal cracking		28	18	8	2	O
Delayed coking		12	14	16	19	T,N
Visbreaking		13	16	18	22	T,N
Fluid coking		2	2	4	6	T,N
Catalytic Cracking	25	51	56	60	65	
Fluid catalytic cracking		39	45	50	60	T,N
Thermofor catalytic cracking		13	12	10	6	O
Houdriflow		3	3	2	0	O
Hydrocracking	0	2	8	25	34	
Isomax			4	11	15	N
Unicracking			2	8	12	N
H-G hydrocracking		0.3	0.8	3	3	N
H-oil			0.4	1	1	N
Reforming		62	67	74	79	
Platforming		37	40	44	47	O,T,N
Catalytic reforming-Engelhard		5	9	11	12	O,T
Powerforming		1	2	3	3	T,N
Ultraforming		6	6	7	8	T,N
Polymerization	25	42	33	26	7	
Bulk acid polymerization						T,N
Solid phosphoric acid condensation						T
Sulfuric acid polymerization						O
Thermal polymerization		1	0.4			O
Alkylation	10	38	47	54	62	
Sulfuric acid alkylation		22	26	32	38	T,N
HF alkylation		16	21	22	25	O,T,N
DIP alkylation						N
Thermal alkylation						O
Isomerization		5	7	10	15	
Isomerate		1	1.5	3	6	N
Liquid-phase isomerization		2	3	4	5	N
Butamer		1	1	2	2	N
Penex		0.7	1	1	2	N

Table 67, continued.

	1950	1963	1967	1972	1977	Technology[a]
Solvent Refining		25%	29%	30%	32%	
Furfural refining		14	15	16	16	O,T,N
Duo-Sol		2	3	3	3	T,N
Phenol extraction		10	10	11	11	O,T,N
Udex		3	5	8	8	T,N
Dewaxing		11	11	11	11	
Solvent dewaxing (MEK)		8	8	9	9	O,T,N
Propane dewaxing		2	2	2	2	O,T
Pressing and sweating		1	1	0	0	O
Hydrotreating		47	56	70	80	
Unifining		22	23	30	35	T,N
Hydrofining		3	3	5	8	T,N
Trickle hydrodesulfurization		0.3	2	4	5	T,N
Ultrafining		3	5	8	10	T,N
Deasphalting		20	23	25	27	
Propane deasphalting and fractionation		15	18	20	21	O,T,N
Solvent decarbonizing		4	5	5	6	T,N
Drying and Sweetening		80	80	80	80	
Copper sweetening						O,T
Doctor sweetening						O
Merox						N
Girbotol						O,T,N
Wax Finishing		11	11	11	11	
Wax fractionation		10	9	6	5	O,T
Wax manufacturing, MIBK		1	1	1	1	O,T
Hydrotreating			1	4	5	N
Grease Manufacture		12	12	10	10	O,T,N
Lube Oil Finishing		19	19	20	20	
Percolation filtration		11	7	5	2	O,T
Continuous contact filtration		6	7	7	7	O,T
Hydrotreating		2	5	8	11	N
Hydrogen Manufacture		2	8	25	34	
Hydrogen partial oxidation		1	3	10	12	N
Hydrogen, steam-reforming		1	5	15	22	N
Total Number of Refineries	346	293	261	236	211	

[a]O, older; T, typical; N, newer.

Table 68. Frequent Pollutant Emissions from some Commonly Encountered Hydrocarbon Processing Units[78]

Process Unit	Source of Pollutant	Pollutant	Possible Terminal Remedy
Crude Running	Electrical desalting Steam distillation	Brine Hydrocarbon emulsions	Deep-well injection Settling and flocculation
Catalytic Cracking	Catalyst regeneration Reactor effluent Reactor effluent	Catalyst fines Hydrogen sulfide Phenolic compounds	Cyclones and electrical precipitators Scrubbing with absorbent Bioxidation
Polymerization, Alkylation, Isomerization	Catalysts	Spent acid catalysts Aluminum chloride	Regeneration Dispersion
Product Treating	Sweetening agents	Spent caustic Spent acid Salts	Regeneration with steam, air Co-product sales Dispersion
Catalytic Reforming	Reactor effluent	Hydrogen sulfide, ammonia	Scrubbing with absorbent
Aromatics Extraction	Extract water washing Solvent regeneration	Aromatic hydrocarbons Solvents	Settling and flocculation Bioxidation
Steam Cracking for Olefins	Furnace effluent	Acid gases Phenolic compounds Spent caustic Polymerization products	Scrubbing with absorbent Bioxidation Regeneration, deep well injection Combustion
Styrene from Benzene	Reactor effluents Waste catalyst	Aromatic hydrocarbons Aluminum chloride	Settling and flocculation Decomposition
Thermal Cracking for Acetylene	Furnace effluent	Soot Acid gases	Combustion, filtration Scrubbing with absorbent
Oxychlorination Units	Reactor effluent	Chlorine, hydrogen chloride By-products	Scrubbing with absorbent Co-product sales, recycling
Hydrocarbon Oxidation for Aldehydes, Alcohols	Process slops	Acetone, formaldehyde, acetaldehyde, and soluble products	Bioxidation
Ethylene Oxide, Ethylene Glycol, Propylene Oxide, and Propylene Glycol Manufacture	Process slops	Calcium chloride, hydrocarbon polymers, and soluble products	Co-product sales, bioxidation
Acrylonitrile Manufacture	Wastewater	Cyanides	Oxidation with chlorine
Phenol Manufacture	Wastewater	Phenolics	Bioxidation, deep-well injection
Ammonia Manufacture	Anion, cation demineralization Regeneration, process condensate	Acids, bases Ammonia	Ponding for neutralization Steam-air stripping

rinsing processes, blending processes, cleaning tank cars and tankers before loading, and the processing of additives. For example, tetraethyl lead (TEL) and tetramethyl lead (TML) are gasoline additives which are anti-knock agents. These compounds are extremely toxic and can gain entrance to wastewaters via two avenues:

1. TEL and TML are separated from other compounds by a steam distillation and purification process. Water is then contaminated by the condensing steam.

2. TEL and TML are present in tank bottom sludges and contaminate waters through washings and other maintenance.

Other wastewaters are produced during hydrogen manufacturing, power generation, refrigeration (cooling towers) and recirculation operations. Table 69 lists the various petrochemical processes as wastewater sources.

SOLID WASTES

Typical solid wastes generated at a refinery include process sludges, spent catalysts, waste materials and various sediments. Along with identification on the waste stream a flow diagram should provide a quantitative and qualitative description of its characteristics. Refinery solid wastes are grouped into three general categories: process solids, effluent treatment solids, and general waste. Tables 70, 71, and 72 give the sources, characteristics and quantities of solid wastes generated. Recovery should always be considered to reduce volumes. Disposal practices can involve separation into landfill wastes and incineration wastes. Wastes which are innocuous or have been neutralized can be landfilled. Hazardous wastes may require further processing including incineration. At this time energy recovery has become cost-effective and must be considered.

AIR EMISSIONS

All refinery air emissions must be identified, described and quantified. Interim heat releases, start-up, shut-down, safety valve releases, leaks and other potential sources are to be assessed. The refining operation includes all facilities related to the oil processing project such as marine terminals, storage facilities and distribution network. Table 73 identifies potential sources of specific emissions from oil refineries. The major sources of air emissions from a petroleum refinery are listed.[13,80]

Table 69. Petrochemical Processes as Waste Sources[79]

Process	Source	Pollutants
Alkylation: Ethylbenzene		Tar, hydrochloric acid, caustic soda, fuel oil
Ammonia Production	Demineralization	Acids, bases
	Regeneration, process condensates	Ammonia
	Furnace effluents	Carbon dioxide, carbon monoxide
Aromatics Recovery	Extract water	Aromatic hydrocarbons
	Solvent purification	Solvents—sulfur dioxide, diethylene glycol
Cyanide Production	Water slops	Hydrogen cyanide, unreacted so soluble hydrocarbons
Dehydrogenation		
Butadiene production from n-butane and butylene	Quench waters	Residue gas, tars, oils, soluble hydrocarbons
Ketone production	Distillation slops	Hydrocarbon polymers, chlorinated hydrocarbons, glycerol, sodium chloride
Styrene from ethylbenzene	Catalyst	Spent catalyst (Fe, Mg, K, Cu, Cr, Zn)
	Condensates from spray tower	Aromatic hydrocarbons, including styrene, ethylbenzene, and toluene, tars
Desulfurization		Hydrogen sulfide, mercaptans
Extraction and Purification		
Isobutylene	Acid and caustic wastes	Sulfuric acid, C_4 hydrocarbon, caustic soda
Butylene	Solvent and caustic wash	Acetone, oils, C_4 hydrocarbon, caustic soda, sulfuric acid
Styrene	Still bottoms	Heavy tars
Butadiene absorption	Solvent	Cuprous ammonium acetate, C_4 hydrocarbons, oils
Extractive distillation	Solvent	Furfural, C_4 hydrocarbons
Halogenation (Principally Chlorination)		
Addition to olefins	Separator	Spent caustic
Substitution	HCl absorber, scrubber	Chlorine, hydrogen chloride, spent caustic, hydrocarbon isomers and chlorinated products, oils
	Dehydrohalogenation	Dilute salt solution
Hypochlorination	Hydrolysis	Calcium chloride, soluble organics, tars
Hydrochlorination	Surge tank	Tars, spent catalyst, alkyl halides
Hydrocarboxylation (OXO Process)	Still slops	Soluble hydrocarbons, aldehydes
Hydrocyanation (for Acrylonitrile, Adipic Acid)	Process effluents	Cyanides, organic and inorganic
Isomerization in General	Process wastes	Hydrocarbons; aliphatic, aromatic and derivative tars

Table 69, Continued

Process	Source	Pollutants
Nitration Paraffins		By-product aldehydes, ketones, acids, alcohols, olefins, carbon dioxide
Aromatics		Sulfuric acid, nitric acid, aromatics
Oxidation Ethylene oxide and glycol manufacture	Process slops	Calcium chloride, spent lime, hydrocarbon polymers, ethylene oxide, glycols, dichloride
Aldehydes, alcohols, and acids from hydrocarbons	Process slops	Acetone, formaldehyde, acetaldehyde, methanol, higher alcohols, organic acids
Acids and anhydrides from aromatic oxidation	Condensates Still slops	Anhydrides, aromatics, acids Pitch
Phenol and acetone from cumene oxidation	Decanter	Formic acid, hydrocarbons
Carbon black manufacture	Cooling, quenching	Carbon black, particulates, dissolved solids
Polymerization, Alkylation	Catalysts	Spent acid catalysts (phosphoric acid), aluminum chloride
Polymerization Polyethylene	Catalysts	Chromium, nickel, cobalt, molybdenum
Butyl rubber	Process wastes	Scrap butyl, oil, light hydrocarbons
Copolymer rubber	Process wastes	Butadiene, styrene serum, softener sludge
Nylon 66	Process wastes	Cyclohexane oxidation products, succinic acid, adipic acid, glutaric acid, hexamethylene diamides, adiponitrile, acetone, methyl ethyl ketone
Sulfation of Olefins		Alcohols, polymerized hydrocarbons, sodium sulfate, ethers
Sulfonation of Aromatics	Caustic wash	Spent caustic
Thermal Cracking for Olefin Production (Including Fractionation and Purification	Furnace effluent and caustic treating	Acids, hydrogen sulfide, mercaptans, soluble hydrocarbons, polymerization products, spent caustic, phenolic compounds, residue gases, tars and heavy oils
Utilities	Boiler blowdown	Phosphates, lignins, heat, total dissolved solids, tannins
	Cooling system blowdown	Chromates, phosphates, algacides, heat
	Water treatment	Calcium and magnesium chlorides, sulfates, carbonates

Table 70. Sources and Characteristics of Refinery Solid Waste[13]

Type of Waste	Sources	Description	Characteristics
Process Solids	Crude oil storage, desalter	Basic sediment and water	Iron rust, iron sulfides, clay, sand, water, oil
	Catalytic cracking	Catalyst fines	Inert solids, catalyst particles, carbon
	Coker	Coker fines	Carbon particles, hydrocarbons
	Alkylation	Spent sludges	Calcium flouride, bauxite, aluminum chloride
	Lube oil treatment	Spent clay sludges, press dumps	Clay, acid sludges, oil
	Drying and sweetening	Copper sweetening residues	Copper compounds, sulfides, hydrocarbons
	Storage tanks	Tank bottoms	Oil, water, solids
	Slop oil treatment	Precoat vacuum filter sludges	Oil, diatomaceous earth, solids
Effluent Treatment Solids	API separator	Separator sludge	Oil, sand and various process solids
	Chemical treatment	Flocculant aided precipitates	Aluminum or ferric hydroxides, calcium carbonate
	Air flotation	Scums or froth	Oil, solids, flocculants (if used)
	Biological treatment	Waste sludges	Water, biological solids, inerts
	Water treatment plant	Water treatment sludges	Calcium carbonate, alumina, ferric oxide, silica
General Waste	Office	Waste paper	Paper, cardboard
	Cafeteria	Food wastes (garbage)	Putrescible matter, paper
	Shipping and receiving	Packaging materials, strapping pallets, cartons, returned products, cans, drums	Paper, wood, some metal, wire
	Boiler plant	Ashes, dust	Inert solids
	Laboratory	Used samples, bottles, cans	Glass, metals, waste products
	Plant expansion	Construction and demolition	Dirt, building materials, insulation, scrap metal
	Maintenance	General refuse	Insulation, dirt, scrapped materials—valves, hoses, pipe

Table 71. Characteristics of Refinery Solid Wastes[13] (all values are percentages and approximate)

Waste Type	Oil or Hydrocarbon	Water	Volatile Solids	Inert Solids	Characteristics
API Separator Sludge	15	66	6	13	Fluid slurry of oil, water and sand
Tank Bottoms	48	40	4	8	Oil-water mixture
Chemical Treatment Sludge	5	90	–	5	Slightly viscous fluid
Air Flotation Froth	22	75	–	3	Thick, oily fluid
Precoat Vacuum Filter Sludges	22	29	–	49	Stiff material, semi-solid at ambient temperatures
Biological Treatment Sludges					
Raw	0	98	1.5	0.5	Water consistency
Mechanically-thickened	0	94	4	2	Thick, but pumpable
Centrifuged	0	85	10	5	Viscous–peanut butter consistency
Vacuum-filtered	0	7515	15	10	Wet, crumbly solid
Screw pressed	0	40	40	20	Intact, solid cake
Water Treatment Sludge	0	95	–	5	Pumpable fluid, sometimes gelatinous

Table 72. Estimated Refinery Solid Waste Quantities[13]

Waste Types	Unit Loads	Factors	Quantities (lb/day)
API Separator Sludges	200 mg/l suspended solids	5.0 mgd	8,320
Chemical Treatment (API separator effluent)	50 mg/l suspended solids removed only	5.0 mgd	2,080 (dry solids)
Biological Sludges	0.7 lb dry solids per lb BOD removed	4,500 lb/BOD$_5$/day	3,150
Water Treatment Sludge			
Lime soda ash	2 parts dry sludge per 1 part hardness removed	200 ppm hardness removed	16,700
Ion exchange	0.4 lb salt per 1000 grains hardness	200 ppm hardness removed	560 (dry salt)
Office Wastes	1.0 ft^3 per employee/month	120 employees	1,200
Cafeteria	0.6 lb per meal	100 meals/day	60

Table 73. Potential Sources of Specific Air Emissions from Petroleum Refineries[80]

Emission	Potential Sources
Oxides of Sulfur	Boilers, process heaters, catalytic cracking unit regenerators, treating units, H_2S flares, decoking operations
Hydrocarbons	Loading facilities, turnarounds, sampling, storage tanks, wastewater separators, blow-down systems, catalyst regenerators, pumps, valves, blind changing, cooling towers, vacuum jets, barometric condensers, air-blowing, high pressure equipment handling volatile hydrocarbons, process heaters, boilers, compressor engines
Oxides of Nitrogen	Process heaters, boilers, compressor engines, catalyst regenerators, flares
Particulate Matter	Catalyst regenerators, boilers, process heaters, decoking operations, incinerators
Aldehydes	Catalyst regenerators
Ammonia	Catalyst regenerators
Odors	Treating units (air-blowing, steam-blowing), drains, tank vents, barometric condenser sumps, wastewater separators
Carbon Monoxide	Catalyst regeneration, decoking, compressor engines, incinerators

storage tanks	cooling towers
catalyst regeneration units	loading facilities
pipeline valves and flanges	blowdown systems
pressure relief valves	pipeline blind-flange changing
pumps and compressors	boilers and process heaters
compressor engines	vacuum jets
acid treating	sampling
wastewater separators and process drains	air blowing

Hydrocarbon emission regulations require control at the source. Storage tanks are a major source of hydrocarbons emitted to the atmosphere. Figure 87 illustrates the various types of storage tanks. The regulations suggest floating roofs for new storage facilities. Table 74 presents hydrocarbon emission factors for evaporation losses from storage tanks. It compares fixed roof losses to floating roof losses. Table 75 estimates total hydrocarbon losses from tankers. The cost-effectiveness and economic analysis of a vapor recovery system are outlined in Tables 76A and 76B.

Emission factors for crude oil separation units are listed in Table 77 for the various pollutants. Factors are given for pounds of pollutant per cubic foot of air emission and pounds of pollutant per 1000 barrels of oil. Similar emission factors are given for the catalyst regenerator unit in Table 78.

The sulfur recovery unit of the refinery should be described in detail with particular consideration given to tail-gas emissions. Other economic data pertaining to air emissions should be given.

FACILITY IMPACTS

All facilities, including ports, pipelines, storage areas, loading areas and extraction points must be assessed as to their environmental consequences. Detail of the impact evaluation is at the discretion of the EPA regional administrator. The EIS must provide a detailed description depending on whether the offsite facility is wholly or partially owned by the builder of the refinery, or depending on the degree of expansion necessary to accommodate the new refinery.

Marine terminals and deep-water port facilities are discussed if they are constructed or expanded as a result of the proposed refinery. Several aspects of this construction must be considered in an EIS.[13]

1. site location in relationship to transportation; and industrial, commercial and residential areas;
2. aesthetics;
3. local ordinances and rights-of-way;
4. utility services;
5. regional land use plans;

ORDINARY CONE ROOF TANK
Roof may be wood or steel

WATER ROOF TANK
Roof is flat with shallow layer of water

FLOATING ROOF TANK
Roof deck rests upon liquid and moves

EXPANSION ROOF TANK
Designed for light pressure flexes upward or downward
with vapor volume changes. Upper position indicated by
solid line, lower position by dotted line

LIFTER ROOF TANK
Liquid-sealed roof moves upward and downward
with vapor volume changes

LIFTER ROOF TANK FOR REFRIGERATED CONTENTS
Insulated tank holds refrigerated liquids. Vapor from above
liquid is compressed and cooled to furnish refrigeration

VAPORDOME ROOF TANK
Flexible diaphram in hemispherical roof moves in
accordance with vapor volume changes

Figure 87. Various types of atmospheric pressure storage tanks.[81]

Table 74. Hydrocarbon Emission Factors for Evaporation Losses from
the Storage of Petroleum Products[82]

| Type of Tank | Units | Type of Material Stored | |
		Gasoline or Finished Petroleum Product	Crude Oil
Fixed Roof			
Breathing Loss	lb/day-1000-gal storage capacity	0.4	0.3
	kg/day-1000-liter storage capacity	0.05	0.04
Working Loss	lb/1000-gal throughput	11	8
	kg/1000-liter throughput	1.32	0.96
Floating Roof			
Breathing Loss	lb/day-tank	140(40-210)	100(30-160)
	kg/day-tank	63.5	45.4
Working Loss	lb/1000-gal throughput	Negative	Negative
	kg/1000-liter throughput	Negative	Negative

Table 75. Estimated Total Hydrocarbon Losses from Tankers[83]
(high vapor pressure, nonboiling cargoes)

Capacity of Tanker	dwt	40,000	100,000	200,000	300,000
Tank Depth	ft	50	70	80	95
Final Oil Depth	ft	49	68.6	78.4	93.1
Mass of Oil (per ft^2 of surface in full tanks, SG = 0.85)	ton	1.17	1.62	1.87	2.22
Gas Vented (per ft^2 of oil surface)	ft^3	1.25	1.25	1.25	1.25
Hence, Hydrocarbon Gas Vented (per ton of oil)	ft^3/ton	1.07	0.772	0.667	0.563
Total Volume Hydrocarbon Gas Vented	ft^3	42,800	77,200	133,400	168,900
Mass of Hydrocarbon Gas Vented (mol wt = 44, temp = $100°F$)	ton	2.05	3.70	6.38	8.10

Table 76A. An Economic Analysis of a Vapor Recovery System[83]

Refinery Throughput (gal/month)	Invested Capital	Annual Expenditure[a]	Depreciation ($/yr)	Annual Savings[b] (costs)
1 x 10⁶	$ 88,200 (15 M ft³ vapor holder)	$ 4,150	$ 5,850	22,200-SL 7,200-SSL
5 x 10⁶	92,200 (25 M ft³ vapor holder)	4,350	6,150	111,000-SL 36,000-SSL
10 x 10⁶	120,000 (30 M ft³ vapor holder)	5,052	8,020	222,000-SL 72,000-SSL
20 x 10⁶	133,200 (30 M ft³ vapor holder)	5,360	8,850	444,000-SL 144,000-SSL
90 x 10⁶	200,000 (40 M ft³ vapor holder)	6,700	13,340	1,995,000-SL 648,000-SSL

Table 76B. The Cost Effectiveness of a Vapor Recovery System[83]

Refinery Throughput (gal/yr)	Investment Cost	Cost ($/T/yr)	Savings Vapor Recovery vs Splash (tons/yr)	Savings Vapor Recovery vs SSL (tons/yr)	Cost ($/T/yr)
12,000,000	$ 88,200	$1,278	69.0	22.3	$3,995
60,000,000	92,200	265	344.0	111.6	826
120,000,000	120,200	175	688.0	223.2	539
240,000,000	133,200	97	1376.0	446.0	298
1,080,000,000	200,000	32	6192.0	2008.0	100

[a]Annual expenditures for operation can be classified into three categories:
1. Proportional costs: includes expenditures which vary with length of time system is in operation (e.g., electric power).
2. Fixed costs: expenditures which tend to remain more or less constant on a yearly basis (e.g., maintenance, operations and overhead).
3. Variable costs: expenditures which either come up unexpectedly, such as major breakdowns which may require the presence of factory or distributor service personnel or costs which are incurred on a yearly basis but differ from year to year. These include taxes which are paid based on the assessed equipment value and the outstanding insurance rates which are also based on assessed value.

[b]SL = splash loading; SSL = subsurface loading.

Table 77. Pollutants from Crude-Oil Separation Units[82]

Source	Pollutant	Emission Factor	
		lb/1000 ft	lb/1000 bbl oil
Combustion	Sulfur oxides	Negative	8400 x % S
Process Heaters	Hydrocarbons	0.03	140
Boilers	Particulates	0.02	800
	Carbon monoxide	–	2
	Nitrogen oxides	0.02	2900
Barometric Condensers	Hydrocarbons	–	130 lb/1000 bbl charge to vacuum distillation tower
	Odors	–	Odors caused by noncondensables and light hydrocarbons
Miscellaneous (sampling, spillage, leaks, drains and blowdown)	Hydrocarbons	–	150 lb/1000 bbl crude

Table 78. Emission Factors for Pollutants from Catalyst Regenerator[82]

Pollutant	Source	Emission Factor (lb/1000 bbl charge)
Sulfur Dioxide	Regenerator	500
Particulates	Regenerator (with precipitator)	61
Hydrocarbons	Regenerator	220
Carbon Monoxide	Regenerator	13,700
Nitrogen Oxides	Regenerator	63

6. pier structure description;
7. piping design and contents;
8. storage tanks;
9. cathodic protection; and
10. dredging operations.

Dredging operations can result in serious environmental impacts. Therefore, the description should consider physiography, dredging method, volume to be dredged, disposal options, effects on marine flora and fauna, sediments and site hydrography.

Tanker casualties, if attributable to the refinery operation, should be considered and the possibility and effects of an oil spill described (see Chapter 8). Leak detection equipment, emergency shut-down procedures, and oil spill contingency plans should be discussed in the EIS. The impact sources of a refining operation are summarized in Table 79. Factors are listed per 1000 barrels of crude oil processed. In this manner total emissions and effluent quantities can be estimated.

Table 79. Summary of Refinery Impact Sources[13]

	BOD lb/1000 bbl	COD lb/1000 bbl	Particulates lb/1000 bbl	NO_x lb/1000 bbl	SO_x lb/1000 bbl	Hydrocarbons lb/1000 bbl	Solids lb/1000 bbl
Transport (crude and product)							
Pipeline	–	–	1.9	55	4.0	5.5	–
Tankers	–	–	0.4	38	5.8	0.2	–
Supertankers	–	–	0.42	38	5.8	0.2	–
Barges	–	–	34	24	26	16	–
Tank trucks	–	–	7	210	15	21	–
Tank cars	–	–	14	43	37	29	–
Processing							
Crude desalting	1	–	0	0	0	0	0
Crude fractionation	0.2	5	2.3	27	0.1	34	–
Cracking	15	18	17	102	300	44	48
Hydrocarbon rebuilding	12	142	7	112	0.1	77	–
Hydrocarbon rearrangement	–	40	5	144	0.05	34	–
Solvent refining	–	–	–	–	–	90	–
Hydrotreating	95	100	5	50	–	120	–
Grease manufacturing	–	–	–	–	–	23	–
Asphalt production	–	32	6	70	0.2	35	–
Storage							
Crude	1	–	0	0	0	62	–
Product	–	–	–	–	–	21	–

POLLUTION CONTROL

Pollution control measures on effluents can effectively reduce the environmental impact on a region. Initial discussions of pollution control should consider reduction of effluents and emissions at the source. Further, reuse and recycling of materials should be maximized. All reuse and recycling measures should be discussed including:

- the use of catalytic cracker accumulator wastewaters rich in H_2S (sour waters) for makeup to crude desalters;
- the use of blowdown condensate from high-pressure boilers for makeup to low-pressure boilers;
- the reuse of waters that have been treated for closed cooling systems, fire mains, and everyday washing operations;
- stormwater use for routine water applications;
- blowdown waters from cooling towers for use as water seals on high-temperature pumps;
- the recirculation of steam condensate;
- the recycling of cooling waters.

General maintenance of the plant should be discussed because such measures can effectively reduce emissions. Good maintenance measures would include:

- the recovery of oil spills and hydrocarbons with vacuum trucks to reduce emissions and water effluents;
- a reduction of leaks and accidents through preventive maintenance;
- the separation of hazardous wastes, concentrated wastes, and other process wastes from general effluents for more effective treatment;
- the diking of process unit areas to control and treat spills, oily stormwater runoff, or periodic washes;
- the reduction of shock pollutant loads on treatment facilities through the periodic flushing of process sewers to prevent contaminant build-up;
- a specialized program for handling hazardous wastes, sludges, washwaters, and other effluents;
- a system to minimize wastes from monitoring stations;
- personnel awareness that the waste treatment is initiated at the process unit.

Actual process changes can in many instances reduce pollution significantly while returning a value through recovery. Technology changes that reduce pollution may not be as cost-effective during process cycles, but may prove to be highly beneficial when waste treatment costs have been reduced. Depending on the feasibility and suitability of a particular project, such process technology changes may include:

- catalyst switching to one of longer life and greater activity reducing regeneration rates;
- a reduction in cooling water usage through the implementation of air-fin coolers
- a reduction in spent caustic and sulfides loadings by including hydrocracking and hydrotreating processes;
- the inclusion of process control instrumentation to employ emergency shut-downs or control upset conditions;

the minimization of filter solids, water washes, and spent caustics and acids through the optimization of drying, sweetening and finishing processes.

COOLING TOWERS

A description of the cooling system is necessary including possible alternatives, *i.e.*, nonevaporative devices. The evaporating cooling systems include spray ponds, mechanical-draft cooling towers, atmospheric cooling towers, and natural-draft cooling towers. Treated wastewater should be considered for makeup purposes. The cooling water blowdown composition is dependent on the composition of the original water used, the operation method, and cooling water treatment. Chromates, zinc, polyphosphates, dust, microorganisms and other corrosion inhibitors are constituents of the cooling treatment wastewaters. A discussion of alternate treatment methods, process operations and piping materials may be in order. A dry cooling system consisting of air-fins to dissipate the undesired heat directly to the atmosphere should be discussed.

WATER TREATMENT

The in-process physical/chemical pretreatment methods should be described. Flow equalization neutralization of spent acid and spent caustic wastewaters, oil separators and slop oil recovery systems, and clarifiers and chemical coagulation and precipitation are common pretreatment methods for petroleum and petrochemical wastewaters. Table 80 identifies and estimates the various wastewater treatment processes used by petroleum refineries. It illustrates the impact of the recent considerations of the environment on the increased usage of wastewater pollution control devices.

Depending on refinery location, refinery plant size, the refining process (degree of crude finishing), and wastewater characteristics (defined in the material balance—see Chapter 3), the wastewater treatment facilities are designed based on the processes in Table 80. Table 81 gives wastewater characteristics and quantities for the various petroleum unit operations. One of the fundamental aspects of the EIS is the consideration and implementation of new technology for pollution curtailment.

Included in the EIS is a complete description of the waste treatment plant. A process flow diagram is necessary to illustrate each step of the treatment process. Figure 88 shows the numerous combinations of treatment operations that can be used to process refinery wastewaters. This diagram might be used in conjunction with Table 82 which estimates the

Table 80. Estimated Percentage of Petroleum Refineries Using Various
Wastewater Treatment Processes[77]

Processes and Subprocesses	1950	1963	1967	1972	1977
API Separators	40%	50%	60%	70%	80%
Earthen Basin Separators	60	50	40	30	20
Evaporation	0-1	0-1	1	1-2	2-5
Air Flotation	0-1	10	15	18	20
Neutralization (total wastewater)	0-1	0-1	0-1	0-1	0-1
Chemical Coagulation and Precipitation	1-5	1-5	5-10	10-15	10-15
Activated Sludge	0	5	10	40	55
Aerated Lagoons	0	5	10	25	30
Trickling Filters	1-2	7	10	10	10
Oxidation Ponds	10	25	25	25	20
Activated Carbon	0	0.5	0.5	3	5
Ozonation	0	1	1	3	5
Ballast Water Treatment (Physical)	9	9	8	5	5
Ballast Water Treatment (Chemical)	1	1	2	5	5
Slop Oil-Vacuum Filtration	0	5	7	12	15
Slop Oil-Centrifugation	0	2	3	10	15
Slop Oil-Separation	100	93	90	80	70
Sour Water-Steam Stripping					
Flue Gas Strippers	60	70	85	90	90
Natural Gas					
Sour Water-Air Oxidation	0	3	3-5	7	10
Sour Water-Vaporization	1	1-2	1	0	0
Sour Water-Incineration[a]	35-40	40	50	30	20
Neutralization of Spent Caustics					
Flue gas	20	30	35	20	20
Spent acid (including springing and stripping)	15	25	30	25	20
Oxidation	0	3	5	5	5
Incineration	25	40	50	20	15

[a]Incineration includes flaring, boiler furnaces, and separate incinerators used only in conjunction with stripping and vaporization.

efficiencies of the various treatment practices on refinery wastewaters. In this manner a first-order estimation of treatment efficiency for a particular refinery operation can be obtained.

Several factors should be considered when selecting a wastewater treatment system. These factors include the system's reliability and susceptibility to upset; its ability to be expanded in the event of the promulgation of more stringent regulations and the effects of that expansion; the flexibility to accommodate different types of wastewater constituents; the capability of modification to remove substances that presently are allowed to enter

Table 81. Waste Loadings and Volumes per Unit of Fundamental Process Throughput in Older, Typical, and Newer Technologies[77]

Fundamental Process	Older Technology				Typical Technology				Newer Technology			
	Flow (gal/bbl)	BOD (lb/bbl)	Phenol (lb/bbl)	Mercaptans & Sulfides (lb/bbl)	Flow (gal/bbl)	BOD (lb/bbl)	Phenol (lb/bbl)	Mercaptans & Sulfides (lb/bbl)	Flow (gal/bbl)	BOD (lb/bbl)	Phenol (lb/bbl)	Mercaptans & Sulfides (lb/bbl)
Crude Oil and Product Storage	4	0.001	–	–	4	0.001	–	–	4	0.001	–	–
Crude Desalting	2	0.002	0.20	0.002	2	0.002	0.10	0.002	2	0.002	0.05	0.002
Crude Fractionation	100	0.020	3.0	0.001	50	0.0002	1.0	0.001	10	0.0002	1.0	0.001
Thermal Cracking	66	0.001	7.0	0.002	2	0.001	0.2	0.001	1.5	0.001	0.2	0.001
Catalytic Cracking	85	0.062	50.0	0.03	30	0.010	20	0.003	5	0.010	5	0.003
Hydrocracking	not in this technology				not in this technology				5	–	–	–
Reforming	9	tr	0.7	tr	6	tr	0.7	0.001	6	tr	0.7	0.001
Polymerization	300	0.003	1.4	0.22	140	0.003	0.4	0.010	not in this technology			
Alkylation	173	0.001	0.1	0.005	60	0.001	0.1	0.010	20	0.001	0.1	0.020
Isomerization	not in this technology				not in this technology				–	–	–	–
Solvent Refining	8	–	3	tr	8	–	3	tr	8	–	3	tr
Dewaxing	247	0.52	2	tr	23	0.50	1.5	tr	20	0.25	1.5	tr
Hydrotreating	1	0.002	0.6	0.007	1	0.002	0.01	0.002	8	0.002	0.01	0.002
Deasphalting	–	–	–	–	–	–	–	–	–	–	–	–
Drying and Sweetening	100	0.10	10	–	40	0.05	10	–	40	0.05	10	–
Wax Finishing	–	–	–	–	not in this technology				–	–	–	–
Grease Manufacture	–	–	–	–					–	–	–	–
Lube Oil Finishing	–	–	–	–					–	–	–	–
Hydrogen Manufacture	not in this technology								–	–	–	–
Blending and Packaging	–	–	–	–	–	–	–	–	–	–	–	–

aData not available for reasonable estimate.

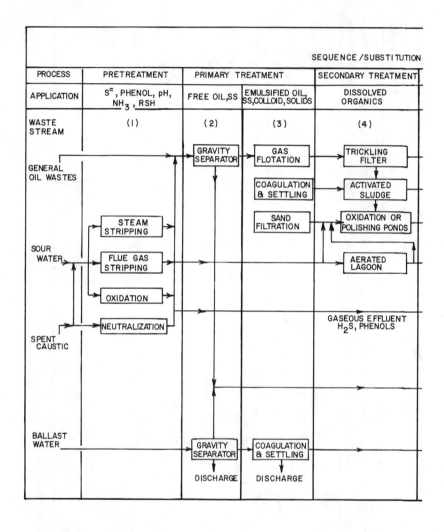

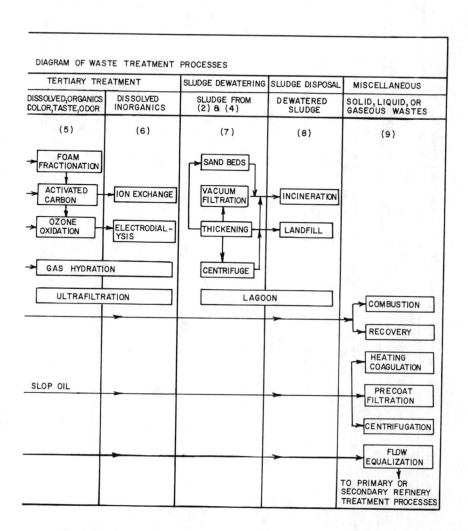

Figure 88. Sequence/substitution diagram of waste treatment process.[77]

Table 82. Efficiency of Oil Refinery Waste Treatment Practices Based on Effluent Quality

Process	Influent[a]	BOD	COD	Separable Oil	Emulsified Oil	Phenol	Sulfide S⁻	Suspended Solids	Chloride	Ammonia	Cyanide	pH	Toxicity	Temp. (°F)
Physical Treatment														
API separators	Raw waste	5-35[b]	5-30[b]	60-99	n.a.	Reduced	n.a.	10-50	n.a.	n.a.	n.a.	n.a.	n.a.	n.a.
Earthen separators	Raw waste	5-50	5-40	50-99	n.a.	Reduced	n.a.	10-85	n.a.	n.a.	n.a.	n.a.	n.a.	n.a.
Evaporation	API effluent	100	100	n.a.	100	100	100	100	100	100	100	n.a.	n.a.	n.a.
Air flotation without chemicals	API effluent	5-25	5-20	70-95	10-40	n.a.	Reduced	10-40	n.a.	n.a.	n.a.	n.a.	n.a.	n.a.
Chemical Treatment														
Air flotation with chemicals	API effluent	10-60	10-50	75-95	50-90	n.a.	Reduced	50-90	n.a.	Reduced	n.a.	n.a.	n.a.	n.a.
Chemical coagulation and precipitation	API effluent	10-70	10-50	60-95	50-90	n.a.	n.a.	50-90	n.a.	n.a.	Altered	Altered	n.a.	n.a.
Biological Treatment														
Activated sludge	API effluent	70-95	30-70	n.a.	50-80	65-99	90-99	60-85	n.a.	50-95	65-99	Altered	Reduced	10-60
Aerated lagoons	API effluent	50-90	25-60	n.a.	50-80	65-99	90-99	0-40	n.a.	0-45	65-99	Altered	Reduced	10-90
Trickling filters	API effluent	50-90	25-60	n.a.	50-80	65-99	80-99	60-85	n.a.	50-95	65-99	Altered	Reduced	10-60
Oxidation ponds	API effluent	40-80	20-50	n.a.	40-70	65-99	70-90	20-70	n.a.	20-90	65-99	Altered	Reduced	10-90
Tertiary Treatment														
Activated carbon	Secondary[c] effluent	50-90	50-90	n.a.	50-90	80-99	80-99	n.a.	n.a.	10-30	80-99	n.a.	Reduced	n.a.
Ozonation	Secondary effluent	50-90	50-90	n.a.	n.a.	80-99	80-99	n.a.	n.a.	10-30	80-99	n.a.	Reduced	n.a.

[a] Most probable process influent—indicates the kind or extent of prior treatment required for efficient utilization of the specific process under consideration.
[b] BOD and COD from separable oil not included.
[c] Chemical or biological treatment.

waterways but in the future may be listed as toxic substances; and the complexity of the system, its monitoring requirements, and maintenance operations.

The EIS should consider the possibility of the construction of a dam across receiving waters. Thus a swiftly moving river can be turned into a reservoir reducing its pollutant assimilation capacity. More stringent effluent water quality criteria may then have to be considered.

SOLID WASTE CONTROL

The EIS should consider all methods for solid waste reuse, recyclability or disposal. Based on Tables 71 and 72 each waste is evaluated with regard to the adequacy of the treatment method. Disposal sites, leachate problems, and long-term effects should be discussed. Oily sludges and spent caustic wastes should be assessed.

Incineration without proper air pollution control devices would be considered an inadequate disposal method as would landfilling materials that are potential leachate problems. The suitability of incineration with controls, pyrolysis and heat recovery should be determined. The generation of sludges, their handling and disposal, and ultimate environmental effects should be described.

AIR EMISSIONS CONTROL

Specific air emission control measures for each pollutant should be described:

Hydrocarbon emissions can be limited through the use of floating-roof tops; manifolding purge lines to a recovery system (condenser or carbon adsorber) or to a flare (see Figure 89); vapor recovery systems on loading facilities; preventive maintenance; enclosed waste treatment plant; the use of a carbon monoxide boiler; mechanical seals on compressors and pumps, and trained and cognizant personnel. A typical scrubbing system for emissions from air-blown asphalt stills is given in Figure 90.

Particulates are controlled with the use of wet scrubbers and high-efficiency mechanical collectors (cyclones, bag houses); electrostatic precipitators on catalyst regenerators and power plant stacks; controlled combustion to reduce smoke; controlled stack and flame temperatures, and improved burner and incinerator design.

Carbon monoxide emissions are controlled at the catalytic cracker and fluid coker units with a CO boiler and at other sites through proper furnace and burner design.

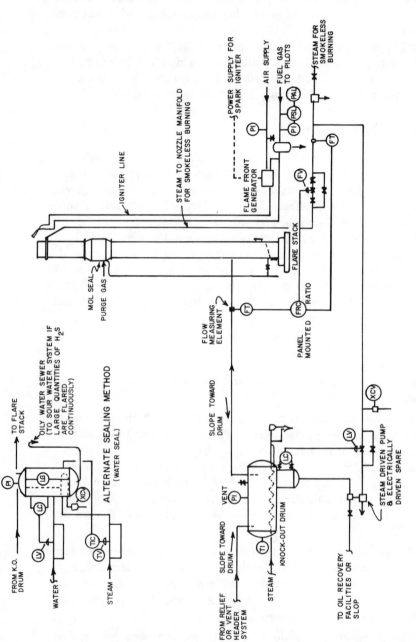

Figure 89. Typical flare installation. [83] Note: This represents an operable system arrangement and its components. Arrangement of the system will vary with the performance required. Correspondingly, the selection of types and quantities of components, as well as their applications, must match the needs of the particular plant and its specifications.

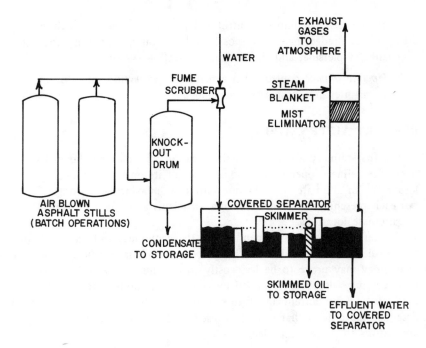

Figure 90. Flow diagram of scrubbing system.[83]

Odors are controlled with a good preventive maintenance program; governmental regulation of hydrocarbon and sulfur emissions; the treatment of H_2S-rich wastewater streams evolving from the catalytic crackers; gas-processing units and vacuum distillation towers; and the flaring of H_2S, mercaptans, other sulfides and other odor-producing compounds.

Sulfur dioxide emissions are controlled primarily through the burning of low-sulfur fuels in furnaces and boilers, the wet scrubbing of high-sulfur dioxide flue gases, and the desulfurization of fuels before use.

Nitrogen oxide emissions can be controlled through an improved combustion process (*i.e.,* lower flame temperature, less excess air), low-nitrogen fuel burning, and good stack dispersion.

OTHER REFINERY IMPACT CONTROLS

All other refinery impact controls should be discussed in detail including:

Aesthetic impact control such as camouflouging stacks and towers through site orientation, directional lighting, and the expanding and modification of the region's green belts.

Marine terminal impact control and preventive measures such as dredging and dredge material disposal methods, oil spill contaminant, oil slop tanks at pierside, and the drainage collection system.

Pipeline impact control such as cathodic protection, monitoring systems, maintenance programs, and land use.

REFINERY ALTERNATIVES

The first refinery alternative is site location. It must be shown that the entire refinery operation or some facet of its operation could be located at an alternate site. Marine terminals, pipelines and storage facilities each present unique environmental problems and might be considered for alternate locations.

The refinery process is dependent on the economics and marketing of the finished product, and the refinery feedstock characteristics. New technology may prove to be less costly than pollution control. Other processes may also be determined by considering alternate refinery finished products and the current and future demand for those products. Chapter 3 describes the criteria that should be used in determining and evaluating alternatives to proposed projects.

ORGANIC CHEMICAL INDUSTRY

The environmental impact of a proposed organic chemical plant can be described in a similar manner to that of the petroleum refinery. The parameters to be considered are in some ways similar, but in many respects different. An organic chemical synthesizing plant may consist of several interrelated processes each with its own unique waste streams. A process flow diagram is required for each process identifying raw materials, raw water, feedstock pretreatment, storage and handling, and product storage and distribution. All process water inputs and outputs and all sources of leaks and spillage are defined. The interconnections between processes should also be presented. Table 83 gives wastewater characteristics associated with some chemical products.

Processes, applications, and other considerations of a proposed organic chemical plant that should be discussed include:

1. all vessels and piping that require cleaning and washing;
2. all steps of liquid, solid and vapor recovery;
3. all process mix variations attributable to market demand, raw material availability and other factors;
4. with respect to 3, all waste load variations;

Table 83. Wastewater Characteristics Associated with Some Chemical Products[84]

Product	Flow (gal/ton) Low	High	BOD (mg/l) Low	High	COD (mg/l) Low	High	Other Important Characteristics
Dyes and Pigments	50,000	250,000	200	400	500	2,000	Heavy-metals and solids contents, color, pH
Primary Petrochemicals							
Ethylene	50	1,500	100	1,000	500	3,000	Phenol and oil contents, pH
Propylene	100	2,000	100	1,000	500	3,000	Phenol content, pH
Primary Intermediates							
Toluene	300	3,000	300	2,500	1,000	5,000	
Xylene	200	3,000	500	4,000	1,000	8,000	
Ammonia	300	4,000	25	100	50	250	Oil and nitrogen contents
Methanol	300	4,000	300	1,000	500	2,000	Oil and solids contents
Ethanol	300	4,000	300	3,000	1,000	4,000	Oil and solids contents
Butanol	200	2,000	500	4,000	1,000	8,000	Heavy-metals content
Ethylbenzene	300	3,000	500	3,000	1,000	7,000	Heavy-metals content
Chlorinated hydrocarbons	50	1,000	50	150	100	500	Oil and solids contents, pH
Secondary Intermediates							
Phenol, cumene	500	2,500	1,200	10,000	2,000	15,000	Phenol and solids content
Acetone	500	1,500	1,000	5,000	2,000	10,000	
Glycerin, glycols	1,000	5,000	500	3,500	1,000	7,000	
Urea	100	2,000	50	300	100	500	
Acetic anhydride	1,000	8,000	300	5,000	500	8,000	pH
Terephthalic acid	1,000	3,000	1,000	3,000	500	4,000	Heavy-metals content
Acrylates	1,000	3,000	500	5,000	2,000	15,000	Solids and cyanide contents, color
Acrylonitrile	1,000	10,000	200	700	500	1,500	Cyanide content, color, pH
Butadiene	100	2,000	25	200	100	400	Oil and solids contents
Styrene	1,000	10,000	300	3,000	1,000	6,000	
Vinyl chloride	10	200	200	2,000	500	5,000	
Primary Polymers							
Polyethylene	400	1,600	—	—	200	4,000	Solids content
Polypropylene	400	1,600	—	—	200	4,000	
Polystyrene	500	1,000	—	—	1,000	3,000	Solids content
Polyvinyl chloride	1,500	3,000	50	500	1,000	2,000	
Cellulose acetate	10	200	500	2,000	1,000	5,000	
Butyl rubber	2,000	6,000	800	2,000	2,500	5,000	
Miscellaneous Organics							
Isocyanate	5,000	10,000	1,000	2,500	4,000	8,000	Nitrogen content
Phenyl glycine	5,000	10,000	1,000	2,500	4,000	8,000	Phenol content
Parathion	3,000	8,000	1,500	3,500	3,000	6,000	Solids content, pH
Tributyl phosphate	1,000	4,000	500	2,000	1,000	3,000	Phosphorus content

5. feedstock and raw material substitutions;
6. all intermediate steps producing chemicals and wastes, and the effects of an accident (if intermediate compound is highly toxic);
7. all water effluents produced by steam stripping and washing and cleaning operations for solvent and catalyst treatment and maintenance;
8. all water effluents produced by process startup, shutdown, or upset conditions; steam and cooling water blowdown; and stormwater runoff. Estimated flow rates and waste loadings should be included. Table 84 lists some chemical waste characteristics.

Table 84. Chemical Waste Characteristics[84]

Principal Products	Flow (mgd)	BOD$_5$ (mg/l)	COD (mg/l)	SS (mg/l)
Phthalic anhydride, maleic anhydride plasticizers, H$_2$SO$_4$	0.002		200	24
Chemical warfare gas, chromium plating	0.002	200	1,100	
Terephthalic acid, isophthalic acid, dimethyl teraphthalate	5.36		9,800	10,600
Butadiene, styrene, polyethylene, olefins	1.68			
Phenol, ethylene	2.0	300	1,200	300
Acrylonitrile	0.302		1,200	239
Fatty acids, esters, glycerol	0.10	10,000	14,000	
Regenerated cellulose	1.41			
Acetylene				
Dyes, pigments, inks	0.452	227		93
Azo and anthraquinine dyes	0.94	352	1,760	152
Anthraquinine vat dyes	5.0	300	1,160	
Ethylene, alcohols, phenol	5.9	1,700	3,600	610
Benzene, ethylene, butyl rubber, butadiene, xylene, isoprene	14.7	91	273	
Acrylonitrile, acetonitrile, hydrogen cyanide	3.9	390	830	106
Terephthalic acid	0.335		4,160	
Glycerine, various glycols	0.49	2,810		
Methyl and ethyl parathion	0.075	3,100	5,000	80
Methyl isocyanate, phosgene diphenol glycine	0.543	1,146	3,420	
Urea, ammonia, nitric acid, NH$_4$NO$_3$	0.65	105	140	
Butadiene, styrene, propylene, polyolefin, adipic acid	1.38	5,630	1,230	225
Butadiene, alkylate, methyl ethyl ketone, styrene, maleic anhydride	2.0	1,870		10
Butadiene, maleic acid, fumaric acid, tetrahydrophthalic anhydride	3.605	959	1,525	
Diphenyl carbonate, dinitrophenol, benzene, quinolin, H$_2$SO$_4$, tear gas, dinitrobenzoic acid	0.098	650	1,380	

Table 84, continued.

Principal Products	Flow (mgd)	BOD$_5$ (mg/l)	COD (mg/l)	SS (mg/l)
Organophosphates, esters, resins, phosphorus chlorides	1.2	845	2,040	322
Phenols	0.215	6,600	13,200	
500 different products	3.2	360	500	673
Organic and inorganic chemicals	2.1	100		
Phenols	0.22	6,600	13,200	
Additives for lubricating oils	0.20	465	1,050	250
Polyethylene, ethylene oxide and polypropylene	2.1	1,385	2,842	
Acrylates, insecticides, enzymes, formaldehydes, amines	1.06	1,960	2,660	80
Ethylene, propylene, butadiene, crude benzene, toluene	0.228	500		
Acids, formaldehyde, acetone, methanol, ketones, nitric acid, nylon salt, vinyl acetate, acetaldehyde	3.46	530	10,130	160
Isocyanates, polyols, urethane foam	0.57	421	1,200	50
Acetaldehyde	1.15	20,000	50,000	200
Acrylonitrile, phenol, butadiene	1.817		29,100	
Ethylene, propylene, toluene, xylene	0.43	1,300	1,500	
Acids, formaldehyde, acetone, methanol, ketones, alcohols, acetaldehyde	1.15	15,000	30,000	
Petrochemicals	15.2	177		
Acrylonitrile, butadiene, styrene, impact polystyrene, crystal polystyrene	0.085	200		60
Ethylene, propylene, butadiene, alpha olefins, polyethylenes	0.750	155	380	120
Organic chemicals	1.4	2,000	4,800	900
Pharmaceuticals (Dallas WPCF 66)	0.037	14,000		500
Organic chemicals	0.077	850	1,700	300
2,4,5-T	0.005	15,000	21,000	700
2-4-D	0.005	15,000	23,000	348
Butadiene	0.288		350	300
Organic chemicals	0.288		750	120
Olefins	0.288		320	400
Adipic acid	0.13		35,000	180
Hexamethylenediamine	0.13		113,000	1,200
Petrochemicals	0.02		40,000	30
Petrochemicals	0.17	24,000	39,000	

9. all wastewater treatment methods emphasizing recycling and reuse of effluents. Table 85 identifies the use of wastewater treatment processes in the various chemical industries.

10. all solid wastes produced, giving a quantitative and qualitative analysis describing storage and handling, treatment and disposal practices.

11. identification of all materials classified hazardous wastes and their waste management procedures. Table 86 lists hazardous organic chemical compounds.

Table 85. Use of Wastewater Treatment Processes in Chemical Industries[84]

Product Type	Oil Separation	Equalization	Coagulation and Sedimentation	Neutralization	Aerated Lagoon	Activated Sludge	Sludge Treatment Yes	Sludge Treatment No
Dye chemicals	X		X			X	X	
Dye chemicals		X	X			X		X
Dye chemicals		X	X	X				
Petrochemicals	X			X		X	X	
Petrochemicals	X			X		X	X	
Organic chemicals					X			
Organic chemicals		X						
Organic chemicals	X					X	X	
Organic chemicals				X				
Organic chemicals		X						
Organic chemicals		X		X		X		X
Organic chemicals	X			X		X	X	
Organic and inorganic chemicals		X	X					
Petrochemicals	X							
Organic and inorganic chemicals		X		X		X		X
Petrochemicals	X					X		X
Monomers-polymers				X	X			
Petrochemicals	X		X	X				
Petrochemicals								
Organic chemicals								
Organic chemicals			X	X	X			
Petrochemicals	X	X	X			X	X	X
Petrochemicals	X	X						
Phenol	X					X		X
Petrochemicals	X							
Misc. chemicals			X			X	X	
Petrochemicals	X					X		X
Petrochemicals	X		X	X				
Petrochemicals	X				X			
Petrochemicals	X		X	X	X			
Misc. chemicals			X		X			
Petrochemicals	X							
Cyclic chemicals	X					X	X	
Organic chemicals			X					
Organic chemicals	X					X		X
Cyclic chemicals			X					
Petrochemicals	X		X					
Organic chemicals			X		X			
Organic chemicals	X		X	X				
Organic chemicals				X				

Table 86. Hazardous Organic Compounds[13]

1. Acetaldehyde	49. Tripropane (Norene)
2. Acetone	50. Turpentine
3. Butyraldehyde	51. Anthracene
4. Camphor	52. Benzene
5. Crotonaldehyde	53. Creosote (coal tar)
6. Cyclohexanone	54. Cumene
7. Diisobutyl Ketone	55. Dodecylbenzene
8. Ethyl Methyl Ketone	56. Ethylbenzene
9. Furfural	57. Naphthalene
10. Isophorone	58. Styrene
11. Mesityl Oxide	59. Styrene Polymers
12. Methyl Isobutyl Ketone	60. Toluene
13. Paraformaldehyde	61. Xylene
14. Propionaldehyde	62. Acrylic Acid
15. Acetic Acid	63. Adipic Acid
16. Acetic Anhydride	64. Benzoic Acid
17. Acetyl Chloride	65. Fatty Acids
18. Formaldehyde	66. Formic Acid
19. Oleic Acid	67. Propionic Acid
20. Phthalic Anhydride	68. Salicylic Acid
21. Sodium Formate	69. Allyl Alcohol
22. Sodium Oxalate	70. Amyl Alcohol
23. Acetone Cyanohydrin	71. Butanols
24. Acetonitrile	72. Cyclohexanol
25. Acrylonitrile	73. Decyl Alcohol
26. Cyanoacetic Acid	74. Diethylene Glycol
27. Ethylene Cyanohydrin	75. Furfural Alcohol
28. Toluene Diisocyanate	76. Glycerol
29. Acetylene	77. Isopropanol
30. Butadiene	78. Methanol
31. Butane	79. Methyl Amyl Alcohol
32. 1-Butene	80. Octyl Alcohol
33. Butylene	81. n-Propyl Alcohol
34. Cyclohexane	82. Propylene Glycol
35. Dicyclopentadiene	83. Sorbitol
36. Diisobutylene	84. Triethylene Glycol
37. Ethane	85. Allyl Chloride
38. Ethylene	86. Aminoethylethanol Amine
39. n-Heptane	87. n-Butylamine
40. 1-Heptene	88. Cyclohexylamine
41. Hexane	89. Diethanolamine
42. Isopentane	90. Diethylamine
43. Isoprene	91. Diethylene Trimine
44. Naphtha (crude)	92. Diisopropanolamine
45. n-Pentane	93. Dimethylamine
46. Propane	94. Ethanolamine (Monoethanolamine)
47. Propylene	95. Ethylamine (Monoethylamine)
48. Tetrapropylene	96. Ethylene Diamine

Table 86, continued

97. Hexamethylene Diamine	145. Dichlorofluoromethane
98. Isopropyl Amine	146. Dichloromethane
99. Methylamine	147. 2,3-Dichloropropane
100. Morpholine	148. 1,3-Dichloropropene
101. Propylamine	149. Dichlorotetrafluoroethane
102. Triethanolamine	150. Epichlorohydrin
103. Triethylamine	151. Ethyl Chloride
104. Triethylene Tetramine	152. Ethylene Dichloride
105. Trimethylamine	153. Methyl Chloroformate
106. Urea	154. Perchloroethylene
107. Amyl Acetate	155. Polyvinyl Chloride
108. Butyl Acetate	156. Tetrachloroethane
109. Butyl Acrylate	157. Trichloroethane
110. Di-n-butyl Phthalate	158. Trichlorofluoromethane
111. Ethyl Acetate	159. Vinyl Chloride
112. Ethyl Acrylate	160. Chlorobenzene (Chlorobenzol)
113. Ethyl Phthalate	161. o-Dichlorobenzene
114. Isobutyl Acetate	162. p-Dichlorobenzene
115. Methyl Acetate	163. Hexachlorophene
116. Methyl Acrylate	164. Trichlorobenzene
117. Methyl Formate	165. Benzoyl Peroxide
118. Methyl Methacrylate	166. 1,2,4-Butanetrioyl Trinitrate (BTTN)
119. n-Propyl Acetate	167. Chloropicrin
120. Vinyl Acetate	168. Cyanuric Triazide
121. Acridine	169. Diethylether
122. Aniline	170. Dioxane
123. 2,4-Dinitroaniline	171. Ethers
124. m-Methylaniline	172. Ethylene Glycol Monoethyl Ether
125. B-Naphthylamine	173. Ethylene Glycol Monoethyl Ether Acetate
126. Phenylhydrazine Hydrochloride	174. Isopropyl Ether
127. Pyridine	175. Polypropylene Glycol Methyl Ether
128. o-Toluidine	176. Propylene Oxide
129. Benzene Sulfonic Acid	177. Tetrahydrofuran
130. Benzyl Chloride	178. Dinitrobenzene
131. o-Butyl Phenol	179. Dinitrophenol
132. p-Butyl Phenol	180. Diphenylamine
133. Carbolic Acids (Phenol)	181. Ethylene Bromide
134. Cresol (Cresylic Acid)	182. Methyl Bromide
135. Diethylstibestrol	183. Methyl Chloride
136. Ethyl Phenol	184. Ethyleneimine
137. Nonyl Phenol	185. Glycerol Monolactate Trinitrate (GLNT)
138. Xylenol	186. Hydrazine Azide/Hydrazine
139. Butyl Mercaptan	187. Hydroquinone
140. Carbon Disulfide	188. Maleic Anhydride
141. Carbon Tetrachloride	189. Manganese Methylcyclopentadienyl-tricarbonyl
142. Chloral Hydrate	
143. Chloroform	
144. Dichloroethyl Ether	

Table 86, continued

190. Nitroaniline	200. Polychlorinated Biphenyls
191. Nitrobenzene	201. Polyvinyl Nitrate (PVN)
192. *m*- and *p*-Nitrochlorobenzene	202. Quinone
193. Nitroethane	203. Tetraethyl- and Tetramethyllead
194. Nitromethane	204. Tetranitromethane
195. 1-Nitropropane	205. Tricresyl Phosphate
196. 4-Nitrophenol	206. Acrolein
197. 4-Nitrotoluene	207. Dimethyl Sulfate
198. Oxalic Acid	208. Pentachlorophenol
199. Phosgene (Carbonyl Chloride)	

12.   all air emissions from stacks and vents; storage and waste treatment; and leaks and spills.

13.   air emissions due to combustion processes, chemical synthesizing processes, upsets, startups, shutdowns, and waste treatment.

14.   all air emission control devices and related equipment included along with descriptions of feedstock substitutions and process technology changes for pollution control.

15.   all other impacts unique to the organic chemical industry.

To the best of his ability the writer of the EIS should project changes in technology and market demands as they would affect organic chemical production and synthesization. The organic chemical industry can have an unknown impact upon the environment due to thousands of presently available compounds and the new compounds constantly being produced.

CHAPTER 12

ENVIRONMENTAL ASSESSMENT
OF THE RUBBER INDUSTRY

Rubber consumption has continuously risen over the last 150 years. At first natural rubber was utilized exclusively. However, around the turn of the century the advent of the straight-chained polymer structure revolutionized the industry and resulted in the common use of synthetic rubber. Various processes exist to produce numerous types of rubber, and thus different wastes are generated. An EIS for the rubber industry would follow the general procedures outlined in previous chapters, but specific impact analyses are often necessary, and they will be described for selected areas of the rubber industry.[13]

NATURAL RUBBER

The impact identification of natural rubber processing begins with a process flow diagram, given as Figure 91. Process wastes from crude latex extraction, coagulation, purification, and additive-mixing operations should be discussed. Natural rubber production generates large amounts of wastes, nearly 100-200 liters per kilogram of final product. Much of this waste is in the form of wastewater. Cooling waters, washing and rinsing waters, rubber cutting and finishing waters, and other auxiliary water use operations are discussed in the environmental assessment.

Characteristic of natural rubber wastewaters are the high BOD levels of around 1500-2500 mg/l and the relatively low suspended solids content of about 400 mg/l. The process wastes can be controlled with a combination physical/chemical and biological wastewater treatment plant. Figure 92 is a diagram of a treatment plant for rubber wastes. In Table 87 the efficiencies of the various waste treatment processes for the combined industrial and municipal treatment facility are given.

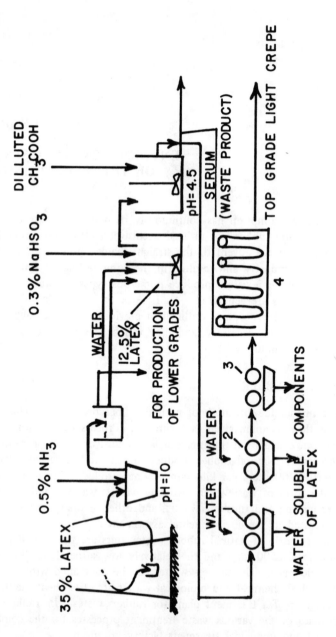

Figure 91. Natural rubber process flow diagram.[76]

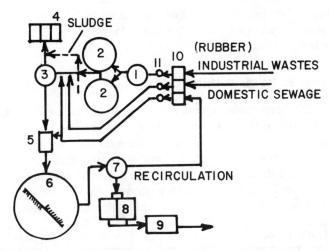

1–EQUALIZING TANK, 2–DIGESTERS, 3–PRELIMINARY SETTLING
TANK, 4–SLUDGE DRYING BEDS, 5–MIXING CHAMBER, 6–HIGH
RATE TRICKLING FILTER, 7–SECONDARY SETTLING TANK,
8–SAND FILTERS, 9–CONTACT TANK, 10–CONTAINER CHAMBER,
11–EJECTORS

Figure 92. Flow diagram for a natural rubber processing and municipal waste
treatment plant.[76]

Table 87. Efficiencies of a Combined Natural Rubber Processing and
Municipal Waste Treatment Facility[85]

	Average Over a 6-Month Period
BOD_5 of wastes from the manufacturing plant (mg O_2/l)	1838
BOD_5 of effluent from the digesters (mg O_2/l)	741
Reduction in BOD_5 due to digestion (%)	60
BOD_5 of domestic sewage (mg O_2/l)	240
BOD_5 of wastes fed to the settling tank (mg O_2/l)[a]	582
BOD_5 of effluent from the settling tank (mg O_2/l)	327
Reduction in BOD_5 in the settling tank and the trickling filter (%)	44
BOD_5 of effluent from the sand filters (mg O_2/l)	129
Reduction in BOD_5 on the sand filters (%)	60
Reduction in BOD_5 for the total wastes over the whole treatment (%)	90.1

[a]Effluent from the digesters and domestic sewage.

Alternatives to a natural rubber processing facility should consider synthetic or reclaimed rubber factories. Previously described alternative procedures should be followed.

SYNTHETIC RUBBER INDUSTRY

The synthetic rubber industry impacts should be identified based on the following parameters. A flow diagram of a typical synthetic (butadiene-styrene) rubber process is shown in Figure 93, and the EIS would include a similar diagram for the specific process under consideration. Impacts are then identified. Figures 94, 95 and Figure 96 are flow diagrams for three different rubber production processes. Each has its own characteristic wastes which are considered in the environmental assessment.

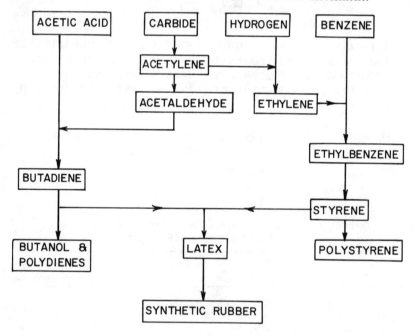

Figure 93. Flow diagram of a butadiene-styrene rubber production process.[84]

Wastewater Sources

Wastewater discussions of emulsion-polymerized crumb rubber processes should include each unit operation: the caustic soda scrubber unit and monomer recovery unit and strippers. Coagulation and crumb dewatering processes as well as all tank and reactor washings and rinses are assessed.

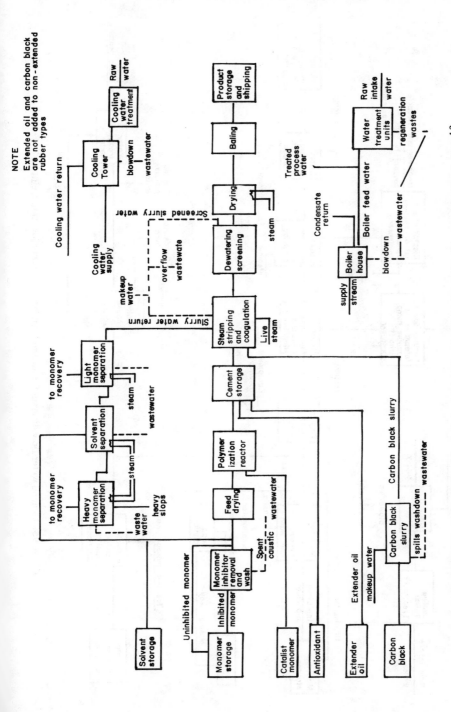

Figure 94. General water flow diagram for a solution-polymerized crumb rubber production facility.[13]

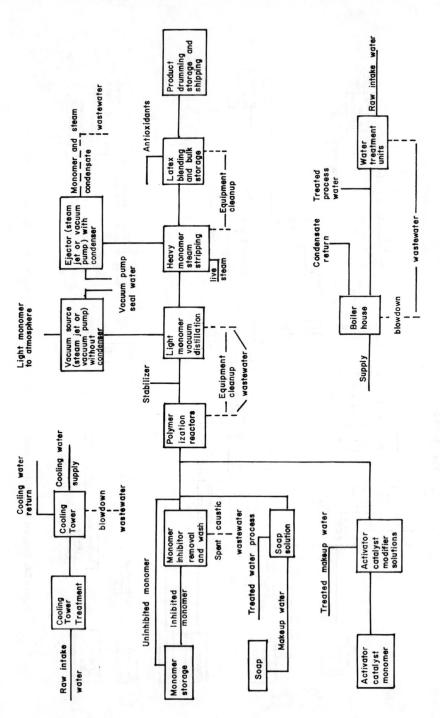

Figure 95. General water flow diagram for an emulsion latex rubber production facility.[13]

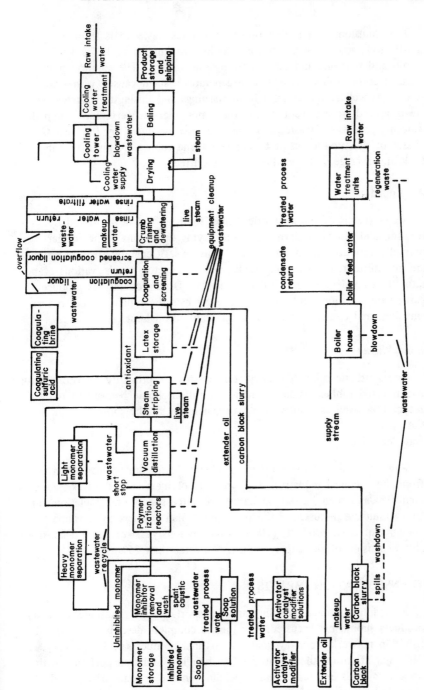

Figure 96. General water flow diagram for an emulsion-polymerized crumb rubber production facility.[13]

The emulsion latex rubber process generates wastewaters from its caustic soda scrubbers; monomer stripping; washing of tanks, tank cars, vessels and reactors; and other washings and rinses. The solution polymerized crumb rubber process also generates a spent caustic wastewater stream, as well as wastewaters from washings and rinsings. Wastewaters are produced from solvent purification, monomer recovery, and crumb dewatering processes. Wastewater characteristics produced by the various phases of the production of butadiene-styrene rubber are given in Table 88. Peak and average waste loadings are also listed.

Air Emissions

Air emissions of the synthetic rubber industry can be separated by: reaction and blowdown processes, and drying operations. The former normally produce organic compound emissions. Generally these emissions are intermittent, from purging tanks, venting reactor vessels, various evaporators and condensers, and leaks and spills. Drying operations produce particulate emissions and odors. Table 89 lists the typical rubber industry emissions and emission rates based on plant throughput.

Solid Wastes

The synthetic rubber industry solid wastes are primarily soapstone sludges and rejected rubber products. Major attention should be paid to reuse and recycling as well as control of waste generation at the source.

Raw Materials

Raw materials for the synthetic rubber industry are normally derived from the chemical industry. The original polymers and additives should be identified and their impact on natural resource supply and depletion industries should be discussed if the proposed rubber plant is to consume large amounts of these products (100,000 metric tons/year) or cause a considerable expansion of the supply industry (greater than 25%).

TRANSPORTATION IMPACTS

All loading, storage, handling and transportation of raw materials and products should be discussed. Further, a complete description of the distribution network from raw material to final product sale is necessary. Particular attention should be given to handling and transportation with respect to spills and leaks, and depending on the product or material characteristics (flammability, toxicity, persistence).

Table 88. Wastewater Characteristics of the Butadiene-Styrene Rubber Process[85]

1	Plant Sections								Combined Wastes		
	Production and Purification of Butadiene			Separation & Purification of Styrene	Dehydration of Pseudo-butylene	Polymerization	Separation and Processing of Rubber		Minimum	Maximum	Average
	Fuel Waters from Fraction-ating Columns	Fuel Waters from the Columns After Discharge of Aldehyde Water	Hydrogenation of Ether and Polymerization of Hydrocarbons				1st Stage	2nd Stage			
	2	3	4	5	6	7	8	9	10	11	12
Temperature ($^\circ$C)	70-80	45	50	30	29	30	25	10	26	30	28
Permanganate value (mg O_2/l)	5.0	12.3	—	78.4	—	—	752	221	288	475	370.5
pH	—	—	5.5	6.7	8.0	8.85	5.75	6.95	6.5	8.0	8.0
COD (iodate method) (mg O_2/l)	1640	18,015	993	1115	2734	2512	2844	876	918	2186	1529
BOD$_5$ (mg O_2/l)	1247	10,360	487	377	865	750	200	275	215	345	303
BOD, total (mg O_2/l)	1588	11,360	792	480	1560	1498	5262	1060	440	535	498
Total dry residue (mg/l)	1024	2694	98	448	38	726	3700	592	1270	2337	2100
Residue on ignition (mg/l)	422	1092	31	256	—	380	198	90	1250	1731	1562
Organic acids[a] (CH_3COOH) (mg/l)	483	99	335.5	4.5	18	16.5	b	b	60	90	75
Aldehydes[a] (CH_3COOH) (mg/l)	b	413.7	b	11	b	—	b	b	14	37.5	27.7
Styrene (mg/l)	b	b	b	180	b	397	b	b	57	180	109
Ethyl alcohol (mg/l)	520	7300	36	b	b	b	—	—	b	b	b
Nekal (mg/l)	b	b	b	b	b	b	1312	565.3	333	528	437
Ammonia nitrogen, N (mg/l)	4.5	7.4	1.4	0.7	46.9	53.7	5.4	3.3	7	28	14.7
Chlorides, Cl (mg/l)	—	—	—	—	—	—	—	—	186.2	404	321.3
Sulfates, SO_4 (mg/l)	—	—	—	—	—	—	—	—	520	619	565
Suspended solids (mg/l)	16	34	16	11	b	b	40	13.4	40	72	60
Sludge volume after 2-hr settling (cm^3/l)	—	—	—	—	—	—	—	—	0.28	0.69	0.44
Moisture content of sludge (%)	—	—	—	—	—	—	—	—	—	—	98.8

[a] The content of organic acids and aldehydes has been given in terms of acetic acid.

[b] Not detected.

Table 89. Emission Factors Based on Plant Throughput for Typical
Synthetic Rubber Industry Emissions[13]

Compound	Emissions	
	(lb/ton)	(kg/MT)
Alkenes		
Butadiene	40	20
Methylpropene	15	7.5
Butyne	3	1.5
Pentadiene	1	0.5
Alkanes		
Dimethylheptane	1	0.5
Pentane	2	1
Ethanonitrile	1	0.5
Carbonyls		
Acrylonitrile	17	8.5
Acrolein	3	1.5

PROCESS IMPACT CONTROL

In-process control of wastes should be discussed. New technology must be considered to alleviate some problems. Source control of wastes should include discussions for control of crumb rinses and coagulation liquor overflows, elimination of steam-driven vacuum systems, equalization of scrubber solutions, and control of spills.

End of process wastes control should also be discussed. The waste control measures may involve primary, secondary and tertiary treatments such as physical/chemical or biological methods. Specifically, these methods would include: fine rubber solids removal in primary clarification; coagulation and settling of latex; separator, skimmer, and disposal of floating oils and materials; biological treatment through the use of nutrient additives; and carbon adsorption.

Molded, extruded, and fabricated rubber products generate large amounts of solid wastes. A process flow diagram is given as Figure 97. All process wastes should be described.

The large amounts of solid wastes generated are illustrated in Table 90. Those wastes specific to the proposed plant are described. Methods for minimizing these wastes and effecting their recycle are considered.

Air emissions are primarily the result of the mixing of raw materials, drying operations, venting of organic solvents from reactor vessels or storage tanks, and onsite steam and heating generation effecting SO_2, NO_x, and particulates.

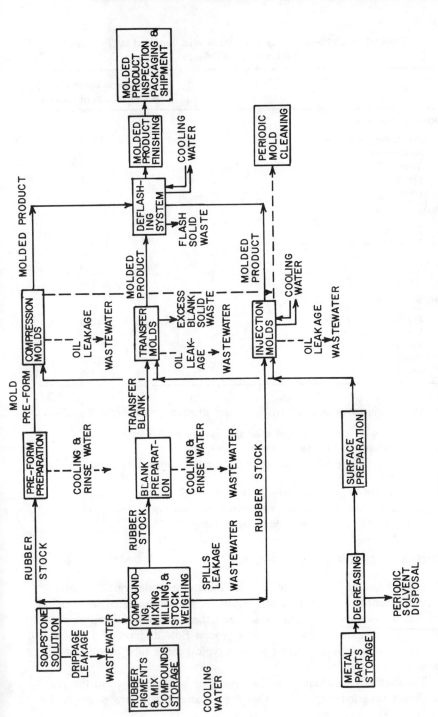

Figure 97. Flow diagram for the production of a typical molded item.[13]

Table 90. Solid Wastes Generated by the Molded, Extruded and Fabricated
Rubber Products Industry[13]

	Millions of Pounds Per Million Pounds of Product
Rubber Footwear	
Paper, cardboard, wood	17,000
Rubber compound	90,000
Textile	66,000
Metal	1,000
Other	35,000
Belts	
Paper, cardboard, wood	27,300
Rubber compound	41,000
Textile	25,200
Metal	2,300
Other	30,200
Hose	
Paper, cardboard, wood	33,000
Rubber compound	54,700
Textile	18,300
Metal	10,200
Other	49,300

At the source process control of waste must be considered, including:

* more effective control of spills and leaks in process lines, and at storage and handling terminals;
* separate disposal systems for oily and greasy wastes, soapstone, and other solutions;
* separation of metals from solutions before processing;
* preventive measures to keep wastes from plant sewers; and
* closed-loop water and process liquor systems.

TIRE AND INNER TUBE INDUSTRY

For the tire and inner tube areas of the rubber industry all impacts must be identified and evaluated. Again, as previously stated, each industry has its own unique waste streams and these are first defined in process flow diagrams.[86] Figure 98 is a process flow diagram of a typical tire production facility. Figure 99 is a typical inner tube production facility.

The diagrams given illustrate the origins of wastewaters from the tire and inner tube industry. Note that these waste streams are generated by

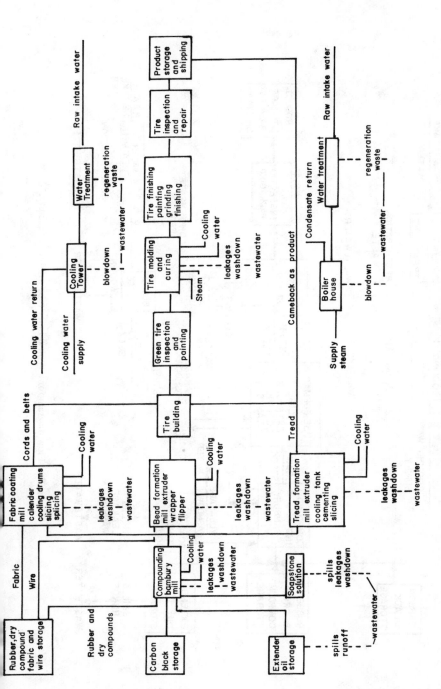

Figure 98. Flow diagram for a typical tire and camelback production facility. 86.

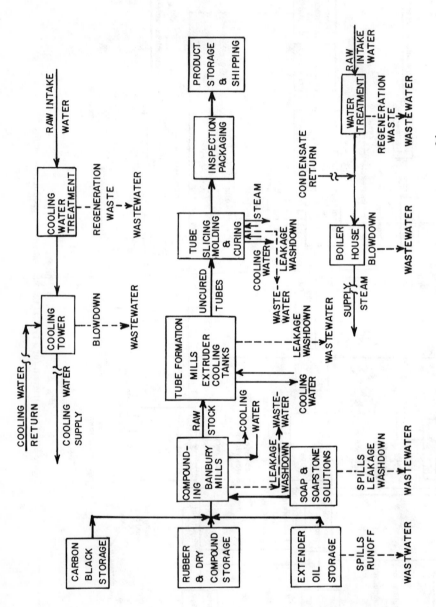

Figure 99. Flow diagram for a typical inner tube production facility.[86]

spills and leaks, cooling tower blowdowns, storm runoff, and washes and rinses. Regeneration wastes are common water contaminants of the inner tube manufacturing process.

Solid wastes for the tire and inner tube industry are generated from packaging, processes, and defective products and materials. These wastes can be summarized as in Table 91. Specifically, these wastes are derived as follows:

- Packaging of raw materials, products, and by-products produces large amounts of paper wastes.
- Cuttings, trimmings and edgings of the rubber product produce large amounts of rubber wastes.
- Cuttings, trimmings and edgings of textiles and tire cords are another source of solid wastes.
- Scrap metal wastes from tire beads, steel belts, and defects contribute to the industry's solid wastes.

Table 91. Summary of the Generator Rates of the Tire and Inner Tube Industry Solid Wastes[87]

Type of Solid Wastes	Pounds of Waste per Million Pounds of Product Shipped
Paper, cardboard, and wood	13,400
Rubber compound	11,900
Textile material	5,900
Metal	9,700
Other	14,000

All solid wastes should be considered for recycle and methods of minimizing their production should be discussed.

AIR EMISSIONS

Air emissions should be discussed also. The emissions are primarily the result of: mixing and compounding raw materials; burning rubber wastes producing odors, particulates, and sulfurous gases; drying and storage operations effecting organic solids; leaks, spills, and safety valve releases; and products of combustion (SO_2 NO_x, particulates) used to generate onsite steam and power. All air pollution control measures including equipment and process design are discussed. These would include:

- dust collectors (baghouses) in the mixing and compounding areas to reduce particulate levels;
- collection systems in the sidewall-grinding balancing machines, and tire repair areas;

* collection and control systems in painting areas; and
* enclosures on oil/water separators.

Wastewater Control

The control of wastewaters should be discussed in detail and an explanation of preventive measures should be provided. Specifically, these would include:

* Control system for soapstone discharge. Closed-loop recirculation systems have been utilized to prevent continuous soapstone discharges. A description of the workings of the recycle system would be necessary.
* Preventive and safety measures for leaks and spills. A description of dikes, sump pumps, drip pans, etc., should be included. The amount of reduction in spills and leaks should be discussed with respect to control measures.
* Control of washing and rinsing effluents with special areas and slop tanks designated for this use. A description of these facilities and their utilization is included.
* Separate sewers to prevent storm water pollution. Stormwater runoff effects should be discussed. Roofs and other protective measures may be required.
* Wastewater control incorporated into plant design and operation. Special systems and new technology to be used by the proposed plant should be described in the EIS.

Solid Waste Control

The measures of solid waste control in the proposed tire and inner tube plant should be discussed. These would include: efforts towards solid waste reduction through recycling and reuse; disposal methods of reject and test tires; and comparison of the alternate disposal methods for all the solid wastes of the plant (e.g., incineration vs sanitary landfill) and their environmental effects.

IMPACT CONTROL THROUGH DESIGN FOR POLLUTION CONTROL

One of the basic elements of the EIS is to encourage proper plant design for reduction of pollutants. The basic order of pollution abatement design in a plant should be as follows:[88] in-process solutions; in-plant solutions; end of pipe treatment. Figure 100 outlines a six-step procedure to abatement.[89]

Step 1. Examine the basic process and consider alternatives that can give the same results but with less pollution. If the alternative process is more costly, these cost figures should be compared to the added-on expense of end of pipe treatment (both capital and operating costs).

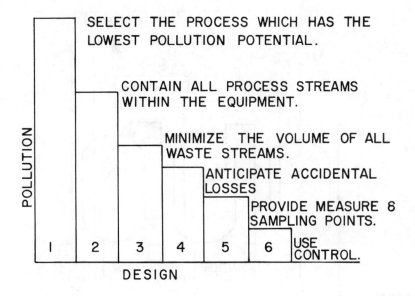

Figure 100. Pollution abatement in six steps.[89]

Step 2. Contain and control all raw materials, products, and byproducts. Spills and leaks should be kept to a minimum.

Step 3. Minimize waste streams by reducing dilution requirements if possible. Separation of wastes from high concentration waste streams is more easily obtained than low concentration streams. Further, larger volumes are more costly and difficult to handle and process.

Step 4. Preventive maintenance and safety precautions should be designed into the plant. Pollution abatement through avoiding accidents and minimizing their effects is imperative.

Step 5. Monitoring and sampling stations should be designed into the plant with provisions for good instrumentation and trained personnel. All waste streams should be measured to obtain the best possible control techniques.

Step 6. Utilize good end-of-pipe pollution control processes and practices in the project's design. New technology can be evaluated as well as specialty systems to eliminate specific pollutants.

A poor example of a pollution abatement design follows. A specified process requires three vessels to produce completion of the reaction. Figure 101 is a flow diagram of the process. Note the several pumps that

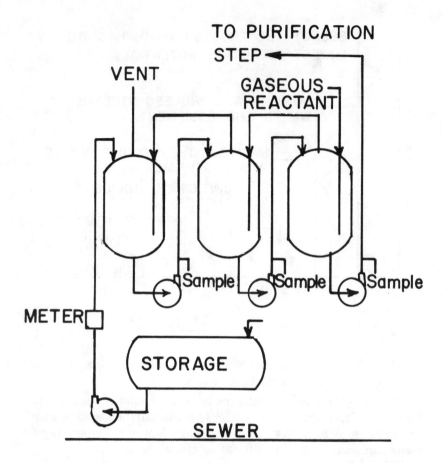

Figure 101. Poor arrangement of reactor vessels for pollution control.[89]

could leak or rupture. The first tank could overflow if the in-line control meter failed. Sampling points provide sources for leaks and spills, and even if collected in dikes or dry pans they must then be disposed of properly. Further, the storage tank is above ground and must be pump fed, providing the means for an overflow. After the reaction is completed the product must then be transported to the purification step.

The alternate arrangement for the vessels of this process is given in Figure 102. In this design most pump leakage and rupture problems have been eliminated. Gravity flow feeds the below-ground storage tank and most of the vessels, and in-the-tank vertical pump feeds the initial reactor. Sampling ports are provided with catch basins and drains to an in-process

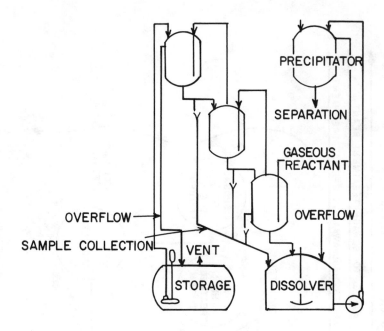

Figure 102. An arrangement of process vessels that provides for pollution control.[89]

destination. Vessels and tanks include overflow pipes and the final purification step is adjacent to the reactors to minimize transporting or conveying pollution risks.

A similar concept of design for pollution control is illustrated in Figure 103. The original process was analyzed as generating wastes in unacceptable quantities. The first step was to consider a source connection, a process modification and/or resource recovery that could reduce waste outputs. Another important aspect in pollution control is the segregation of wastes. Some wastes are more easily treated than others and should be separated from the problem streams. Each, in being treated with a method more suitable to its disposal, can be dealt with more effectively. The industrial wastestream is assessed to determine the most efficient treatment method. A recycling of water is always considered. Finally, the effluent is disposed in a receiving body of water in a manner least detrimental to its environment. For example, a plant might only be able to discharge into a tidal stream twice a day during high tides. An assessment of the receiving waters is therefore essential. Similar principles for the above can be applied to air, solid waste, and radioactive sources of pollution.

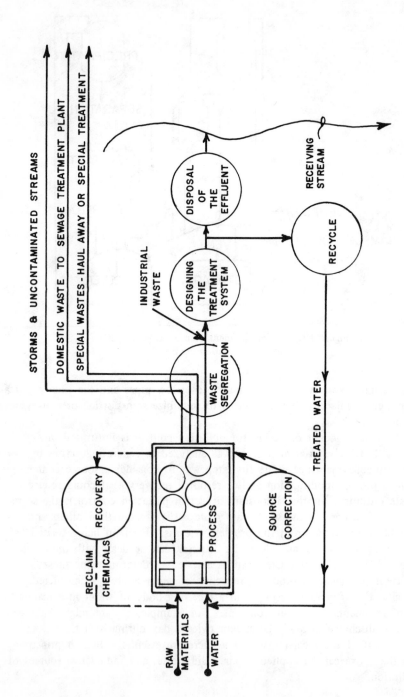

Figure 103.   A flow diagram for analysis of waste treatment design.[90]

APPENDICES

NATIONAL ENVIRONMENTAL POLICY ACT OF 1969

SUBCHAPTER I.–POLICIES AND GOALS

§ 4321. Congressional declaration of purposes

The purposes of this chapter are: To declare a national policy which will encourage productive and enjoyable harmony between man and his environment; to promote efforts which will prevent or eliminate damage to the environment and biosphere and stimulate the health and welfare of man; to enrich the understanding of the ecological systems and natural resources important to the Nation: and to establish a Council on Environmental Quality.
Pub.L. 91–190, § 2, Jan. 1, 1970, 83 Stat. 852.

SUBCHAPTER I.–POLICIES AND GOALS

§ 4331. Congressional declaration of national environmental policy

(a) The Congress, recognizing the profound impact of man's activity on the interrelations of all components of the natural environment, particularly the profound influences of population growth, high-density urbanization, industrial expansion, resource exploitation, and new and expanding technological advances and recognizing further the critical importance of restoring and maintaining environmental quality to the overall welfare and development of man, declares that it is the continuing policy of the Federal Government, in cooperation with State and local governments, and other concerned public and private organizations, to use all practicable means and measures, including financial and technical assistance, in a manner calculated to foster and promote the general welfare, to create and maintain conditions under which man and nature can exist in productive harmony, and fulfill the social, economic, and other requirements of present and future generations of Americans.

(b) In order to carry out the policy set forth in this chapter, it is the continuing responsibility of the Federal Government to use all practicable means, consistent with other essential considerations of national policy, to improve and coordinate Federal plans, functions, programs, and resources to the end that the Nation may—

(1) fulfill the responsibilities of each generation as trustee of the environment for succeeding generations;

(2) assure for all Americans safe, healthful, productive, and esthetically and culturally pleasing surroundings;

355

(3) attain the widest range of beneficial uses of the environment without degradation, risk to health or safety, or other undesirable and unintended consequences;

(4) preserve important historic, cultural, and natural aspects of our national heritage, and maintain, wherever possible, an environment which supports diversity and variety of individual choice;

(5) achieve a balance between population and resource use which will permit high standards of living and a wide sharing of life's amenities; and

(6) enhance the quality of renewable resources and approach the maximum attainable recycling of depletable resources.

(c) The Congress recognizes that each person should enjoy a healthful environment and that each person has a responsibility to contribute to the preservation and enhancement of the environment.

Pub.L. 91–190, Title I, § 101, Jan. 1, 1970, 83 Stat. 852.

§ 4332. Cooperation of agencies; reports; availability of information; recommendations; international and national coordination of efforts

The Congress authorizes and directs that, to the fullest extent possible: (1) the policies, regulations, and public laws of the United States shall be interpreted and administered in accordance with the policies set forth in this chapter, and (2) all agencies of the Federal Government shall—

(A) utilize a systematic, interdisciplinary approach which will insure the integrated use of the natural and social sciences and the environmental design arts in planning and in decisionmaking which may have an impact on man's environment;

(B) identify and develop methods and procedures, in consultation with the Council on Environmental Quality established by subchapter II of this chapter, which will insure that presently unquantified environmental amenities and values may be given appropriate consideration in decisionmaking along with economic and technical considerations;

(C) include in every recommendation or report on proposals for legislation and other major Federal actions significantly affecting the quality of the human environment, a detailed statement by the responsible official on—

(i) the environmental impact of the proposed action,

(ii) any adverse environmental effects which cannot be avoided should the proposal be implemented,

(iii) alternatives to the proposed action,

(iv) the relationship between local short-term uses of man's environment and the maintenance and enhancement of long-term productivity, and

(v) any irreversible and irretrievable commitments of resources which would be involved in the proposed action should it be implemented.

Prior to making any detailed statement, the responsible Federal official shall consult with and obtain the comments of any Federal agency which has jurisdiction by law or special expertise with respect to any environmental impact involved. Copies of such statement and the comments and views of the appropriate Federal, State, and local agencies, which are authorized to develop and enforce environmental standards, shall be made available to the President, the Council on Environmental Quality and to the public as provided by section 552 of Title 5, and shall accompany the proposal through the existing agency review processes;

(D) study, develop, and describe appropriate alternatives to recommended courses of action in any proposal which involves unresolved conflicts concerning alternative uses of available resources;

(E) recognize the worldwide and long-range character of environmental problems and, where consistent with the foreign policy of the United States, lend appropriate support to initiatives, resolutions, and programs designed to maximize international cooperation in anticipating and preventing a decline in the quality of mankind's world environment;

(F) make available to States, counties, municipalities, institutions, and individuals, advice and information useful in restoring, maintaining, and enhancing the quality of the environment;

(G) initiate and utilize ecological information in the planning and development of resource-oriented projects; and

(H) assist the Council on Environmental Quality established by subchapter II of this chapter.

Pub.L. 91–190, Title I, § 102, Jan. 1, 1970, 83 Stat. 853.

§ 4333. Conformity of administrative procedures to national environmental policy

All agencies of the Federal Government shall review their present statutory authority, administrative regulations, and current policies and procedures for the purpose of determining whether there are any deficiencies or inconsistencies therein which prohibit full compliance with the purposes and provisions of this chapter and shall propose to the President not later than July 1, 1971, such measures as may be necessary to bring their authority and policies into conformity with the intent, purposes, and procedures set forth in this chapter.

Pub.L. 91–190, Title I, § 103, Jan. 1, 1970, 83 Stat. 854.

§ 4334. Other statutory obligations of agencies

Nothing in section 4332 or 4333 of this title shall in any way affect the specific statutory obligations of any Federal agency (1) to comply with criteria or standards of environmental quality, (2) to coordinate or consult with any other Federal or State agency, or (3) to act, or refrain from acting contingent upon the recommendations or certification of any other Federal or State agency.

Pub.L. 91–190, Title I, § 104, Jan. 1, 1970, 83 Stat. 854.

§ 4335. Efforts supplemental to existing authorizations

The policies and goals set forth in this chapter are supplementary to those set forth in existing authorizations of Federal agencies.

Pub.L. 91–190, Title I, § 105, Jan. 1, 1970, 83 Stat. 854.

SUBCHAPTER II.–COUNCIL ON ENVIRONMENTAL QUALITY

§ 4341. Reports to Congress; recommendations for legislation

The President shall transmit to the Congress annually beginning July 1, 1970, an Environmental Quality Report (hereinafter referred to as the "report") which shall set forth (1) the status and condition of the major natural, manmade, or altered environmental classes of the Nation, including, but not limited to, the air, the aquatic, including marine, estuarine, and fresh water, and the terrestrial environment, including, but not limited to, the forest, dryland, wetland, range, urban, suburban, and rural environment; (2) current and foreseeable trends in the quality, management and utilization of such environments and the effects of those trends on the social, economic, and other requirements of the Nation; (3) the adequacy of available natural resources for fulfilling human and economic requirements of the Nation in the light of expected population pressures; (4) a review of the programs and activities (including regulatory activities) of the Federal Government, the State and local governments, and nongovernmental entities or individuals, with particular reference to their effect on the environment and on the conservation, development and utilization of natural resources; and (5) a program for remedying the deficiencies of existing programs and activities, together with recommendations for legislation.

Pub.L. 91–190, Title II, § 201, Jan. 1, 1970, 83 Stat. 854.

§ 4342. Establishment; membership; Chairman; appointments

There is created in the Executive Office of the President a Council on Environmental Quality (hereinafter referred to as the "Council"). The Council shall be composed of three members who shall be appointed by the President to serve at his pleasure, by and with the advice and consent of the Senate. The President shall designate one of the members of the Council to serve as Chairman. Each member shall be a person who, as a result of his training, experience, and attainments, is exceptionally well qualified to analyze and interpret environmental trends and information of all kinds; to appraise programs and activities of the Federal Government in the light of the policy set forth in subchapter I of this chapter; to be conscious of and responsive to the scientific, economic, social, esthetic, and cultural needs and interests of the Nation; and to formulate and recommend national policies to promote the improvement of the quality of the environment.
Pub.L. 91–190, Title II, § 202, Jan. 1, 1970, 83 Stat. 854.

§ 4343. Employment of personnel, experts and consultants

The Council may employ such officers and employees as may be necessary to carry out its functions under this chapter. In addition, the Council may employ and fix the compensation of such experts and consultants as may be necessary for the carrying out of its functions under this chapter, in accordance with section 3109 of Title 5 (but without regard to the last sentence thereof).
Pub.L. 91–190, Title II, § 203, |Jan. 1, 1970, 83 Stat. 855.

§ 4344. Duties and functions

It shall be the duty and function of the Council—
1 (1) to assist and advise the President in the preparation of the Environmental Quality Report required by section 4341 of this title;
(2) to gather timely and authoritative information concerning the conditions and trends in the quality of the environment both current and prospective, to analyze and interpret such information for the purpose of determining whether such conditions and trends are interfering, or are likely to interfere, with the achievement of the policy set forth in subchapter I of this chapter, and to compile and submit to the President studies relating to such conditions and trends;
(3) to review and appraise the various programs and activities of the Federal Government in the light of the policy set forth in subchapter I of this chapter for the purpose of determining the extent to which such programs and activities are contributing to the achievement of such policy, and to make recommendations to the President with respect thereto;
(4) to develop and recommend to the President national policies to foster and promote the improvement of environmental quality to meet the conservation, social, economic, health, and other requirements and goals of the Nation;
(5) to conduct investigations, studies, surveys, research, and analyses relating to ecological systems and environmental quality;
(6) to document and define changes in the natural environment, including the plant and animal systems, and to accumulate necessary data and other information for a continuing analysis of these changes or trends and an interpretation of their underlying causes;
(7) to report at least once each year to the President on the state and condition of the environment; and
(8) to make and furnish such studies, reports thereon, and recommendations with respect to matters of policy and legislation as the President may request.
Pub.L. 91–190, Title II, § 204, Jan. 1, 1970, 83 Stat. 855.

§ 4345. Consultation with the Citizen's Advisory Committee on Environmental Quality and other representatives

In exercising its powers, functions and duties under this chapter, the Council shall—
  (1) consult with the Citizens' Advisory Committee on Environmental Quality established by Executive Order numbered 11472, dated May 29, 1969, and with such representatives of science, industry, agriculture, labor, conservation organizations, State and local governments and other groups, as it deems advisable; and
  (2) utilize, to the fullest extent possible, the services, facilities, and information (including statistical information) of public and private agencies and organizations, and individuals, in order that duplication of effort and expense may be avoided, thus assuring that the Council's activities will not unnecessarily overlap or conflict with similar activities authorized by law and performed by established agencies.
Pub.L. 91–190, Title II, § 205, Jan. 1, 1970, 83 Stat. 855.

§ 4346. Tenure and compensation of members

Members of the Council shall serve full time and the Chairman of the Council shall be compensated at the rate provided for Level II of the Executive Schedule Pay Rates. The other members of the Council shall be compensated at the rate provided for Level IV or the Executive Schedule Pay Rates.
Pub.L. 91–190, Title II, § 206, Jan. 1, 1970, 83 Stat. 856.

§ 4347. Authorization of appropriations

There are authorized to be appropriated to carry out the provisions of this chapter not to exceed $300,000 for fiscal year 1970, $700,000 for fiscal year 1971, and $1,000,000 for each fiscal year thereafter.
Pub.L. 91–190, Title II, § 207, Jan. 1, 1970, 83 Stat. 856.

CHAPTER V—COUNCIL ON ENVIRONMENTAL QUALITY

PART 1500—PREPARATION OF ENVIRONMENTAL IMPACT STATEMENTS: GUIDELINES

On May 2, 1973, the Council on Environmental Quality published in the FEDERAL REGISTER, for public comment, a proposed revision of its guidelines for the preparation of environmental impact statements. Pursuant to the National Environmental Policy Act (P.L. 91-190, 42 U.S.C. 4321 et seq.) and Executive Order 11514 (35 FR 4247) all Federal departments, agencies, and establishments are required to prepare such statements in connection with their proposals for legislation and other major Federal actions significantly affecting the quality of the human environment. The authority for the Council's guidelines is set forth below in § 1500.1. The specific policies to be implemented by the guidelines is set forth below in § 1500.2.

The Council received numerous comments on its proposed guidelines from environmental groups, Federal, State, and local agencies, industry, and private individuals. Two general themes were presented in the majority of the comments. First, the Council should increase the opportunity for public involvement in the impact statement process. Second, the Council should provide more detailed guidance on the responsibilities of Federal agencies in light of recent court decisions interpreting the Act. The proposed guidelines have been revised in light of the specific comments relating to these general themes, as well as other comments received, and are now being issued in final form.

The guidelines will appear in the Code of Federal Regulations in Title 40, Chapter V, at Part 1500. They are being codified, in part, because they affect State and local governmental agencies, environmental groups, industry, and private individuals, in addition to Federal agencies, to which they are specifically directed, and the resultant need to make them widely and readily available.

Sec.
1500.1 Purpose and authority.
1500.2 Policy.
1500.3 Agency and OMB procedures.
1500.4 Federal agencies included; effect of the act on existing agency mandates.
1500.5 Types of actions covered by the act.
1500.6 Identifying major actions significantly affecting the environment.

1500.7 Preparing draft environmental statements; public hearings.
1500.8 Content of environmental statements.
1500.9 Review of draft environmental statements by Federal, Federal-State, and local agencies and by the public.
1500.10 Preparation and circulation of final environmental statements.
1500.11 Transmittal of statements to the Council; minimum periods for review; requests by the Council.
1500.12 Legislative actions.
1500.13 Application of section 102(2)(C) procedure to existing projects and programs.
1500.14 Supplementary guidelines; evaluation of procedures.

Appendix I Summary to accompany draft and final statements.
Appendix II Areas of environmental impact and Federal agencies and Federal State agencies with jurisdiction by law or special expertise to comment thereon.
Appendix III Offices within Federal agencies and Federal-State agencies for information regarding the agencies' NEPA activities and for receiving other agencies' impact statements for which comments are requested.
Appendix IV State and local agency review of impact statements.

AUTHORITY: National Environmental Act (P.L. 91-190, 42 U.S.C. 4321 et seq.) and Executive Order 11514.

§ 1500.1 Purpose and authority.

(a) This directive provides guidelines to Federal departments, agencies, and establishments for preparing detailed environmental statements on proposals for legislation and other major Federal actions significantly affecting the quality of the human environment as required by section 102(2)(C) of the National Environmental Policy Act (P.L. 91-190, 42 U.S.C. 4321 et. seq.) (hereafter "the Act"). Underlying the preparation of such environmental statements is the mandate of both the Act and Executive Order 11514 (35 FR 4247) of March 5, 1970, that all Federal agencies, to the fullest extent possible, direct their policies, plans and programs to protect and enhance environmental quality. Agencies are required to view their actions in a manner calculated to encourage productive and enjoyable harmony between man and his environment, to promote efforts preventing or eliminating damage to the environment and biosphere and stimulating the health and welfare of man, and to enrich the understanding of the ecological systems and natural resources important to the Nation. The objective of section 102(2) (C) of the Act and of these guidelines is to assist agencies in implementing these policies. This requires agencies to build into their decisionmaking process, beginning at the earliest possible point, an appropriate and careful consideration of the environmental aspects of proposed action in order that adverse environmental effects may be avoided or minimized and environmental quality previously lost may be restored. This directive also provides guidance to Federal, State, and local agencies and the public in commenting on statements prepared under these guidelines.

(b) Pursuant to section 204(3) of the Act the Council on Environmental Quality (hereafter "the Council") is assigned the duty and function of reviewing and appraising the programs and activities of the Federal Government, in the light of the Act's policy, for the purpose of determining the extent to which such programs and activities are contributing to the achievement of such policy, and to make recommendations to the President with respect thereto. Section 102(2) (B) of the Act directs all Federal agencies to identify and develop methods and procedures, in consultation with the Council, to insure that unquantified environmental values be given appropriate consideration in decisionmaking along with economic and technical considerations; section 102(2) (C) of the Act directs that copies of all environmental impact statements be filed with the Council; and section 102(2) (H) directs all Federal agencies to assist the Council in the performance of its functions. These provisions have been supplemented in sections 3(h) and (i) of Executive Order 11514 by directions that the Council issue guidelines to Federal agencies for preparation of environmental impact statements and such other instructions to agencies and requests for reports and information as may be required to carry out the Council's responsibilities under the Act.

§ 1500.2 Policy.

(a) As early as possible and in all cases prior to agency decision concerning recommendations or favorable reports on proposals for (1) legislation significantly affecting the quality of the human environment (see §§ 1500.5(i) and 1500.12) (hereafter "legislative actions") and (2) all other major Federal actions significantly affecting the quality of the human environment (hereafter "administrative actions"), Federal agencies will, in consultation with other appropriate Federal, State and local agencies and the public assess in detail the potential environmental impact.

(b) Initial assessments of the environmental impacts of proposed action should be undertaken concurrently with initial technical and economic studies and, where required, a draft environmental impact statement prepared and circulated for comment in time to accompany the proposal through the existing agency review processes for such action. In this process, Federal agencies shall: (1) Provide for circulation of draft environmental statements to other Federal, State, and local agencies and for their availability to the public in accordance with the provisions of these guidelines; (2) consider the comments of the agencies and the public; and (3) issue final environmental impact statements responsive to the comments received. The purpose of this assessment and consultation process is to provide agencies and other decisionmakers as well as members of the public with an understanding of the potential environmental effects of proposed actions, to avoid or minimize adverse effects wherever possible, and to restore or enhance environmental quality to the fullest extent practicable. In particular, agencies should use the environmental impact statement process to explore alternative actions that will avoid or minimize adverse impacts and to evaluate both the long- and short-range implications of proposed actions to man, his physical and social surroundings, and to nature. Agencies should consider the results of their environmental assessments along with their assessments of the net economic, technical and other benefits of proposed actions and use all practicable means, consistent with other essential considerations of national policy, to restore environmental quality as well as to avoid or minimize undesirable consequences for the environment.

§ 1500.3 Agency and OMB procedures.

(a) Pursuant to section 2(f) of Executive Order 11514, the heads of Federal agencies have been directed to proceed with measures required by section 102 (2) (C) of the Act. Previous guidelines of the Council directed each agency to establish its own formal procedures for (1) identifying those agency actions requiring environmental statements, the appropriate time prior to decision for the consultations required by section 102

362

(2)(C) and the agency review process for which environmental statements are to be available, (2) obtaining information required in their preparation, (3) designating the officials who are to be responsible for the statements, (4) consulting with and taking account of the comments of appropriate Federal, State and local agencies and the public, including obtaining the comment of the Administrator of the Environmental Protection Agency when required under section 309 of the Clean Air Act, as amended, and (5) meeting the requirements of section 2(b) of Executive Order 11514 for providing timely public information on Federal plans and programs with environmental impact. Each agency, including both departmental and subdepartmental components having such procedures, shall review its procedures and shall revise them, in consultation with the Council, as may be necessary in order to respond to requirements imposed by these revised guidelines as well as by such previous directives. After such consultation, proposed revisions of such agency procedures shall be published in the FEDERAL REGISTER no later than October 30, 1973. A minimum 45-day period for public comment shall be provided, followed by publication of final procedures no later than forty-five (45) days after the conclusion of the comment period. Each agency shall submit seven (7) copies of all such procedures to the Council. Any future revision of such agency procedures shall similarly be proposed and adopted only after prior consultation with the Council and, in the case of substantial revision, opportunity for public comment. All revisions shall be published in the FEDERAL REGISTER,

(b) Each Federal agency should consult, with the assistance of the Council and the Office of Management and Budget if desired, with other appropriate Federal agencies in the development and revision of the above procedures so as to achieve consistency in dealing with similar activities and to assure effective coordination among agencies in their review of proposed activities. Where applicable, State and local review of such agency procedures should be conducted pursuant to procedures established by Office of Management and Budget Circular No. A-85.

(c) Existing mechanisms for obtaining the views of Federal, State, and local agencies on proposed Federal actions should be utilized to the maximum extent practicable in dealing with environmental matters. The Office of Management and Budget will issue instructions, as necessary, to take full advantage of such existing mechanisms.

§ 1500.4 Federal agencies included; effect of the Act on existing agency mandates.

(a) Section 102(2)(C) of the Act applies to all agencies of the Federal Government. Section 102 of the Act provides that "to the fullest extent possible: (1) The policies, regulations, and public laws of the United States shall be interpreted and administered in accordance with the policies set forth in this Act," and section 105 of the Act provides that "the policies and goals set forth in this Act are supplementary to those set forth in existing authorizations of Federal agencies." This means that each agency shall interpret the provisions of the Act as a supplement to its existing authority and as a mandate to view traditional policies and missions in the light of the Act's national environmental objectives. In accordance with this purpose, agencies should continue to review their policies, procedures, and regulations and to revise them as necessary to ensure full compliance with the purposes and provisions of the Act. The phrase "to the fullest extent possible" in section 102 is meant to make clear that each agency of the Federal Government shall comply with that section unless existing law applicable to the agency's operations expressly prohibits or makes compliance impossible.

§ 1500.5 Types of actions covered by the Act.

(a) "Actions" include but are not limited to:

(1) Recommendations or favorable reports relating to legislation including requests for appropriations. The requirement for following the section 102 (2)(C) procedure as elaborated in these guidelines applies to both (i) agency recommendations on their own proposals for legislation (see § 1500.12); and (ii) agency reports on legislation initiated elsewhere. In the latter case only the agency which has primary responsibility for the subject matter involved will prepare an environmental statement.

(2) New and continuing projects and program activities: directly undertaken by Federal agencies; or supported in whole or in part through Federal contracts, grants, subsidies, loans, or other

forms of funding assistance (except where such assistance is solely in the form of general revenue sharing funds, distributed under the State and Local Fiscal Assistance Act of 1972, 31 U.S.C. 1221 et. seq. with no Federal agency control over the subsequent use of such funds); or involving a Federal lease, permit, license certificate or other entitlement for use.

(3) The making, modification, or establishment of regulations, rules, procedures, and policy.

§ 1500.6 Identifying major actions significantly affecting the environment.

(a) The statutory clause "major Federal actions significantly affecting the quality of the human environment" is to be construed by agencies with a view to the overall, cumulative impact of the action proposed, related Federal actions and projects in the area, and further actions contemplated. Such actions may be localized in their impact, but if there is potential that the environment may be significantly affected, the statement is to be prepared. Proposed major actions, the enviornmental impact of which is likely to be highly controversial, should be covered in all cases. In considering what constitutes major action significantly affecting the environment, agencies should bear in mind that the effect of many Federal decisions about a project or complex of projects can be individually limited but cumulatively considerable. This can occur when one or more agencies over a period of years puts into a project individually minor but collectively major resources, when one decision involving a limited amount of money is a precedent for action in much larger cases or represents a decision in principle about a future major course of action, or when several Government agencies individually make decisions about partial aspects of a major action. In all such cases, an environmental statement should be prepared if it is reasonable to anticipate a cumulatively significant impact on the environment from Federal action. The Council, on the basis of a written assessment of the impacts involved, is available to assist agencies in determining whether specific actions require impact statements.

(b) Section 101(b) of the Act indicates the broad range of aspects of the environment to be surveyed in any assessment of significant effect. The Act also indicates that adverse significant effects include those that degrade the quality of the environment, curtail the range of beneficial uses of the environment, and serve short-term, to the disadvantage of long-term, environmental goals. Significant effects can also include actions which may have both beneficial and detrimental effects, even if on balance the agency believes that the effect will be beneficial. Significant effects also include secondary effects, as described more fully, for example, in § 1500.8(a) (iii)(B). The significance of a proposed action may also vary with the setting, with the result that an action that would have little impact in an urban area may be significant in a rural setting or vice versa. While a precise definition of environmental "significance," valid in all contexts, is not possible, effects to be considered in assessing significance include, but are not limited to, those outlined in Appendix II of these guidelines.

(c) Each of the provisions of the Act, except section 102(2)(C), applies to all Federal agency actions. Section 102(2)(C) requires the preparation of a detailed environmental impact statement in the case of "major Federal actions significantly affecting the quality of the human environment." The identification of major actions significantly affecting environment is the responsibility of each Federal agency, to be carried out against the background of its own particular operations. The action must be a (1) "major" action, (2) which is a "Federal action," (3) which has a "significant" effect, and (4) which involves the "quality of the human environment." The words "major" and "significantly" are intended to imply thresholds of importance and impact that must be met before a statement is required. The action causing the impact must also be one where there is sufficient Federal control and responsibility to constitute "Federal action" in contrast to cases where such Federal control and responsibility are not present as, for example, when Federal funds are distributed in the form of general revenue sharing to be used by State and local governments (see § 1500.5(ii)). Finally, the action must be one that significantly affects the quality of the human environment either by directly affecting human beings or by indirectly affecting human beings through adverse effects on the environment. Each agency should review the typical classes of actions that it undertakes and, in consultation with the Council, should develop specific criteria and methods for identifying those actions likely to require environmental statements and those actions likely not

364

to require environmental statements. Normally this will involve:

(i) Making an initial assessment of the environmental impacts typically associated with principal types of agency action.

(ii) Identifying on the basis of this assessment, types of actions which normally do, and types of actions which normally do not, require statements.

(iii) With respect to remaining actions that may require statements depending on the circumstances, and those actions determined under the preceding paragraph (C) (4) (ii) of this section as likely to require statements, identifying: (a) what basic information needs to be gathered; (b) how and when such information is to be assembled and analyzed; and (c) on what bases environmental assessments and decisions to prepare impact statements will be made. Agencies may either include this substantive guidance in the procedures issued pursuant to § 1500.3(a) of these guidelines, or issue such guidance as supplemental instructions to aid relevant agency personnel in implementing the impact statement process. Pursuant to § 1500.14 of these guidelines, agencies shall report to the Council by June 30, 1974, on the progress made in developing such substantive guidance.

(d) (1) Agencies should give careful attention to identifying and defining the purpose and scope of the action which would most appropriately serve as the subject of the statement. In many cases, broad program statements will be required in order to assess the environmental effects of a number of individual actions on a given geographical area (e.g., coal leases), or environmental impacts that are generic or common to a series of agency actions (e.g., maintenance or waste handling practices), or the overall impact of a large-scale program or chain of contemplated projects (e.g., major lengths of highway as opposed to small segments). Subsequent statements on major individual actions will be necessary where such actions have significant environmental impacts not adequately evaluated in the program statement.

(2) Agencies engaging in major technology research and development programs should develop procedures for periodic evaluation to determine when a program statement is required for such programs. Factors to be considered in making this determination include the magnitude of Federal investment in the program, the likelihood of widespread application of the technology, the degree of environmental impact which would occur if the technology were widely applied, and the extent to which continued investment in the new technology is likely to restrict future alternatives. Statements must be written late enough in the development process to contain meaningful information, but early enough so that this information can practically serve as an input in the decision-making process. Where it is anticipated that a statement may ultimately be required but that its preparation is still premature, the agency should prepare an evaluation briefly setting forth the reasons for its determination that a statement is not yet necessary. This evaluation should be periodically updated, particularly when significant new information becomes available concerning the potential environmental impact of the program. In any case, a statement must be prepared before research activities have reached a stage of investment or commitment to implementation likely to determine subsequent development or restrict later alternatives. Statements on technology research and development programs should include an analysis not only of alternative forms of the same technology that might reduce any adverse environmental impacts but also of alternative technologies that would serve the same function as the technology under consideration. Efforts should be made to involve other Federal agencies and interested groups with relevant expertise in the preparation of such statements because the impacts and alternatives to be considered are likely to be less well defined than in other types of statements.

(e) In accordance with the policy of the Act and Executive Order 11514 agencies have a responsibility to develop procedures to insure the fullest practicable provision of timely public information and understanding of Federal plans and programs with environmental impact in order to obtain the views of interested parties. In furtherance of this policy, agency procedures should include an appropriate early notice system for informing the public of the decision to prepare a draft environmental statement on proposed administrative actions (and for soliciting comments that may be helpful in preparing the statement) as soon as is practicable after the decision to prepare the statement is made. In this connection, agencies should: (1) maintain a list of administrative actions for which environmental statements are being prepared; (2) revise the list at regular intervals specified in the agency's procedures developed pursuant to § 1500.3(a) of these guidelines (but not less than quarterly) and transmit each such revision to the Council; and (3) make the list available for public inspection on request. The Council will periodically publish such lists in the FEDERAL REGISTER. If an agency decides that an environmental statement is not necessary for a proposed action (i) which the agency has identified pursuant to § 1500.6(c)(4) (ii) as normally requiring preparation of a statement, (ii) which is similar to actions for which the agency has prepared a significant number of statements, (iii) which the agency has previously announced would be the subject of a statement, or (iv) for which the agency has made a negative determination in response to a request from the Council pursuant to § 1500.11(f), the agency shall prepare a publicly available record briefly setting forth the agency's decision and the reasons for that determination. Lists of such negative determinations, and any evaluations made pursuant to § 1500.6 which conclude that preparation of a statement is not yet timely, shall be prepared and made available in the same manner as provided in this subsection for lists of statements under preparation.

§ 1500.7 Preparing draft environmental statements; public hearings.

(a) Each environmental impact statement shall be prepared and circulated in draft form for comment in accordance with the provisions of these guidelines. The draft statement must fulfill and satisfy to the fullest extent possible at the time the draft is prepared the requirements established for final statements by section 102(2)(C). (Where an agency has an established practice of declining to favor an alternative until public comments on a proposed action have been received, the draft environmental statement may indicate that two or more alternatives are under consideration.) Comments received shall be carefully evaluated and considered in the decision process. A final statement with substantive comments attached shall then be issued and circulated in accordance with applicable provisions of §§ 1500.10, 1500.11, or 1500.12. It is important that draft environmental statements be prepared and circulated for comment and furnished to the Council as early as possible in the agency review process in order to permit agency decisionmakers and outside reviewers to give meaningful consideration to the environmental issues involved. In particular, agencies should keep in mind that such statements are to serve as the means of assessing the environmental impact of proposed agency actions, rather than as a justification for decisions already made. This means that draft statements on administrative actions should be prepared and circulated for comment prior to the first significant point of decision in the agency review process. For major categories of agency action, this point should be identified in the procedures issued pursuant to § 1500.3(a). For major categories of projects involving an applicant and identified pursuant to § 1500.6 (c)(c)(ii) as normally requiring the preparation of a statement, agencies should include in their procedures provisions limiting actions which an applicant is permitted to take prior to completion and review of the final statement with respect to his application.

(b) Where more than one agency (1) directly sponsors an action, or is directly involved in an action through funding, licenses, or permits, or (2) is involved in a group of actions directly related to each other because of their functional interdependence and geographical proximity, consideration should be given to preparing one statement for all the Federal actions involved (see § 1500.6(d)(1)). Agencies in such cases should consider the possibility of joint preparation of a statement by all agencies concerned, or designation of a single "lead agency" to assume supervisory responsibility for preparation of the statement. Where a lead agency prepares the statement, the other agencies involved should provide assistance with respect to their areas of jurisdiction and expertise. In either case, the statement should contain an environmental assessment of the full range of Federal actions involved. should reflect the views of all participating agencies, and should be prepared before major or irreversible actions have been taken by any of the participating agencies. Factors relevant in determining an appropriate lead agency include the time sequence in which the agencies become involved, the magnitude of their respective involvement, and their relative expertise with respect to the project's environmental effects. As necessary, the Council will assist in resolving questions of responsibility for statement preparation in the case of multi-agency actions. Federal Regional Councils, agencies and the public are encouraged to bring to the attention of the Council and other relevant agencies appropriate situations where a geographic or regionally focused statement would be desirable because of the cumulative environmental effects likely to result from multi-agency actions in the area.

(c) Where an agency relies on an applicant to submit initial environmental information, the agency should assist the applicant by outlining the types of information required. In all cases, the agency should make its own evaluation of the environmental issues and take responsibility for the scope and content of draft and final environmental statements.

(d) Agency procedures developed pursuant to § 1500.3(a) of these guidelines should indicate as explicitly as possible those types of agency decisions or actions which utilize hearings as part of the normal agency review process, either as a result of statutory requirement or agency practice. To the fullest extent possible, all such hearings shall include consideration of the environmental aspects of the proposed action. Agency procedures shall also specifically include provision for public hearings on major actions with environmental impact, whenever appropriate, and for providing the public with relevant information, including information on alternative courses of action. In deciding whether a public hearing is appropriate, an agency should consider: (1) The magnitude of the proposal in terms of economic costs, the geographic area involved, and the uniqueness or size of commitment of the resources involved; (2) the degree of interest in the proposal, as evidenced by requests from the public and from Federal, State and local authorities that a hearing be held; (3) the complexity of the issue and the likelihood that information will be presented at the hearing which will be of assistance to the agency in fulfilling its responsibilities under the Act; and (4) the extent to which public involvement already has been achieved through other means, such as earlier public hearings, meetings with citizen representatives, and/or written comments on the proposed action. Agencies should make any draft environmental statements to be issued available to the public at least fifteen (15) days prior to the time of such hearings.

§ 1500.8 Content of environmental statements.

(a) The following points are to be covered:

(1) A description of the proposed action, a statement of its purposes, and a description of the environment affected, including information, summary technical data, and maps and diagrams where relevant, adequate to permit an assessment of potential environmental impact by commenting agencies and the public. Highly technical and specialized analyses and data should be avoided in the body of the draft impact statement. Such materials should be attached as appendices or footnoted with adequate bibliographic references. The statement should also succinctly describe the environment of the area affected as it exists prior to a proposed action, including other Federal activities in the area affected by the proposed action which are related to the proposed action. The interrelationships and cumulative environmental impacts of the proposed action and other related Federal projects shall be presented in the statement. The amount of detail provided in such descriptions should be commensurate with the extent and expected impact of the action, and with the amount of information required at the particular level of decisionmaking (planning, feasibility, design, etc.). In order to ensure accurate descriptions and environmental assessments, site visits should be made where feasible. Agencies should also take care to identify, as appropriate, population and growth characteristics of the affected area and any population and growth assumptions used to justify the project or program or to determine secondary population and growth impacts resulting from the proposed action and its alternatives (see paragraph (a)(1)(3)(ii), of this section). In discussing these population aspects, agencies should give consideration to using the rates of growth in the region of the project contained in the projection compiled for the Water Resources Council by the Bureau of Economic Analysis of the Department of Commerce and the Economic Research Service of the Department of Agriculture (the "OBERS" projection). In any event it is essential that the sources of data used to identify, quantify or evaluate any and all environmental consequences be expressly noted.

(2) The relationship of the proposed action to land use plans, policies, and controls for the affected area. This requires a discussion of how the proposed action may conform or conflict with the objectives and specific terms of approved or proposed Federal, State, and local land use plans, policies, and controls, if

any, for the area affected including those developed in response to the Clean Air Act or the Federal Water Pollution Control Act Amendments of 1972. Where a conflict or inconsistency exists, the statement should describe the extent to which the agency has reconciled its proposed action with the plan, policy or control, and the reasons why the agency has decided to proceed notwithstanding the absence of full reconciliation.

(3) The probable impact of the proposed action on the environment.

(i) This requires agencies to assess the positive and negative effects of the proposed action as it affects both the national and international environment. The attention given to different environmental factors will vary according to the nature, scale, and location of proposed actions. Among factors to consider should be the potential effect of the action on such aspects of the environment as those listed in Appendix II of these guidelines. Primary attention should be given in the statement to discussing those factors most evidently impacted by the proposed action.

(ii) Secondary or indirect, as well as primary or direct, consequences for the environment should be included in the analysis. Many major Federal actions, in particular those that involve the construction or licensing of infrastructure investments (e.g., highways, airports, sewer systems, water resource projects, etc.), stimulate or induce secondary effects in the form of associated investments and changed patterns of social and economic activities. Such secondary effects, through their impacts on existing community facilities and activities, through inducing new facilities and activities, or through changes in natural conditions, may often be even more substantial than the primary effects of the original action itself. For example, the effects of the proposed action on population and growth may be among the more significant secondary effects. Such population and growth impacts should be estimated if expected to be significant (using data identified as indicated in § 1500.8(a)(1)) and an assessment made of the effect of any possible change in population patterns or growth upon the resource base, including land use, water, and public services, of the area in question.

(4) Alternatives to the proposed action, including, where relevant, those not within the existing authority of the responsible agency. (Section 102(2)(D) of

the Act requires the responsible agency to "study, develop, and describe appropriate alternatives to recommended courses of action in any proposal which involves unresolved conflicts concerning alternative uses of available resources"). A rigorous exploration and objective evaluation of the environmental impacts of all reasonable alternative actions, particularly those that might enhance environmental quality or avoid some or all of the adverse environmental effects, is essential. Sufficient analysis of such alternatives and their environmental benefits, costs and risks should accompany the proposed action through the agency review process in order not to foreclose prematurely options which might enhance environmental quality or have less detrimental effects. Examples of such alternatives include: the alternative of taking no action or of postponing action pending further study; alternatives requiring actions of a significantly different nature which would provide similar benefits with different environmental impacts (e.g., nonstructural alternatives to flood control programs, or mass transit alternatives to highway construction); alternatives related to different designs or details of the proposed action which would present different environmental impacts (e.g., cooling ponds vs. cooling towers for a power plant or alternatives that will significantly conserve energy); alternative measures to provide for compensation of fish and wildlife losses, including the acquisition of land, waters, and interests therein. In each case, the analysis should be sufficiently detailed to reveal the agency's comparative evaluation of the environmental benefits, costs and risks of the proposed action and each reasonable alternative. Where an existing impact statement already contains such an analysis, its treatment of alternatives may be incorporated provided that such treatment is current and relevant to the precise purpose of the proposed action.

(5.) Any probable adverse environmental effects which cannot be avoided (such as water or air pollution, undesirable land use patterns, damage to life systems, urban congestion, threats to health or other consequences adverse to the environmental goals set out in section 101 (b) of the Act). This should be a brief section summarizing in one place those effects discussed in paragraph (a)(3) of this section that are adverse and unavoidable under the proposed action. Included for purposes of contrast should

368

be a clear statement of how other avoidable adverse effects discussed in paragraph (a)(2) of this section will be mitigated.

(6) The relationship between local short-term uses of man's environment and the maintenance and enhancement of long-term productivity. This section should contain a brief discussion of the extent to which the proposed action involves tradeoffs between short-term environmental gains at the expense of long-term losses, or vice versa, and a discussion of the extent to which the proposed action forecloses future options. In this context short-term and long-term do not refer to any fixed time periods, but should be viewed in terms of the environmentally significant consequences of the proposed action.

(7) Any irreversible and irretrievable commitments of resources that would be involved in the proposed action should it be implemented. This requires the agency to identify from its survey of unavoidable impacts in paragraph (a)(5) of this section the extent to which the action irreversibly curtails the range of potential uses of the environment. Agencies should avoid construing the term "resources" to mean only the labor and materials devoted to an action. "Resources" also means the natural and cultural resources committed to loss or destruction by the action.

(8) An indication of what other interests and considerations of Federal policy are thought to offset the adverse environmental effects of the proposed action identified pursuant to paragraphs (a)(3) and (5) of this section. The statement should also indicate the extent to which these stated countervailing benefits could be realized by following reasonable alternatives to the proposed action (as identified in paragraph (a)(4) of this section) that would avoid some or all of the adverse environmental effects. In this connection, agencies that prepare cost-benefit analyses of proposed actions should attach such analyses, or summaries thereof, to the environmental impact statement, and should clearly indicate the extent to which environmental costs have not been reflected in such analyses.

(b) In developing the above points agencies should make every effort to convey the required information succinctly in a form easily understood, both by members of the public and by public decisionmakers, giving attention to the substance of the information conveyed

rather than to the particular form, or length, or detail of the statement. Each of the above points, for example, need not always occupy a distinct section of the statement if it is otherwise adequately covered in discussing the impact of the proposed action and its alternatives—which items should normally be the focus of the statement. Draft statements should indicate at appropriate points in the text any underlying studies, reports, and other information obtained and considered by the agency in preparing the statement including any cost-benefit analyses prepared by the agency, and reports of consulting agencies under the Fish and Wildlife Coordination Act, 16 U.S.C. 661 et seq., and the National Historic Preservation Act of 1966, 16 U.S.C. 470 et seq., where such consultation has taken place. In the case of documents not likely to be easily accessible (such as internal studies or reports), the agency should indicate how such information may be obtained. If such information is attached to the statement, care should be taken to ensure that the statement remains an essentially self-contained instrument, capable of being understood by the reader without the need for undue cross reference.

(c) Each environmental statement should be prepared in accordance with the precept in section 102(2)(A) of the Act that all agencies of the Federal Government "utilize a systematic, interdisciplinary approach which will insure the integrated use of the natural and social sciences and the environmental design arts in planning and decisionmaking which may have an impact on man's environment." Agencies should attempt to have relevant disciplines represented on their own staffs; where this is not feasible they should make appropriate use of relevant Federal, State, and local agencies or the professional services of universities and outside consultants. The interdisciplinary approach should not be limited to the preparation of the environmental impact statement, but should also be used in the early planning stages of the proposed action. Early application of such an approach should help assure a systematic evaluation of reasonable alternative courses of action and their potential social, economic, and environmental consequences.

(d) Appendix I prescribes the form of the summary sheet which should accompany each draft and final environmental statement.

§ 1500.9 Review of draft environmental statements by Federal, Federal-State, State, and local agencies and by the public.

(a) *Federal agency review.* (1) *In general.* A Federal agency considering an action requiring an environmental statement should consult with, and (on the basis of a draft environmental statement for which the agency takes responsibility) obtain the comment on the environmental impact of the action of Federal and Federal-State agencies with jurisdiction by law or special expertise with respect to any environmental impact involved. These Federal and Federal-State agencies and their relevant areas of expertise include those identified in Appendices II and III to these guidelines. It is recommended that the listed departments and agencies establish contact points, which may be regional offices, for providing comments on the environmental statements. The requirement in section 102(2)(C) to obtain comment from Federal agencies having jurisdiction or special expertise is in addition to any specific statutory obligation of any Federal agency to coordinate or consult with any other Federal or State agency. Agencies should, for example, be alert to consultation requirements of the Fish and Wildlife Coordination Act, 16 U.S.C. 661 et seq., and the National Historic Preservation Act of 1966, 16 U.S.C. 470 et seq. To the extent possible, statements or findings concerning environmental impact required by other statutes, such as section 4(f) of the Department of Transportation Act of 1966, 49 U.S.C. 1653(f), or section 106 of the National Historic Preservation Act of 1966, should be combined with compliance with the environmental impact statement requirements of section 102(2)(C) of the Act to yield a single document which meets all applicable requirements. The Advisory Council on Historic Preservation, the Department of Transportation, and the Department of the Interior, in consultation with the Council, will issue any necessary supplementing instructions for furnishing information or findings not forthcoming under the environmental impact statement process.

(b) *EPA review.* Section 309 of the Clean Air Act, as amended (42 U.S.C. § 1857h-7), provides that the Administrator of the Environmental Protection Agency shall comment in writing on the environmental impact of any matter relating to his duties and responsibilities, and shall refer to the Council any matter that the Administrator determines is unsatisfactory from the standpoint of public health or welfare or environmental quality. Accordingly, wherever an agency action related to air or water quality, noise abatement and control, pesticide regulation, solid waste disposal, generally applicable environmental radiation criteria and standards, or other provision of the authority of the Administrator is involved, Federal agencies are required to submit such proposed actions and their environmental impact statements, if such have been prepared, to the Administrator for review and comment in writing. In all cases where EPA determines that proposed agency action is environmentally unsatisfactory, or where EPA determines that an environmental statement is so inadequate that such a determination cannot be made, EPA shall publish its determination and notify the Council as soon as practicable. The Administrator's comments shall constitute his comments for the purposes of both section 309 of the Clean Air Act and section 102(2)(C) of the National Environmental Policy Act.

(c) State and local review. Office of Management and Budget Circular No. A–95 (Revised) through its system of State and areawide clearinghouses provides a means for securnig the views of State and local environmental agencies, which can assist in the preparation and review of environmental impact statements. Current instructions for obtaining the views of such agencies are contained in the joint OMB–CEQ memorandum attached to these guidelines as Appendix IV. A current listing of clearinghouses is issued periodically by the Office of Management and Budget.

(d) *Public review.* The procedures established by these guidelines are designed to encourage public participation in the impact statement process at the earliest possible time. Agency procedures should make provision for facilitating the comment of public and private organizations and individuals by announcing the availability of draft environmental statements and by making copies available to organizations and individuals that request an opportunity to comment. Agencies should devise methods for publicizing the existence of draft statements, for example, by publication of notices in local newspapers or by maintaining a list of groups, including relevant conservation commissions, known to be interested in the agency's activities and directly notifying such groups of the existence of a draft statement, or sending them a copy, as soon as it has been prepared. A copy of the draft statement should in all cases be sent to any applicant whose project is the subject of the statement. Materials to be made available to the public shall be provided without charge to the extent practicable, or at a fee which is not more than the actual cost of reproducing copies required to be sent to other Federal agencies, including the Council.

(e) *Responsibilities of commenting entities.* (1) Agencies and members of the public submitting comments on proposed actions on the basis of draft environmental statements should endeavor to make their comments as specific, substantive, and factual as possible without undue attention to matters of form in the impact statement. Although the comments need not conform to any particular format, it would assist agencies reviewing comments if the comments were organized in a manner consistent with the structure of the draft statement. Emphasis should be placed on the assessment of the environmental impacts of the proposed action, and the acceptability of those impacts on the quality of the environment, particularly as contrasted with the impacts of reasonable alternatives to the action. Commenting entities may recommend modifications to the proposed action and/or new alternatives that will enhance environmental quality and avoid or minimize adverse environmental impacts.

(2) Commenting agencies should indicate whether any of their projects not identified in the draft statement are sufficiently advanced in planning and related environmentally to the proposed action so that a discussion of the environmental interrelationships should be included in the final statement (see § 1500.8(a)(1)). The Council is available to assist agencies in making such determinations.

(3) Agencies and members of the public should indicate in their comments the nature of any monitoring of the environmental effects of the proposed project that appears particularly appropriate. Such monitoring may be necessary during the construction, startup, or operation phases of the project. Agencies with special expertise with respect to the environmental impacts involved are encouraged to assist the sponsoring agency in the establishment and operation of appropriate environmental monitoring.

(f) Agencies seeking comment shall establish time limits of not less than forty-five (45) days for reply, after which it may be presumed, unless the agency or party consulted requests a specified extension of time, that the agency or party consulted has no comment to make. Agencies seeking comment should endeavor to comply with requests for extensions of time of up to fifteen (15) days. In determining an appropriate period for comment, agencies should consider the magnitude and complexity of the statement and the extent of citizen interest in the proposed action.

§ 1500.10 Preparation and circulation of final environmental statements.

(a) Agencies should make every effort to discover and discuss all major points of view on the environmental effects of the proposed action and its alternatives in the draft statement itself. However, where opposing professional views and responsible opinion have been overlooked in the draft statement and are brought to the agency's attention through the commenting process, the agency should review the environmental effects of the action in light of those views and should make a meaningful reference in the final statement to the existence of any responsible opposing view not adequately discussed in the draft statement, indicating the agency's response to the issues raised. All substantive comments received on the draft (or summaries thereof where response has been exceptionally voluminous) should be attached to the final statement, whether or not each such comment is thought to merit individual discussion by the agency in the text of the statement.

(b) Copies of final statements, with comments attached, shall be sent to all Federal, State, and local agencies and private organizations that made substantive comments on the draft statement and to individuals who requested a copy of the final statement, as well as any applicant whose project is the subject of the statement. Copies of final statements shall in all cases be sent to the Environmental Protection Agency to assist it in carrying out its responsibilities under section 309 of the Clean Air Act. Where the number of comments on a draft statement is such that distribution of the final statement to all commenting entities appears impracticable, the agency shall consult with the Council concerning alternative arrangements for distribution of the statement.

§ 1500.11 Transmittal of statements to the Council; minimum periods for review; requests by the Council.

(a) As soon as they have been prepared, ten (10) copies of draft environmental statements, five (5) copies of all comments made thereon (to be forwarded to the Council by the entity making comment at the time comment is forwarded to the responsible agency), and ten (10) copies of the final text of environmental statements (together with the substance of all comments received by the responsible agency from Federal, State, and local agencies and from private organizations and individuals) shall be supplied to the Council. This will serve to meet the statutory requirement to make environmental statements available to the President. At the same time that copies of draft and final statements are sent to the Council, copies should also be sent to relevant commenting entities as set forth in §§ 1500.9 and 1500.10(b) of these guidelines.

(b) To the maximum extent practicable no administrative action subject to section 102(2)(C) is to be taken sooner than ninety (90) days after a draft environmental statement has been circulated for comment, furnished to the Council and, except where advance public disclosure will result in significantly increased costs of procurement to the Government, made available to the public pursuant to these guidelines; neither should such administrative action be taken sooner than thirty (30) days after the final text of an environmental statement (together with comments) has been made available to the Council, commenting agencies, and the public. In all cases, agencies should allot a sufficient review period for the final statement so as to comply with the statutory requirement that the "statement and the comments and views of appropriate Federal, State, and local agencies * * * accompany the proposal through the existing agency review processes." If the final text of an environmental statement is filed within ninety (90) days after a draft statement has been circulated for comment, furnished to the Council and made public pursuant to this section of these guidelines, the minimum thirty (30) day period and the ninety (90) day period may run concurrently to the extent that they overlap. An agency may at any time supplement or amend a draft or final environmental statement, particularly when substantial changes are made in the proposed action, or signifi-

cant new information becomes available concerning its environmental aspects. In such cases the agency should consult with the Council with respect to the possible need for or desirability of recirculation of the statement for the appropriate period.

(c) The Council will publish weekly in the FEDERAL REGISTER lists of environmental statements received during the preceding week that are available for public comment. The date of publication of such lists shall be the date from which the minimum periods for review and advance availability of statements shall be calculated.

(d) The Council's publication of notice of the availability of statements is in addition to the agency's responsibility, as described in § 1500.9(d) of these guidelines, to insure the fullest practicable provision of timely public information concerning the existence and availability of environmental statements. The agency responsible for the environmental statement is also responsible for making the statement, the comments received, and any underlying documents available to the public pursuant to the provisions of the Freedom of Information Act (5 U.S.C., 552), without regard to the exclusion of intra- or interagency memoranda when such memoranda transmit comments of Federal agencies on the environmental impact of the proposed action pursuant to § 1500.9 of these guidelines. Agency procedures prepared pursuant to § 1500.3(a) of these guidelines shall implement these public information requirements and shall include arrangements for availability of environmental statements and comments at the head and appropriate regional offices of the responsible agency and at appropriate State and areawide clearinghouses unless the Governor of the State involved designates to the Council some other point for receipt of this information. Notice of such designation of an alternate point for receipt of this information will be included in the Office of Management and Budget listing of clearinghouses referred to in § 1500.9(c).

(e) Where emergency circumstances make it necessary to take an action with significant environmental impact without observing the provisions of these guidelines concerning minimum periods for agency review and advance availability of environmental statements, the Federal agency proposing to take the action should consult with the Council about alternative arrangements. Similarly where there are overriding considerations of expense to the Government or impaired program effectiveness, the responsible agency should consult with the Council concerning appropriate modifications of the minimum periods.

(f) In order to assist the Council in fulfilling its responsibilities under the Act and under Executive Order 11514, all agencies shall (as required by section 102(2)(H) of the Act and section 3(i) of Executive Order 11514) be responsive to requests by the Council for reports and other information dealing with issues arising in connection with the implementation of the Act. In particular, agencies shall be responsive to a request by the Council for the preparation and circulation of an environmental statement, unless the agency determines that such a statement is not required, in which case the agency shall prepare an environmental assessment and a publicly available record briefly setting forth the reasons for its determination. In no case, however, shall the Council's silence or failure to comment or request preparation, modification, or recirculation of an environmental statement or to take other action with respect to an environmental statement be construed as bearing in any way on the question of the legal requirement for or the adequacy of such statement under the Act.

§ 1500.12 Legislative actions.

(a) The Council and the Office of Management and Budget will cooperate in giving guidance as needed to assist agencies in identifying legislative items believed to have environmental significance. Agencies should prepare impact statements prior to submission of their legislative proposals to the Office of Management and Budget. In this regard, agencies should identify types of repetitive legislation requiring environmental impact statements (such as certain types of bills affecting transportation policy or annual construction authorizations).

(b) With respect to recommendations or reports on proposals for legislation to which section 102(2)(C) applies, the final text of the environmental statement and comments thereon should be available to the Congress and to the public for consideration in connection with the proposed legislation or report. In cases where the scheduling of congressional hearings on recommendations or reports on proposals for legislation which the Federal agency has forwarded to the Congress does not allow adequate time for the completion of a final text of an environmental statement (together with comments), a draft environmental statement may be furnished to the Congress and made available to the public pending transmittal of the comments as received and the final text.

§ 1500.13 Application of section 102 (2)(C) procedure to existing projects and programs.

Agencies have an obligation to reassess ongoing projects and programs in order to avoid or minimize adverse environmental effects. The section 102(2)(C) procedure shall be applied to further major Federal actions having a significant effect on the environment even though they arise from projects or programs initiated prior to enactment of the Act on January 1, 1970. While the status of the work and degree of completion may be considered in determining whether to proceed with the project, it is essential that the environmental impacts of proceeding are reassessed pursuant to the Act's policies and procedures and, if the project or program is continued, that further incremental major actions be shaped so as to enhance and restore environmental quality as well as to avoid or minimize adverse environmental consequences. It is also important in further action that account be taken of environmental consequences not fully evaluated at the outset of the project or program.

§ 1500.14 Supplementary guidelines; evaluation of procedures.

(a) The Council after examining environmental statements and agency procedures with respect to such statements will issue such supplements to these guidelines as are necessary.

(b) Agencies will continue to assess their experience in the implementation of the section 102(2)(C) provisions of the Act and in conforming with these guidelines and report thereon to the Council by June 30, 1974. Such reports should include an identification of the problem areas and suggestions for revision or clarification of these guidelines to achieve effective coordination of views on environmental aspects (and alternatives, where appropriate) of proposed actions without imposing unproductive administrative procedures. Such reports shall also indicate what progress the agency has made in developing substantive criteria and guidance for making environmental assessments as required by § 1500.6(c) of this directive and by section 102(2)(B) of the Act.

Effective date. The revisions of these guidelines shall apply to all draft and final impact statements filed with the Council after January 28, 1973.

RUSSELL E. TRAIN,
Chairman.

APPENDIX I—SUMMARY TO ACCOMPANY DRAFT AND FINAL STATEMENTS

(Check one) ( ) Draft. ( ) Final Environmental Statement.

Name of responsible Federal agency (with name of operating division where appropriate). Name, address, and telephone number of individual at the agency who can be contacted for additional information about the proposed action or the statement.

1. Name of action (Check one) ( ) Administrative Action. ( ) Legislative Action.

2. Brief description of action and its purpose. Indicate what States (and counties) particularly affected, and what other proposed Federal actions in the area, if any, are discussed in the statement.

3. Summary of environmental impacts and adverse environmental effects.

4. Summary of major alternatives considered.

5. (For draft statements) List all Federal, State, and local agencies and other parties from which comments have been requested. (For final statements) List all Federal, State, and local agencies and other parties from which written comments have been received.

6. Date draft statement (and final environmental statement, if one has been issued) made available to the Council and the public.

APPENDIX II—AREAS OF ENVIRONMENTAL IMPACT AND FEDERAL AGENCIES AND FEDERAL STATE AGENCIES [1] WITH JURISDICTION BY LAW OR SPECIAL EXPERTISE TO COMMENT THEREON [2]

AIR

Air Quality

Department of Agriculture—
Forest Service (effects on vegetation)
Atomic Energy Commission (radioactive substances)
Department of Health, Education, and Welfare
Environmental Protection Agency
Department of the Interior—
Bureau of Mines (fossil and gaseous fuel combustion)

---

[1] River Basin Commissions (Delaware, Great Lakes, Missouri, New England, Ohio, Pacific Northwest, Souris-Red-Rainy, Susquehanna, Upper Mississippi) and similar Federal-State agencies should be consulted on actions affecting the environment of their specific geographic jurisdictions.

[2] In all cases where a proposed action will have significant international environmental effects, the Department of State should be consulted, and should be sent a copy of any draft and final impact statement which covers such action.

Bureau of Sport Fisheries and Wildlife (effect on wildlife)
Bureau of Outdoor Recreation (effects on recreation)
Bureau of Land Management (public lands)
Bureau of Indian Affairs (Indian lands)
National Aeronautics and Space Administration (remote sensing, aircraft emissions)
Department of Transportation—
Assistant Secretary for Systems Development and Technology (auto emissions)
Coast Guard (vessel emissions)
Federal Aviation Administration (aircraft emissions)

Weather Modification

Department of Agriculture—
Forest Service
Department of Commerce—
National Oceanic and Atmospheric Administration
Department of Defense—
Department of the Air Force
Department of the Interior
Bureau of Reclamation

WATER RESOURCES COUNCIL
WATER

Water Quality

Department of Agriculture—
Soil Conservation Service
Forest Service
Atomic Energy Commission (radioactive substances)
Department of the Interior—
Bureau of Reclamation
Bureau of Land Management (public lands)
Bureau of Indian Affairs (Indian lands)
Bureau of Sports Fisheries and Wildlife
Bureau of Outdoor Recreation
Geological Survey
Office of Saline Water
Environmental Protection Agency
Department of Health, Education, and Welfare
Department of Defense—
Army Corps of Engineers
Department of the Navy (ship pollution control)
National Aeronautics and Space Administration (remote sensing)
Department of Transportation—
Coast Guard (oil spills, ship sanitation)
Department of Commerce—
National Oceanic and Atmospheric Administration
Water Resources Council
River Basin Commissions (as geographically appropriate)

Marine Pollution, Commercial Fishery Conservation, and Shellfish Sanitation

Department of Commerce—
National Oceanic and Atmospheric Administration
Department of Defense—
Army Corps of Engineers
Office of the Oceanographer of the Navy
Department of Health, Education, and Welfare

Department of the Interior—
Bureau of Sport Fisheries and Wildlife
Bureau of Outdoor Recreation
Bureau of Land Management (outer continental shelf)
Geological Survey (outer continental shelf)
Department of Transportation—
Coast Guard
Environmental Protection Agency
National Aeronautics and Space Administration (remote sensing)
Water Resources Council
River Basin Commissions (as geographically appropriate)

Waterway Regulation and Stream Modification

Department of Agriculture—
Soil Conservation Service
Department of Defense—
Army Corps of Engineers
Department of the Interior—
Bureau of Reclamation
Bureau of Sport Fisheries and Wildlife
Bureau of Outdoor Recreation
Geological Survey
Department of Transportation—
Coast Guard
Environmental Protection Agency

National Aeronautics and Space Administration (remote sensing)
Water Resources Council
River Basin Commissions (as geographically appropriate)

FISH AND WILDLIFE

Department of Agriculture—
Forest Service
Soil Conservation Service
Department of Commerce—
National Oceanic and Atmospheric Administration (marine species)
Department of the Interior—
Bureau of Sport Fisheries and Wildlife
Bureau of Land Management
Bureau of Outdoor Recreation
Environmental Protection Agency

SOLID WASTE

Atomic Energy Commission (radioactive waste)
Department of Defense—
Army Corps of Engineers
Department of Health, Education, and Welfare
Department of the Interior—
Bureau of Mines (mineral waste, mine acid waste, municipal solid waste, recycling)
Bureau of Land Management (public lands)
Bureau of Indian Affairs (Indian lands)
Geological Survey (geologic and hydrologic effects)
Office of Saline Water (demineralization)
Department of Transportation—
Coast Guard (ship sanitation)
Environmental Protection Agency
River Basin Commissions (as geographically appropriate)
Water Resources Council

NOISE

Department of Commerce—
National Bureau of Standards
Department of Health, Education, and Welfare
Department of Housing and Urban Development (land use and building materials aspects)
Department of Labor—
Occupational Safety and Health Administration
Department of Transportation—
Assistant Secretary for Systems Development and Technology
Federal Aviation Administration, Office of Noise Abatement
Environmental Protection Agency
National Aeronautics and Space Administration

RADIATION

Atomic Energy Commission
Department of Commerce—
National Bureau of Standards
Department of Health, Education, and Welfare
Department of the Interior—
Bureau of Mines (uranium mines)
Mining Enforcement and Safety Administration (uranium mines)
Environmental Protection Agency

HAZARDOUS SUBSTANCES
Toxic Materials

Atomic Energy Commission (radioactive substances)
Department of Agriculture—
Agricultural Research Service
Consumer and Marketing Service
Department of Commerce—
National Oceanic and Atmospheric Administration
Department of Defense
Department of Health, Education, and Welfare
Environmental Protection Agency
Food Additives and Contamination of Foodstuffs
Department of Agriculture—
Consumer and Marketing Service (meat and poultry products)
Department of Health, Education, and Welfare
Environmental Protection Agency

Pesticides
Department of Agriculture—
Agricultural Research Service (biological controls, food and fiber production)
Consumer and Marketing Service
Forest Service
Department of Commerce—
National Oceanic and Atmospheric Administration
Department of Health, Education, and Welfare
Department of the Interior—
Bureau of Sport Fisheries and Wildlife (fish and wildlife effects)
Bureau of Land Management (public lands)

Bureau of Indian Affairs (Indian lands)
Bureau of Reclamation (irrigated lands)
Environmental Protection Agency

Transportation and Handling of Hazardous Materials

Atomic Energy Commission (radioactive substances)
Department of Commerce—
Maritime Administration
National Oceanic and Atmospheric Administration (effects on marine life and the coastal zone)
Department of Defense—
Armed Services Explosive Safety Board
Army Corps of Engineers (navigable waterways)
Department of Transportation—
Federal Highway Administration, Bureau of Motor Carrier Safety
Coast Guard
Federal Railroad Administration
Federal Aviation Administration
Assistant Secretary for Systems Development and Technology
Office of Hazardous Materials
Office of Pipeline Safety
Environmental Protection Agency

ENGERY SUPPLY AND NATURAL RESOURCES DEVELOPMENT

Electric Energy Development, Generation, and Transmission, and Use

Atomic Energy Commission (nuclear)
Department of Agriculture—
Rural Electrification Administration (rural areas)
Department of Defense—
Army Corps of Engineers (hydro)
Department of Health, Education, and Welfare (radiation effects)
Department of Housing and Urban Development (urban areas)
Department of the Interior—
Bureau of Indian Affairs (Indian lands)
Bureau of Land Management (public lands)
Bureau of Reclamation
Power Marketing Administrations
Geological Survey
Bureau of Sport Fisheries and Wildlife
Bureau of Outdoor Recreation
National Park Service
Environmental Protection Agency
Federal Power Commission (hydro, transmission, and supply)
River Basin Commissions (as geographically appropriate)
Tennessee Valley Authority
Water Resources Council

Petroleum Development, Extraction, Refining, Transport, and Use

Department of the Interior—
Office of Oil and Gas
Bureau of Mines
Geological Survey
Bureau of Land Management (public lands and outer continental shelf)
Bureau of Indian Affairs (Indian lands)

Bureau of Sport Fisheries and Wildlife (effects on fish and wildlife)
Bureau of Outdoor Recreation
National Park Service
Department of Transportation (Transport and Pipeline Safety)
Environmental Protection Agency
Interstate Commerce Commission

Natural Gas Development, Production, Transmission, and Use

Department of Housing and Urban Development (urban areas)
Department of the Interior—
Office of Oil and Gas
Geological Survey
Bureau of Mines
Bureau of Land Management (public lands)
Bureau of Indian Affairs (Indian lands)
Bureau of Sport Fisheries and Wildlife
Bureau of Outdoor Recreation
National Park Service
Department of Transportation (transport and safety)
Environmental Protection Agency
Federal Power Commission (production, transmission, and supply)
Interstate Commerce Commission

Coal and Minerals Development, Mining, Conversion, Processing, Transport, and Use

Appalachian Regional Commission
Department of Agriculture—
Forest Service
Department of Commerce
Department of the Interior—
Office of Coal Research
Mining Enforcement and Safety Administration
Bureau of Mines
Geological Survey
Bureau of Indian Affairs (Indian lands)
Bureau of Land Management (public lands)
Bureau of Sport Fisheries and Wildlife
Bureau of Outdoor Recreation
National Park Service
Department of Labor—
Occupational Safety and Health Administration
Department of Transportation
Environmental Protection Agency
Interstate Commerce Commission
Tennessee Valley Authority

Renewable Resource Development, Production, Management, Harvest, Transport, and Use

Department of Agriculture—
Forest Service
Soil Conservation Service
Department of Commerce
Department of Housing and Urban Development (building materials)
Department of the Interior—
Geological Survey
Bureau of Land Management (public lands)
Bureau of Indian Affairs (Indian lands)
Bureau of Sport Fisheries and Wildlife
Bureau of Outdoor Recreation

National Park Service
Department of Transportation
Environmental Protection Agency
Interstate Commerce Commission (freight rates)

Energy and Natural Resources Conservation

Department of Agriculture—
Forest Service
Soil Conservation Service
Department of Commerce—
National Bureau of Standards (energy efficiency)
Department of Housing and Urban Development—
Federal Housing Administration (housing standards)
Department of the Interior—
Office of Energy Conservation
Bureau of Mines
Bureau of Reclamation
Geological Survey
Power Marketing Administration
Department of Transportation
Environmental Protection Agency
Federal Power Commission
General Services Administration (design and operation of buildings)
Tennessee Valley Authority

LAND USE AND MANAGEMENT

Land Use Changes, Planning and Regulation of Land Development

Department of Agriculture—
Forest Service (forest lands)
Agricultural Research Service (agricultural lands)
Department of Housing and Urban Development
Department of the Interior—
Office of Land Use and Water Planning
Bureau of Land Management (public la
Bureau of Land Management (public lands)
Bureau of Indian Affairs (Indian lands)
Bureau of Sport Fisheries and Wildlife (wildlife refuges)
Bureau of Outdoor Recreation (recreation lands)
National Park Service (NPS units)
Department of Transportation
Environmental Protection Agency (pollution effects)
National Aeronautics and Space Administration (remote sensing)
River Basins Commissions (as geographically appropriate).

Public Land Management

Department of Agriculture—
Forest Service (forests)
Department of Defense
Department of the Interior—
Bureau of Land Management
Bureau of Indian Affairs (Indian lands)
Bureau of Sport Fisheries and Wildlife (wildlife refuges)
Bureau of Outdoor Recreation (recreation lands)
National Park Service (NPS units)
Federal Power Commission (project lands)

General Services Administration
National Aeronautics and Space Administration (remote sensing)
Tennessee Valley Authority (project lands)

PROTECTION OF ENVIRONMENTALLY CRITICAL AREAS—FLOODPLAINS, WETLANDS, BEACHES AND DUNES, UNSTABLE SOILS, STEEP SLOPES, AQUIFER RECHARGE AREAS, ETC.

Department of Agriculture—
Agricultural Stabilization and Conservation Service
Soil Conservation Service
Forest Service
Department of Commerce—
National Oceanic and Atmospheric Administration (coastal areas)
Department of Defense—
Army Corps of Engineers
Department of Housing and Urban Development (urban and floodplain areas)
Department of the Interior—
Office of Land Use and Water Planning
Bureau of Outdoor Recreation
Bureau of Reclamation
Bureau of Sport Fisheries and Wildlife
Bureau of Land Management
Geological Survey
Environmental Protection Agency (pollution effects)
National Aeronautics and Space Administration (remote sensing)
River Basins Commissions (as geographically appropriate)
Water Resources Council

LAND USE IN COASTAL AREAS

Department of Agriculture—
Forest Service
Soil Conservation Service (soil stability, hydrology)
Department of Commerce—
National Oceanic and Atmospheric Administration (impact on marine life and coastal zone management)
Department of Defense—
Army Corps of Engineers (beaches, dredge and fill permits, Refuse Act permits)
Department of Housing and Urban Development (urban areas)
Department of the Interior—
Office of Land Use and Water Planning
Bureau of Sport Fisheries and Wildlife
National Park Service
Geological Survey
Bureau of Outdoor Recreation
Bureau of Land Management (public lands)
Department of Transportation—
Coast Guard (bridges, navigation)
Environmental Protection Agency (pollution effects)
National Aeronautics and Space Administration (remote sensing)

REDEVELOPMENT AND CONSTRUCTION IN BUILT-UP AREAS

Department of Commerce—
Economic Development Administration (designated areas)
Department of Housing and Urban Development

Department of the Interior—
Office of Land Use and Water Planning
Department of Transportation
Environmental Protection Agency
General Services Administration
Office of Economic Opportunity

DENSITY AND CONGESTION MITIGATION

Department of Health, Education, and Welfare
Department of Housing and Urban Development
Department of the Interior—
Office of Land Use and Water Planning
Bureau of Outdoor Recreation
Department of Transportation
Environmental Protection Agency

NEIGHBORHOOD CHARACTER AND CONTINUITY

Department of Health, Education, and Welfare
Department of Housing and Urban Development
National Endowment for the Arts
Office of Economic Opportunity

IMPACTS ON LOW-INCOME POPULATIONS

Department of Commerce—
Economic Development Administration (designated areas)
Department of Health, Education, and Welfare
Department of Housing and Urban Development
Office of Economic Opportunity

HISTORIC, ARCHITECTURAL, AND ARCHEOLOGICAL PRESERVATION

Advisory Council on Historic Preservation
Department of Housing and Urban Development
Department of the Interior—
National Park Service
Bureau of Land Management (public lands)
Bureau of Indian Affairs (Indian lands)
General Services Administration
National Endowment for the Arts

SOIL AND PLANT CONSERVATION AND HYDROLOGY

Department of Agriculture—
Soil Conservation Service
Agricultural Service
Forest Service
Department of Commerce—
National Oceanic and Atmospheric Administration
Department of Defense—
Army Corps of Engineers (dredging, aquatic plants)
Department of Health, Education, and Welfare
Department of the Interior—
Bureau of Land Management
Bureau of Sport Fisheries and Wildlife
Geological Survey
Bureau of Reclamation
Environmental Protection Agency
National Aeronautics and Space Administration (remote sensing)

River Basin Commissions (as geographically appropriate)
Water Resources Council

OUTDOOR RECREATION

Department of Agriculture—
Forest Service
Soil Conservation Service
Department of Defense—
Army Corps of Engineers
Department of Housing and Urban Development (urban areas)
Department of the Interior—
Bureau of Land Management
National Park Service
Bureau of Outdoor Recreation
Bureau of Sport Fisheries and Wildlife
Bureau of Indian Affairs
Environmental Protection Agency
National Aeronautics and Space Administration (remote sensing)
River Basin Commissions (as geographically appropriate)
Water Resources Council

APPENDIX III—OFFICES WITHIN FEDERAL AGENCIES AND FEDERAL-STATE AGENCIES FOR INFORMATION REGARDING THE AGENCIES' NEPA ACTIVITIES AND FOR RECEIVING OTHER AGENCIES' IMPACT STATEMENTS FOR WHICH COMMENTS ARE REQUESTED

ADVISORY COUNCIL ON HISTORIC PRESERVATION

Office of Architectural and Environmental Preservation, Advisory Council on Historic Preservation, Suite 430, 1522 K Street, N.W., Washington, D.C. 20005 254–3974

DEPARTMENT OF AGRICULTURE [1]

Office of the Secretary, Attn: Coordinator Environmental Quality Activities, U.S. Department of Agriculture, Washington, D.C. 20250 447–3965

APPALACHIAN REGIONAL COMMISSION

Office of the Alternate Federal Co-Chairman, Appalachian Regional Commission, 1666 Connecticut Avenue, N.W., Washington, D.C. 20235 967–4103

DEPARTMENT OF THE ARMY (CORPS OF ENGINEERS)

Executive Director of Civil Works, Office of the Chief of Engineers, U.S. Army Corps of Engineers, Washington, D.C. 20314 693–7168

ATOMIC ENERGY COMMISSION

For nonregulatory matters: Office of Assistant General Manager for Biomedical and Environmental Research and Safety Programs, Atomic Energy Commission, Washington, D.C. 20545 973–3208

---

[1] Requests for comments or information from individual units of the Department of Agriculture, e.g., Soil Conservation Service, Forest Service, etc. should be sent to the Office of the Secretary, Department of Agriculture, at the address given above.

For regulatory matters: Office of the Assistant Director for Environmental Projects, Atomic Energy Commission, Washington, D.C. 20545 973–7531

DEPARTMENT OF COMMERCE

Office of the Deputy Assistant Secretary for Environmental Affairs, U.S. Department of Commerce, Washington, D.C. 20230 967–4335

DEPARTMENT OF DEFENSE

Office of the Assistant Secretary for Defense (Health and Environment), U.S. Department of Defense, Room 3E172, The Pentagon, Washington, D.C. 20301 697–2111

DELAWARE RIVER BASIN COMMISSION

Office of the Secretary, Delaware River Basin Commission, Post Office Box 360, Trenton, N.J. 08603 (609) 883–9500

ENVIRONMENTAL PROTECTION AGENCY [2]

Director, Office of Federal Activities, Environmental Protection Agency, 401 M Street, S.W., Washington, D.C. 20460 755–0777

FEDERAL POWER COMMISSION

Commission's Advisor on Environmental Quality, Federal Power Commission, 825 N. Capitol Street, N.E., Washington, D.C. 20426 386–6084

GENERAL SERVICES ADMINISTRATION

Office of Environmental Affairs, Office of the Deputy Administrator for Special Projects, General Services Administration, Washington, D.C. 20405 343–4161

GREAT LAKES BASIN COMMISSION

Office of the Chairman, Great Lakes Basin Commission, 3475 Plymouth Road, P.O. Box 999, Ann Arbor, Michigan 48105 (313) 769–7431

DEPARTMENT OF HEALTH, EDUCATION AND WELFARE [3]

Office of Environmental Affairs, Office of the

---

[2] Contact the Office of Federal Activities for environmental statements concerning legislation, regulations, national program proposals or other major policy issues.

For all other EPA consultation, contact the Regional Administrator in whose area the proposed action (e.g., highway or water resource construction projects) will take place. The Regional Administrators will coordinate the EPA review.

[3] Contact the Office of Environmental Affairs for information on HEW's environmental statements concerning legislation, regulations, national program proposals or other major policy issues, and for all requests for HEW comment on impact statements of other agencies.

For information with respect to HEW actions occurring within the jurisdiction of the Departments' Regional Directors, contact the appropriate Regional Environmental Officer.

Assistant Secretary for Administration and Management, Department of Health, Education and Welfare, Washington, D.C. 20202 963–4456

DEPARTMENT OF HOUSING AND URBAN DEVELOPMENT [4]

Director, Office of Community and Environmental Standards, Department of Housing and Urban Development, Room 7206, Washington, D.C. 20410 755–5980

DEPARTMENT OF THE INTERIOR [5]

Director, Office of Environmental Project Review, Department of the Interior, Interior Building, Washington, D.C. 20240 343–3891

INTERSTATE COMMERCE COMMISSION

Office of Proceedings, Interstate Commerce, Commission, Washington, D.C. 20423 343–6167

DEPARTMENT OF LABOR

Assistant Secretary for Occupational Safety and Health, Department of Labor, Washington, D.C. 20210 961–3405

MISSOURI RIVER BASINS COMMISSION

Office of the Chairman, Missouri River Basins Commission, 10050 Regency Circle, Omaha, Nebraska 68114 (402) 397–5714

NATIONAL AERONAUTICS AND SPACE ADMINISTRATION

Office of the Comptroller, National Aeronautics and Space Administration, Washington, D.C. 20546 755–8440

NATIONAL CAPITAL PLANNING COMMISSION

Office of Environmental Affairs, Office of the Executive Director, National Capital Planning Commission, Washington, D.C. 20576 382–7200

NATIONAL ENDOWMENT FOR THE ARTS

Office of Architecture and Environmental Arts Program, National Endowment for the Arts, Washington, D.C. 20506 382–5765

---

[4] Contact the Director with regard to environmental impacts of legislation, policy statements, program regulations and procedures, and precedent-making project decisions. For all other HUD consultation, contact the HUD Regional Administrator in whose jurisdiction the project lies, as follows:

[5] Requests for comments or information from individual units of the Department of the Interior should be sent to the Office of Environmental Project Review at the address given above.

Office of the Chairman, New England River Basins Commission, 55 Court Street, Boston, Mass. 02108 (617) 223-6244

OFFICE OF ECONOMIC OPPORTUNITY

Office of the Director, Office of Economic Opportunity, 1200 19th Street, N.W., Washington, D.C. 20506. 254-6000

OHIO RIVER BASIN COMMISSION

Office of the Chairman, Ohio River Basin Commission, 36 East 4th Street, Suite 208-20, Cincinnati, Ohio 45202 (513) 684-3831

PACIFIC NORTHWEST RIVER BASINS COMMISSION

Office of the Chairman, Pacific Northwest River Basins Commission, 1 Columbia River, Vancouver, Washington 98660 (206) 695-3606

SOURIS-RED-RAINY RIVER BASINS COMMISSION

Office of the Chairman, Souris-Red-Rainy River Basins Commission, Suite 6, Professional Building, Holiday Mall, Moorhead, Minnesota 56560 (701) 237-5227

DEPARTMENT OF STATE

Office of the Special Assistant to the Secretary for Environmental Affairs, Department of State, Washington, D.C. 20520 632-7964

SUSQUEHANNA RIVER BASIN COMMISSION

Office of the Executive Director, Susquehanna River Basin Commission, 5012 Lenker Street, Mechanicsburg, Pa. 17055 (717) 737-0501

TENNESSEE VALLEY AUTHORITY

Office of the Director of Environmental Research and Development, Tennessee Valley Authority, 720 Edney - Building, Chattanooga, Tennessee 37401 (615) 755-2002

DEPARTMENT OF TRANSPORTATION [6]

Director, Office of Environmental Quality, Office of the Assistant Secretary for Environment, Safety, and Consumer Affairs, Department of Transportation, Washington, D.C. 20590 426-4357

---

[6] Contact the Office of Environmental Quality, Department of Transportation, for information on DOT's environmental statements concerning legislation, regulations, national program proposals, or other major policy issues.

For information regarding the Department of Transportation's other environmental statements, contact the national office for the appropriate administration.

U.S. Coast Guard

Office of Marine Environment and Systems, U.S. Coast Guard, 400 7th Street, S.W., Washington, D.C. 20590, 426-2007

Federal Aviation Administration

Office of Environmental Quality, Federal Aviation Administration, 800 Independence Avenue, S.W., Washington, D.C. 20591, 426-8406

Federal Highway Administration

Office of Environmental Policy, Federal Highway Administration, 400 7th Street, S.W., Washington, D.C. 20590, 426-0351

Federal Railroad Administration

Office of Policy and Plans, Federal Railroad Administration, 400 7th Street, S.W., Washington, D.C. 20590, 426-1567

Urban Mass Transportation Administration

Office of Program Operations, Urban Mass Transportation Administration, 400 7th Street, S.W., Washington, D.C. 20590, 426-4020

For other administration's not listed above, contact the Office of Environmental Quality, Department of Transportation, at the address given above.

For comments on other agencies' environmental statements, contact the appropriate administration's regional office. If more than one administration within the Department of Transportation is to be requested to comment, contact the Secretarial Representative in the appropriate Regional Office for coordination of the Department's comments:

SECRETARIAL REPRESENTATIVE

Region I Secretarial Representative, U.S. Department of Transportation, Transportation Systems Center, 55 Broadway, Cambridge, Massachusetts 02142 (617) 494-2709

Region II Secretarial Representative, U.S. Department of Transportation, 26 Federal Plaza, Room 1811, New York, New York 10007 (212) 264-2672

Region III Secretarial Representative, U.S. Department of Transportation, Mall Building, Suite 1214, 325 Chestnut Street, Philadelphia, Pennsylvania 19106 (215) 597-0407

Region IV Secretarial Representative, U.S. Department of Transportation, Suite 515, 1720 Peachtree Rd., N.W. Atlanta, Georgia 30309 (404) 526-3738

Region V Secretarial Representative, U.S. Department of Transportation, 17th Floor, 300 S. Wacker Drive, Chicago, Illinois 60606 (312) 353-4000

Region V Secretarial Representative, U.S. Department of Transportation, 9-C-18 Federal Center, 1100 Commerce Street, Dallas, Texas 75202 (214) 749-1851

Region VII Secretarial Representative, U.S. Department of Transportation, 601 E. 12th Street, Room 634, Kansas City, Missouri 64106 (816) 374-2761

Region VIII Secretarial Representative, U.S. Department of Transportation, Prudential

Plaza, Suite 1822, 1050 17th Street, Denver, Colorado 80275 (303) 837-3242

Region IX Secretarial Representative, U.S. Department of Transportation, 450 Golden Gate Avenue, Box 36133, San Francisco, California 94102 (415) 556-5961

Region X Secretarial Representative, U.S. Department of Transportation, 1321 Second Avenue, Room 507, Seattle, Washington 98101 (206) 442-0590

FEDERAL AVIATION ADMINISTRATION

New England Region, Office of the Regional Director, Federal Aviation Administration, 154 Middlesex Street, Burlington, Massachusetts 01803 (617) 272-2350

Eastern Region, Office of the Regional Director, Federal Aviation Administration, Federal Building, JFK International Airport, Jamaica, New York 11430 (212) 995-3333

Southern Region, Office of the Regional Director, Federal Aviation Administration, P.O. Box 20636, Atlanta, Georgia 30320 (404) 526-7222

Great Lakes Region, Office of the Regional Director, Federal Aviation Administration, 2300 East Devon, Des Plaines, Illinois 60018 (312) 694-4500

Southwest Region, Office of the Regional Director, Federal Aviation Administration, P.O. Box 1689, Fort Worth, Texas 76101 (817) 624-4911

Central Region, Office of the Regional Director, Federal Aviation Administration, 601 E. 12th Street, Kansas City, Missouri 64106 (816) 374-5626

Rocky Mountain Region, Office of the Regional Director, Federal Aviation Administration, Park Hill Station, P.O. Box 7213, Denver, Colorado 80207 (303) 837-3646

Western Region, Office of the Regional Director, Federal Aviation Administration, P.O. Box 92007, WorldWay Postal Center, Los Angeles, California 90009 (213) 536-6427

Northwest Region, Office of the Regional Director, Federal Aviation Administration, FAA Building, Boeing Field, Seattle, Washington 98108 (206) 767-2780

FEDERAL HIGHWAY ADMINISTRATION

Region 1, Regional Administrator, Federal Highway Administration, 4 Normanskill Boulevard, Delmar, New York 12054 (518) 472-6476

Region 3, Regional Administrator, Federal Highway Administration, Room 1621, George H. Fallon Federal Office Building, 31 Hopkins Plaza, Baltimore, Maryland 21201 (301) 962-2361

Region 4, Regional Administrator, Federal Highway Administration, Suite 200, 1720 Peachtree Road, N.W., Atlanta, Georgia 30309 (404) 526-5078

Region 5, Regional Administrator, Federal Highway Administration, Dixie Highway, Homewood, Illinois 60430 (312) 799-6300

Region 6, Regional Administrator, Federal Highway Administration, 819 Taylor Street, Fort Worth, Texas 76102 (817) 334-3232

Region 7, Regional Administrator, Federal Highway Administration, P.O. Box 7186,

Country Club Station, Kansas City, Missouri 64113 (816) 361-7563

Region 8, Regional Administrator, Federal Highway Administration, Room 242, Building 40, Denver Federal Center, Denver, Colorado 80225

Region 9, Regional Administrator, Federal Highway Administration, 450 Golden Gate Avenue, Box 36096, San Francisco, California 94102 (415) 556-3895

Region 10, Regional Administrator, Federal Highway Administration, Room 412, Mohawk Building, 222 S.W. Morrison Street, Portland, Oregon 97204 (503) 221-2065

URBAN MASS TRANSPORTATION ADMINISTRATION

Region I, Office of the UMTA Representative, Urban Mass Transportation Administration, Transportation Systems Center, Technology Building, Room 277, 55 Broadway, Boston, Massachusetts 02142 (617) 494-2055

Region II, Office of the UMTA Representative, Urban Mass Transportation Administration, 26 Federal Plaza, Suite 1809, New York, New York 10007 (212) 264-8162

Region III, Office of the UMTA Representative, Urban Mass Transportation Administration, Mall Building, Suite 1214, 325 Chestnut Street, Philadelphia, Pennsylvania 19106 (215) 597-0407

Region IV, Office of the UMTA Representative, Urban Mass Transportation Administration, 1720 Peachtree Road, Northwest, Suite 501, Atlanta, Georgia 30309 (404) 526-3948

Region V, Office of the UMTA Representative, Urban Mass Transportation Administration, 300 South Wacker Drive, Suite 700, Chicago, Illinois 60606 (312) 353-6005

Region VI, Office of the UMTA Represevative, Urban Mass Transportation Administration, Federal Center, Suite 9E24, 1100 Commerce Street, Dallas, Texas 75202 (214) 749-7322

Region VII, Office of the UMTA Representative, Urban Mass Transportation Administration, c/o FAA Management Systems Division, Room 1564D, 601 East 12th Street, Kansas City, Missouri 64106 (816) 374-5567

Region VIII, Office of the UMTA Representative, Urban Mass Transportation Administration, Prudential Plaza, Suite 1822, 1050 17th Street, Denver, Colorado 80202 (303) 837-3242

Region IX, Office of the UMTA Representative, Urban Mass Transportation Administration, 450 Golden Gate Avenue, Box 36125, San Francisco, California 94102 (415) 556-2884

Region X, Office of the UMTA Representative, Urban Mass Transportation Administration, 1321 Second Avenue, Suite 5079, Seattle, Washington (206) 442-0590

DEPARTMENT OF THE TREASURY

Office of Assistant Secretary for Administration, Department of the Treasury, Washington, D.C. 20220 964-5391

UPPER MISSISSIPPI RIVER BASIN COMMISSION

Office of the Chairman, Upper Mississippi River Basin Commission, Federal Office Building, Fort Snelling, Twin Cities, Minnesota 55111 (612) 725–4690

WATER RESOURCES COUNCIL

Office of the Associate Director, Water Resources Council, 2120 L Street, N.W., Suite 800, Washington, D.C. 20037 254–6442

APPENDIX IV—STATE AND LOCAL AGENCY REVIEW OF IMPACT STATEMENTS

1. OMB Circular No. A–95 through its system of clearinghouses provides a means for securing the views of State and local environmental agencies, which can assist in the preparation of impact statements. Under A–95, review of the proposed project in the case of federally assisted projects (Part I of A–95) generally takes place prior to the preparation of the impact statement. Therefore, comments on the environmental effects of the proposed project that are secured during this stage of the A–95 process represent inputs to the environmental impact statement.

2. In the case of direct Federal development (Part II of A–95), Federal agencies are required to consult with clearinghouses at the earliest practicable time in the planning of the project or activity. Where such consultation occurs prior to completion of the draft impact statement, comments relating to the environmental effects of the proposed action would also represent inputs to the environmental impact statement.

3. In either case, whatever comments are made on environmental effects of proposed Federal or federally assisted projects by clearinghouses, or by State and local environmental agencies through clearinghouses, in the course of the A–95 review should be attached to the draft impact statement when it is circulated for review. Copies of the statement should be sent to the agencies making such comments. Whether those agencies then elect to comment again on the basis of the draft impact statement is a matter to be left to the discretion of the commenting agency depending on its resources, the significance of the project, and the extent to which its earlier comments were considered in preparing the draft statement.

4. The clearinghouses may also be used, by mutual agreement, for securing reviews of the draft environmental impact statement. However, the Federal agency may wish to deal directly with appropriate State or local agencies in the review of impact statements because the clearinghouses may be unwilling or unable to handle this phase of the process. In some cases, the Governor may have designated a specific agency, other than the clearinghouse, for securing reviews of impact statements. In any case, the clearinghouses should be sent copies of the impact statement.

5. To aid clearinghouses in coordinating State and local comments, draft statements should include copies of State and local agency comments made earlier under the A–95 process and should indicate on the summary sheet those other agencies from which comments have been requested, as specified in Appendix I of the CEQ Guidelines.

[FR Doc.73–15783 Filed 7–31–73;8:45 am]

Title 40—Protection of Environment

CHAPTER I—ENVIRONMENTAL PROTECTION AGENCY

[FRL 327-5]

PART 6—PREPARATION OF ENVIRON-MENTAL IMPACT STATEMENTS

Final Regulation

The National Environmental Policy Act of 1969 (NEPA), implemented by Executive Order 11514 of March 5, 1970, and the Council on Environmental Quality's (CEQ's) Guidelines of August 1, 1973, requires that all agencies of the Federal Government prepare detailed environmental impact statements on proposals for legislation and other major Federal actions significantly affecting the quality of the human environment. NEPA requires that agencies include in their decision-making process an appropriate and careful consideration of all environmental aspects of proposed actions, an explanation of potential environmental effects of proposed actions and their alternatives for public understanding, a discussion of ways to avoid or minimize adverse effects of proposed actions and a discussion of how to restore or enhance environmental quality as much as possible.

On January 17, 1973, the Environmental Protection Agency (EPA) published a new Part 6 in interim form in the FEDERAL REGISTER (38 FR 1696), establishing EPA policy and procedures for the identification and analysis of environmental impacts and the preparation of environmental impact statements (EIS's) when significant impacts on the environment are anticipated.

On July 17, 1974, EPA published a notice of proposed rulemaking the FEDERAL REGISTER (39 FR 26254). The rulemaking provided detailed procedures for applying NEPA to EPA's nonregulatory programs only. A separate notice of administrative procedure published in the October 21, 1974, FEDERAL REGISTER (39 FR 37419) gave EPA's procedures for voluntarily preparing EIS's on certain regulatory activities. EIS procedures for another regulatory activity, issuing National Pollutant Discharge Elimination System (NPDES) discharge permits to new sources, will appear in 40 CFR 6. Associated amendments to the NPDES operating regulations, covering permits to new sources, will appear in 40 CFR 125.

The proposed regulation on the preparation of EIS's for nonregulatory programs was published for public review and comment. EPA received comments on this proposed regulation from environmental groups; Federal, State and local governmental agencies; industry; and private individuals. As a result of the comments received, the following changes have been made:

(1) Coastal zones, wild and scenic rivers, prime agricultural land and wildlife habitat were included in the criteria to be considered during the environmental review.

The Coastal Zone Management Act and the Wild and Scenic Rivers Act are intended to protect these environmentally sensitive areas; therefore, EPA should consider the effects of its projects on these areas. Protection of prime agricultural lands and wildlife habitat has become an important concern as a result of the need to further increase food production from domestic sources as well as commercial harvesting of fish and other wildlife resources and from the continuing need to preserve the diversity of natural resources for future generations.

(2) Consideration of the use of floodplains as required by Executive Order 11296 was added to the environmental review process.

Executive Order 11296 requires agencies to consider project alternatives which will preclude the uneconomic, hazardous or unnecessary use of floodplains to minimize the exposure of facilities to potential flood damage, lessen the need for future Federal expenditures for flood protection and flood disaster relief and preserve the unique and significant public value of the floodplain as an environmental resource.

(3) Statutory definitions of coastal zones and wild and scenic rivers were added to § 6.214(b).

These statutes define sensitive areas and require states to designate areas which must be protected.

(4) The review and comment period for negative declarations was extended from 15 days to 15 working days.

Requests for negative declarations and comments on negative declarations are not acted on during weekends and on holidays. In addition, mail requests often take two or three days to reach the appropriate office and several more days for action and delivery of response. Therefore, the new time frame for review and response to a negative declaration is more realistic without adding too much delay to a project.

(5) Requirements for more data in the negative declaration to clarify the proposed action were added in § 6.212(b).

Requiring a summary of the impacts of a project and other data to support the negative declaration in this document improves its usefulness as a tool to review the decision not to prepare a full EIS on a project.

(6) The definitions of primary and secondary impacts in § 6.304 were clarified.

The definitions were made more specific, especially in the issue areas of induced growth and growth rates, to reduce subjectivity in deciding whether an impact is primary or secondary.

(7) Procedures for EPA public hearings in Subpart D were clarified.

Language was added to this subpart to distinguish EPA public hearings from applicant hearings required by statute or regulation, such as the facilities plan hearings.

(8) The discussion of retroactive application (§ 6.504) was clarified and abbreviated.

The new language retains flexibility in decision making for the Regional Administrator while eliminating the ambiguity of the langauge in the interim regulation.

(9) The criteria for writing an EIS if wetlands may be affected were modified in § 6.510(b).

The new language still requires an EIS on a project which will be located on wetlands but limits the requirements for an EIS on secondary wetland effects to those which are significant and adverse.

(10) A more detailed explanation of the data required in environmental assessments (§ 6.512) was added.

Requiring more specific data in several areas, including energy production and consumption as well as land use trends and population projections, from the applicant will provide a more complete data base for the environmental review. Documentation of the applicant's data will allow EPA to evaluate the validity of this data.

(11) Subpart F, Guidelines for Compliance with NEPA in Research and Development Programs and Activities, was revised.

ORD simplified this subpart by removing the internal procedures and assignments of responsibility for circulation in internal memoranda. Only the general application of this regulation to ORD programs was retained.

(12) The discussions of responsibilities and document distribution procedures were moved to appendices attached to the regulations.

These sections were removed from the regulatory language to improve the read-

ability of the regulation and because these discussions are more explanatory and do not need to have the legal force of regulatory language.

(13) Consideration of the Endangered Species Act of 1973 was incorporated into the regulation.

EPA recognizes its responsibility to assist with implementing legislation which will help preserve or improve our natural resources.

The major issues raised on this regulation were on new and proposed criteria for determining when to prepare an EIS and the retroactive application of the criteria to projects started before July 1, 1975. In addition to the new criteria which were added, CEQ requested the addition of several quantitative criteria for which parameters have not been set. These new criteria are being discussed with CEQ and may be added to the regulation at a future date. Changes in the discussion of retroactive application of the criteria are described in item 8 above.

EPA believes that Agency compliance with the regulations of Part 6 will enhance the present quality of human.life without endangering the quality of the natural environment for future generations.

Effective date: This regulation will become effective April 14, 1975.

Dated: April 3, 1975.

RUSSELL E. TRAIN,
Administrator.

Subpart A—General

Sec.	
6.100	Purpose and policy.
6.102	Definitions.
6.104	Summary of procedures for implementing NEPA.
6.106	Applicability.
6.108	Completion of NEPA precedures before start of administrative action.
6.110	Responsibilities.

Subpart B—Procedures

6.200	Criteria for determining when to prepare an environmental impact statement.
6.202	Environmental assessment.
6.204	Environmental review.
6.206	Notice of intent.
6.208	Draft environmental impact statements.
6.210	Final environmental impact statements.
6.212	Negative declarations and environmental impact appraisals.
6.214	Additional procedures.
6.216	Availability of documents.

Authority: Secs. 102, 103 of 83 Stat. 854 (42 U.S.C. 4321 et seq.)

Subpart A—General

§ 6.100   Purpose and policy.

(a) The National Environmental Policy Act (NEPA) of 1969, implemented by Executive Order 11514 and the Council on Environmental Quality's (CEQ's) Guidelines of August 1, 1973 (38 FR 20550), requires that all agencies of the Federal Government prepare detailed environmental impact statements on proposals for legislation and other major Federal actions significantly affecting the quality of the human environment. NEPA requires that agencies include in the decision-making process appropriate and careful consideration of all environmental effects of proposed actions, explain potential environmental effects of proposed actions and their alternatives for public understanding, avoid or minimize adverse effects of proposed actions and restore or enhance environmental quality as much as possible.

(b) This part establishes Environmental Protection Agency (EPA) policy and procedures for the identification and analysis of the environmental impacts of EPA nonregulatory actions and the preparation and processing of environmental impact statements (EIS's) when significant impacts on the environment are anticipated.

§ 6.102   Definitions.

(a) "Environmental assessment" is a written analysis submitted to EPA by its grantees or contractors describing the environmental impacts of proposed actions undertaken with the financial support of EPA. For facilities or section 208 plans as defined in § 6.102 (j) and (k), the assessment must be an integral, though identifiable, part of the plan submitted to EPA for review.

(b) "Environmental review" is a formal evaluation undertaken by EPA to determine whether a proposed EPA action may have a significant impact on the environment. The environmental assessment is one of the major sources of information used in this review.

(c) "Notice of intent" is a memorandum, prepared after the environmental review, announcing to Federal, regional, State, and local agencies, and to interested persons, that a draft EIS will be prepared.

(d) "Environmental impact statement" is a report, prepared by EPA, which identifies and analyzes in detail the environmental impacts of a proposed EPA action and feasible alternatives.

(e) "Negative declaration" is a written announcement, prepared after the environmental review, which states that EPA has decided not to prepare an EIS and summarizes the environmental impact appraisal.

(f) "Environmental impact appraisal" is based on an environmental review and supports a negative declaration. It describes a proposed EPA action, its expected environmental impact, and the basis for the conclusion that no significant impact is anticipated.

(g) "NEPA-associated documents" are any one or combination of: notices of intent, negative declarations, exemption certifications, environmental impact appraisals, news releases, EIS's, and environmental assessments.

(h) "Responsible official" is an Assistant Administrator, Deputy Assistant Administartor, Regional Administrator or their designee.

(i) "Interested persons" are individuals, citizen groups, conservation organizations, corporations, or other nongovernmental units, including applicants for EPA contracts or grants, who may be interested in, affected by, or technically competent to comment on the environmental impacts of the proposed EPA action.

(j) "Section 208 plan" is an areawide waste treatment management plan prepared under section 208 of the Federal Water Pollution Control Act (FWPCA), as amended, under 40 CFR Part 126 and 40 CFR Part 35, Subpart F.

(k) "Facilities plan" is a preliminary plan prepared as the basis for construction of publicly owned waste treatment works under Title II of FWPCA,. as amended, under 40 CFR 35.917.

(l) "Intramural project" is an in-house project undertaken by EPA personnel.

(m) "Extramural project" is a project undertaken by grant or contract.

§ 6.104 Summary of procedures for implementing NEPA.

(a) *Responsible official.* The responsible official shall utilize a systematic, interdisciplinary approach to integrate natural and social sciences as well as environmental design arts in planning programs and making decisions which are subject to NEPA review. His staff may be supplemented by professionals from other agencies, universities or consultants whenever in-house capabilities are insufficiently interdisciplinary.

(b) *Environmental assessment.* Environmental assessments must be submitted to EPA by its grantees and contractors, as required in Subparts E, F, G, and H of this part. The assessment is used by EPA to decide if an EIS is required and to prepare one if necessary.

(c) *Environmental review.* Environmental reviews shall be made of proposed and certain ongoing EPA actions as required in § 6.106(c). This process shall consist of a study of the action to identify and evaluate the environmental impacts of the action. Types of grants, contracts and other actions requiring study are listed in the subparts following Subpart D. The process shall include a review of any environmental assessment received to determine whether any significant impacts are anticipated, whether any changes can be made in the proposed action to eliminate significant adverse impacts, and whether an EIS is required. EPA has overall responsibility for this review, although its grantees and contractors will contribute to the review through their environmental assessments.

(d) *Notice of intent and EIS's.* When an environmental review indicates that a significant environmental impact may occur and the significant adverse impacts cannot be eliminated by making changes in the project, a notice of intent shall be published, and a draft EIS shall be prepared and distributed. After external coordination and evaluation of the comments received, a final EIS shall be prepared and distributed. EIS's should be prepared first on those proposed actions with the most adverse effects which are scheduled for earliest implementation and on other proposed actions according to priorities assigned by the responsible official.

(e) *Negative declaration and environmental impact appraisal.* When the environmental review indicates no significant impacts are anticipated or when the project is changed to eliminate the significant adverse impacts, a negative declaration shall be issued. For the cases in Subparts E, F, G, and H of this part, an environmental impact appraisal shall be prepared which summaries the impacts, alternatives and reasons an EIS was not prepared. It shall remain on file and be available for public inspection.

§ 6.106 Applicability.

(a) *Administrative actions covered.* This part applies to the administrative actions listed below. The subpart referenced with each action lists the detailed

386

NEPA procedures associated with the action. Administrative actions are:

(1) Development of EPA legislative proposals;

(2) Development of favorable reports on legislation initiated elsewhere and not accompanied by an EIS, when they relate to or affect matters within EPA's primary areas of responsibility;

(3) For the programs under Title II of FWPCA, as amended, those administrative actions in § 6.504;

(4) For the Office of Research and Development, those administrative actions in § 6.604;

(5) For the Office of Solid Waste Management Programs, those administrative actions in § 6.702;

(6) For construction of special purpose facilities and facility renovations, those administrative actions in § 6.804; and

(7) Development of an EPA project in conjunction with or located near a project or complex of projects started by one or more Federal agencies when the cumulative effects of all the projects will be major allocations of resources or foreclosures of future land use options.

(b) *Administrative actions excluded.* The requirements of this part do not apply to environmentally protective regulatory activities undertaken by EPA, nor to projects exempted in § 6.504, § 6.604, and § 6.702.

(c) *Application to ongoing actions.* This regulation shall apply to uncompleted and continuing EPA actions initiated before the promulgation of these procedures when modifications of or alternatives to the EPA action are still available, except for the Title II construction grants program. Specific application for the construction grants program is in § 6.504(c). An EIS shall be prepared for each project found to have significant environmental effects as described in § 6.200.

(d) *Application to legislative proposals.* (1) As noted in paragraphs (a) (1) and (2) of this section, EIS's or negative declarations shall be prepared for legislative proposals or favorable reports relating to legislation which may significantly affect the environment. Because of the nature of the legislative process, EIS's for legislation · must be prepared and reviewed according to the procedures followed in the development and review of the legislative matter. These procedures are described in Office of Management and Budget (OMB) Circular No. A–19.

(2) A working draft EIS shall be prepared by the EPA office responsible for preparing the legislative proposal or report on legislation. It shall be prepared concurrently with the development of the legislative proposal or report and shall contain the information required in § 6.304. The EIS shall be circulated for internal EPA review with the legislative proposal or report and other supporting documentation. The working draft EIS shall be modified to correspond with changes made in the proposal or report during the internal review. All major alternatives developed during the formulation and review of the proposal or report should be retained in the working draft EIS.

(i) The working draft EIS shall accompany the legislative proposal or report to OMB. EPA shall revise the working draft EIS to respond to comments from OMB and other Federal agencies.

(ii) Upon transmittal of the legislative proposal or report to Congress, the working draft EIS will be forwarded to CEQ and the Congress as a formal legislative EIS. Copies will be distributed according to procedures described in Appendix C.

(iii) Comments received by EPA on the legislative EIS shall be forwarded to the appropriate Congressional Committees. EPA also may respond to specific comments and forward its responses with the comments. Because legislation undergoes continuous changes in Congress beyond the control of EPA, no final EIS need be prepared by EPA.

§ 6.108   Completion of NEPA procedures before starting administrative action.

(a) No administrative action shall be taken until the environmental review process, resulting in an EIS or a negative declaration with environmental appraisal, has been completed.

(b) *When an EIS will be prepared.* Except when requested by the responsible official in writing and approved by CEQ, no administrative action shall be taken sooner than ninety (90) calendar days after a draft EIS has been distributed or sooner than thirty (30) calendar days after the final EIS has been made public. If the final text of an EIS is filed within ninety (90) days after a draft EIS has been circulated for comment, furnished to CEQ and made public, the minimum thirty (30) day period and the ninety (90) day period may run concurrently if they overlap. The minimum periods for review and advance availability of EIS's shall begin on the date CEQ publishes the notice of receipt of the EIS in the FEDERAL REGISTER. In addition, the proposed action should be modified to conform with any changes EPA considers necessary before the final EIS is published.

(c) *When an EIS will not be prepared.* If EPA decides not to prepare an EIS on any action listed in this part for which a negative declaration with environmental appraisal has been prepared, no administrative action shall be taken for at least fifteen (15) working days after the negative declaration is issued to allow public review of the decision. If significant environmental issues are raised during the review period, the decision may be changed and a new environmental appraisal or an EIS may be prepared.

§ 6.110 Responsibilities.

See Appendix B for responsibilities of this part.

Subpart B—Procedures

§ 6.200 Criteria for determining when to prepare an EIS.

The following general criteria shall be used when reviewing a proposed EPA action to determine if it will have a significant impact on the environment and therefore require an EIS:

(a) *Significant environmental effects.* (1) An action with both beneficial and detrimental effects should be classified as having significant effects on the environment, even if EPA believes that the net effect will be beneficial. However, preference should be given to preparing EIS's on proposed actions which, on balance, have adverse effects.

(2) When determining the significance of a proposed action's impacts, the responsible official shall consider both its short term and long term effects as well as its primary and secondary effects as defined in § 6.304(c). Particular attention should be given to changes in land use patterns; changes in energy supply and demand; increased development in floodplains; significant changes in ambient air and water quality or noise levels; potential violations of air quality, water quality and noise level standards; significant changes in surface or ground-water quality or quantity; and encroachments on wetlands, coatstal zones, or fish and wildlife habitat, especially when threatened or endangered species may be affected.

(3) Minor actions which may set a precedent for future major actions with significant adverse impacts or a number of actions with individually insignificant but cumulatively significant adverse impacts shall be classified as having significant environmental impacts. If EPA is taking a number of minor, environmentally insignificant actions that are similar in execution and purpose, during a limited time span and in the same general geographic area, the cumulative environmental impact of all of these actions shall be evaluated.

(4) In determining the significance of a proposed action's impact, the unique characteristics of the project area should be carefully considered. For example, proximity to historic sites, parklands or wild and scenic rivers may make the impact significant. A project discharging into a drinking water aquifer may make the impact significant.

(5) A proposed EPA action which will have direct and significant adverse effects on a property listed in or eligible for listing in the National Register of Historic Places or will cause irreparable loss or destruction of significant scientific, prehistoric, historic or archaeological data shall be classified as having significant environmental impacts.

(b) *Controversial actions.* An EIS shall be prepared when the environmental impact of a proposed EPA action is likely to be highly controversial.

(c) *Additional criteria for specific programs.* Additional criteria for various EPA programs are in Subpart E (Title II Wastewater Treatment Works Construction Grants Program), Subpart F (Research and Development Programs), Subpart G (Solid Waste Management Programs) and Subpart H (Construction of Special Facilities and Facility Renovations).

§ 6.202 Environmental assessment.

Environmental assessments must be submitted to EPA by its grantees and contractors as required in Subparts E, F, G, and H of this part. The assessment is to ensure that the applicant considers the environmental impacts of the proposed action at the earliest possible point in his planning process. The assessment and other relevant information are used by EPA to decide if an EIS is required. While EPA is responsible for ensuring that EIS's are factual and comprehensive, it expects assessments and other data submitted by grantees and contrac-

388

tors to be accurate and complete. The responsible official may request additional data and analyses from grantees or other sources any time he determines they are needed to comply adequately with NEPA.

§ 6.204 Environmental review.

Proposed EPA actions, as well as ongoing EPA actions listed in § 6.106(c), shall be subjected to an environmental review. This review shall be a continuing one, starting at the earliest possible point in the development of the project. It shall consist of a study of the proposed action, including a review of any environmental assessments received, to identify and evaluate the environmental impacts of the proposed action and feasible alternatives. The review will determine whether significant impacts are anticipated from the proposed action, whether any feasible alternatives can be adopted or changes can be made in project design to eliminate significant adverse impacts, and whether an EIS or a negative declaration is required. The responsible official shall determine the proper scope of the environmental review. The responsible official may delay approval of related projects until the proposals can be reviewed together to allow EPA to properly evaluate their cumulative impacts.

§ 6.206 Notice of intent.

(a) *General.* (1) When an environmental review indicates a significant impact may occur and significant adverse impacts cannot be eliminated by making changes in the project, a notice of intent, announcing the preparation of a draft EIS, shall be issued by the responsible official. The notice shall briefly describe the EPA action, its location, and the issues involved (Exhibit 1).

(2) The purpose of a notice of intent is to involve other government agencies and interested persons as early as possible in the planning and evaluation of EPA actions which may have significant environmental impacts. This notice should encourage agency and public input to a draft EIS and assure that environmental values will be identified and weighed from the outset rather than accommodated by adjustments at the end of the decision-making process.

(b) *Specific actions.* The specific actions to be taken by the responsible official on notices of intent are:

(1) When the review process indicates a significant impact may occur and significant adverse impacts cannot be eliminated by making changes in the project, prepare a notice of intent immediately after the environmental review.

(2) Distribute copies of the notice of intent as required in Appendix C.

(3) Publish in a local newspaper, with adequate circulation to cover the area affected by the project, a brief public notice stating that an EIS will be prepared on a particular project, and the public may participate in preparing the EIS (Exhibit 2). News releases also may be submitted to other media.

(c) *Regional office assistance to program offices.* Regional offices will provide assistance to program offices in taking these specific actions when the EIS originates in a program office.

§ 6.208 Draft EIS's.

(a) *General.* (1) The responsible official shall assure that a draft EIS is prepared as soon as possible after the release of the notice of intent. Before releasing the draft EIS to CEQ, a preliminary version may be circulated for review to other offices within EPA with interest in or technical expertise related to the action. Then the draft EIS shall be sent to CEQ and circulated to Federal, State, regional and local agencies with special expertise or jurisdiction by law, and to interested persons. If the responsible official determines that a public hearing on the proposed action is warranted, the hearing will be held after the draft EIS is prepared, according to the requirements of § 6.402.

(2) Draft EIS's should be prepared at the earliest possible point in the project development. If the project involves a grant applicant or potential contractor, he must submit any data EPA requests for preparing the EIS. Where a plan or program has been developed by EPA or submitted to EPA for approval, the relationship between the plan and the later projects encompassed by its shall be evaluated to determine the best time to prepare an EIS. Whenever possible, an EIS will be drafted for the total program at the initial planning stage. Then later component projects included in the plan will not require individual EIS's unless they differ substantially from the plan, or unless the overall plan did not provide enough detail to fully assess significant impacts of individual projects. Plans shall be reevaluated by the responsible official to monitor the cumulative impact of the component projects and to preclude the plans' obsolescence.

(b) *Specific actions.* The specific actions to be taken by the responsible official on draft EIS's are:

(1) Distribute the draft EIS according to the procedures in Appendix C.

(2) Inform the agencies to reply directly to the originating EPA office. Commenting agencies shall have at least forty-five (45) calendar days to reply, starting from the date of publication in the FEDERAL REGISTER of lists of statements received by CEQ. If no comments are received during the reply period and no time extension has been requested, it shall be presumed that the agency has no comment to make. EPA may grant extensions of fifteen (15) or more calendar days. The time limits for review and extensions for State and local agencies; State, regional, and metropolitan clearinghouses; and interested persons shall be the same as those available to Federal agencies.

(3) Publish a notice in local newspapers stating that the draft EIS is available for comment and listing where copies may be obtained (Exhibit 2), and submit news releases to other media.

(4) Include in the draft EIS a notice stating that only those Federal, State, regional, and local agencies and interested persons who make substantive comments on the draft EIS or request a copy of the final EIS will be sent a copy.

(c) *Regional office assistance to program office.* If requested, regional offices will provide assistance to program offices in taking these specific actions when the EIS originates in a program office.

§ 6.210 Final EIS's.

(a) Final EIS's shall respond to all substantive comments raised through the review of the draft EIS. Special care should be taken to respond fully to comments disagreeing with EPA's position. (See also § 6.304(g).)

(b) Distribution and other specific actions are described in Appendix C. If there is an applicant, he shall be sent a copy. When the number of comments on the draft EIS is so large that distribution of the final EIS to all commenting entities appears impractical, the program or regional office preparing the EIS shall consult with OFA, which will consult with CEQ about alternative arrangements for distribution of the EIS.

§ 6.212 Negative declaration and environmental impact appraisals.

(a) *General.* When an environmental review indicates there will be no significant impact or significant adverse impacts have been eliminated by making changes in the project, the responsible official shall prepare a negative declaration to allow public review of his decision before it becomes final. The negative declaration and news release must state that interested persons disagreeing with the decision may submit comments for consideration by EPA. EPA shall not take administrative action on the project for at least fifteen (15) working days after release of the negative declaration and may allow more time for response. The responsible official shall have an environmental impact appraisal supporting the negative declaration available for public review when the negative declaration is released for those cases given in Subparts E, F, G, and H.

(b) *Specific actions.* The responsible official shall take the following specific actions on those projects for which both a negative declaration and an impact appraisal will be prepared:

(1) *Negative declaration.* (i) Prepare a negative declaration immediately after the environmental review. This document shall briefly summarize the purpose of the project, its location, the nature and extent of the land use changes related to the project, and the major primary and secondary impacts of the project. It shall describe how the more detailed environmental impact appraisal may be obtained at cost. (See Exhibit 3.)

(ii) Distribute the negative declaration according to procedures in Appendix C. In addition, submit to local newspapers and other appropriate media a brief news release with a negative declaration attached, informing the public that a decision not to prepare an EIS has been made and a negative declaration and environmental impact appraisal are available for public review and comment (Exhibit 2).

(2) *Environmental impact appraisal.* (i) Prepare an environmental impact appraisal concurrently with the negative declaration. This document shall briefly describe the proposed action and feasible alternatives, environmental impacts of the proposed action, unavoidable adverse impacts of the proposed action, the relationship between short term uses of man's environment and the maintenance and enhancement of long term productivity, steps to minimize harm to the environment, irreversible and irretrievable commitments of resources to implement the action, comments and consultations

390

on the project, and reasons for concluding there will be no significant impacts. (See Exhibit 4.)

(ii) Distribute the environmental impact appraisal according to procedures in Appendix C.

§ 6.214 Additional procedures.

(a) *Historical and archaeological sites.* EPA is subject to the requirements of section 106 of the National Historic Preservation Act of 1966, 16 U.S.C. 470 *et seq.,* Executive Order 11593, the Archaeological and Historic Preservation Act of 1974, 16 U.S.C. 469 *et seq.,* and the regulations promulgated under this legislation. These statutes and regulations establish environmental review procedures which are independent of NEPA requirements.

(1) If an EPA action may affect properties with historic, architectural, archaeological or cultural value which are listed in the National Register of Historic Places (published in the FEDERAL REGISTER each February with supplements on the first Tuesday of each month), the responsible official shall comply with the procedures of the Advisory Council on Historic Preservation (36 CFR 800), including determining the need for a Memorandum of Agreement among EPA, the State Historic Preservation Officer and the Advisory Council. If a Memordandum of Agreement is executed, it shall be included in an EIS whenever one is prepared on a proposed action. See § 6.512(c) of this part for additional procedures for the construction grants program under Title II of the FWPCA, as amended.

(2) If an EPA action may cause irreparable loss or destruction of significant scientific, prehistoric, historic or archaeological data, the responsible official shall consult with the State Historic Preservation Officer in compliance with the Archaeological and Historic Preservation Act (P.L. 93–291).

(b) *Wetlands, floodplains, coastal zones, wild and scenic rivers, fish and wildlife.* The following procedures shall be applied to all EPA administrative actions covered by this part that may affect these environmentally sensitive resources.

(1) If an EPA action may affect wetlands, the responsible official shall consult with the appropriate offices of the Department of the Interior, Department of Commerce, and the U.S. Army Corps of Engineers during the environmental review to determine the probable impact of the action on the pertinent fish and wildlife resources and land use of these areas.

(2) If an EPA action may directly cause or induce the construction of buildings or other facilities in a floodplain, the responsible official shall evaluate flood hazards in connection with these facilities as required by Executive Order 11296 and shall, as far as practicable, consider alternatives to preclude the uneconomic, hazardous or unnecessary use of floodplains to minimize the exposure of facilities to potential flood damage, lessen the need for future Federal expenditures for flood protection and flood disaster relief and preserve the unique and significant public value of the floodplain as an environmental resource.

(3) If an EPA action may affect coastal zones or coastal waters as defined in Title III of the Costal Zone Management Act of 1972 (Pub. L. 92–583), the responsible official shall consult with the appropriate State offices and with the appropriate office of the Department of Commerce during the environmental review to determine the probable impact of the action on coastal zone or coastal water resources.

(4) If an EPA action may affect portions of rivers designated wild and scenic or being considered for this designation under the Wild and Scenic Rivers Act (Pub. L. 90–542), the responsible official shall consult with appropriate State offices and with the Secretary of the Interior or, where national forest lands are involved, with the Secretary of Agriculture during the environmental review to determine the status of an affected river and the probable impact of the action on eligible rivers.

(5) If an EPA action will result in the control or structural modification of any stream or other body of water for any purpose, including navigation and drainage, the responsible official shall consult with the United States Fish and Wildlife Service (Department of the Interior), the National Marine Fisheries Service of the National Oceanic and Atmospheric Administration (Department of Commerce), the U.S. Army Corps of Engineers and the head of the agency administering the wildlife resources of the particular State in which the action will take place with a view to the conservation of wildlife resources. This consultation shall follow the procedures in the Fish and Wildlife Coordination Act (Pub. L. 85–624) and shall occur during the environmental review of an action.

(6) If an EPA action may affect threatened or endangered species defined under section 4 of the Endangered Species Act of 1973 (Pub. L. 93–205), the responsible official shall consult with the Secretary of the Interior or the Secretary of Commerce, according to the procedures in section 7 of that act.

(7) Requests for consultation and the results of consultation shall be documented in writing. In all cases where consultation has occurred, the agencies consulted should receive copies of either the notice of intent and EIS or the negative declaration and environmental appraisal prepared on the proposed action. If a decision has already been made to prepare an EIS on a project and wetlands, floodplains, coastal zones, wild and scenic rivers, fish or wildlife may be affected, the required consultation may be deferred until the preparation of the draft EIS.

§ 6.216   Availability of documents.

(a) EPA will print copies of draft and final EIS's for agency and public distribution. A nominal fee may be charged for copies requested by the public.

(b) When EPA no longer has copies of an EIS to distribute, copies shall be made available for public inspection at regional and headquarters Offices of Public Affairs. Interested persons also should be advised of the availability (at cost) of the EIS from the Environmental Law Institute, 1356 Connecticut Avenue NW., Washington, D.C. 20036.

(c) Lists of EIS's prepared or under preparation and lists of negative declarations prepared will be available at both the regional and headquarters Offices of Public Affairs.

Subpart C—Content of Environmental Impact Statements

§ 6.300   Cover sheet.

The cover sheet shall indicate the type of EIS (draft or final), the official project name and number, the responsible EPA office, the date, and the signature of the responsible official. The format is shown in Exhibit 5.

§ 6.302   Summary sheet.

The summary sheet shall conform to the format in Exhibit 6, based on Appendix I of the August 1, 1973, CEQ Guidelines, or the latest revision of the CEQ Guidelines.

§ 6.304   Body of EIS.

The body of the EIS shall identify, develop, and analyze the pertinent issues discussed in the seven sections below; each section need not be a separate chapter. This analysis should include, but not be limited to, consideration of the impacts of the proposed project on the environmental areas listed in Appendix A which are relevant to the project. The EIS shall serve as a means for the responsible official and the public to assess the environmental impacts of a proposed EPA action, rather than as a justification for decisions already made. It shall be prepared using a systematic, interdisciplinary approach and shall incorporate all relevant analytical disciplines to provide meaningful and factual data, information, and analyses. The presentation of data should be clear and concise, yet include all facts necessary to permit independent evaluation and appraisal of the beneficial and adverse environmental effects of alternative actions. The amount of detail provided should be commensurate with the extent and expected impact of the action and the amount of information required at the particular level of decision making. To the extent possible, an EIS shall not be drafted in a style which requires extensive scientific or technical expertise to comprehend and evaluate the environmental impact of a proposed EPA action.

(a) *Background and description of the proposed action.* The EIS shall describe the recommended or proposed action, its purpose, where it is located and its time setting. When a decision has been made not to favor an alternative until public comments on a proposed action have been received, the draft EIS may treat all feasible alternatives at similar levels of detail; the final EIS should focus on the alternative the draft EIS and public comments indicate is the best. The relationship of the proposed action to other projects and proposals directly affected by or stemming from it shall be discussed, including not only other EPA activities, but also those of other governmental and private organizations. Land use patterns and population trends in the project area and the assumptions on which they are based also shall be included. Available maps, photos, and artists' sketches should be incorporated when they help depict the environmental setting.

(b) *Alternatives to the proposed action.* The EIS shall develop, describe, and objectively weigh feasible alternatives to any proposed action, including the options of taking no action or postponing action. The analysis should be detailed enough to show EPA's comparative evaluation of the environmental impacts, commitments of resources, costs, and risks of the proposed action and each feasible alternative. For projects involving construction, alternative sites must be analyzed in enough detail for reviewers independently to judge the relative desirability of each site. For alternatives involving regionalization, the effects of varying degrees of regionalization should be addressed. If a cost-benefit analysis is prepared, it should be appended to the EIS and referenced in the body of the EIS. In addition, the reasons why the proposed action is believed by EPA to be the best course of action shall be explained.

(c) *Environmental impacts of the proposed action.* (1) The positive and negative effects of the proposed action as it affects both the national and international environment should be assessed. The attention given to different environmental factors will vary according to the nature, scale, and location of proposed actions. Primary attention should be given to those factors most evidently affected by the proposed action. The factors shall include, where appropriate, the proposed action's effects on the resource base, including land, water quality and quantity, air quality, public services and energy supply. The EIS shall describe primary and secondary environmental impacts, both beneficial and adverse, anticipated from the action. The description shall include short term and long term impacts on both the natural and human environments.

(2) Primary impacts are those that can be attributed directly to the proposed action. If the action is a field experiment, materials introduced into the environment which might damage certain plant communities or wildlife species would be a primary impact. If the action involves construction of a facility, such as a sewage treatment works, an office building or a laboratory, the primary impacts of the action would include the environmental impacts related to construction and operation of the facility and land use changes at the facility site.

(3) Secondary impacts are indirect or induced changes. If the action involves construction of a facility, the secondary impacts would include the environmental impacts related to:

(i) induced changes in the pattern of land use, population density and related effects on air and water quality or other natural resources;

(ii) increased growth at a faster rate than planned for or above the total level planned by the existing community.

(4) A discussion of how socioeconomic activities and land use changes related to the proposed action conform or conflict with the goals and objectives of approved or proposed Federal, regional, State and local land use plans, policies and controls for the project area should be included in the EIS. If a conflict appears to be unresolved in the EIS, EPA should explain why it has decided to proceed without full reconciliation.

(d) *Adverse impacts which cannot be avoided should the proposal be implemented and steps to minimize harm to the environment.* The EIS shall describe the kinds and magnitudes of adverse impacts which cannot be reduced in severity or which can be reduced to an acceptable level but not eliminated. These may include water or air pollution, undesirable land use patterns, damage to fish and wildlife habitats, urban congestion, threats to human health or other consequences adverse to the environmental goals in section 101(b) of NEPA. Protective and mitigative measures to be taken as part of the proposed action shall be identified. These measures to reduce or compensate for any environmentally detrimental aspect of the proposed action may include those of EPA, its contractors and grantees and others involved in the action.

(e) *Relationship between local short term uses of man's environment and the maintenance and enhancement of long term productivity.* The EIS shall describe the extent to which the proposed action involves tradeoffs between short term environmental gains at the expense of long term gains or vice-versa and the extent to which the proposed action forecloses future options. Special attention shall be given to effects which narrow the range of future uses of land and water resources or pose long term risks to health or safety. Consideration should be given to windfall gains or significant decreases in current property values from implementing the proposed action. In addition, the reasons the proposed action is believed by EPA to be justified now, rather than reserving a long term option for other alternatives, including no action, shall be explained.

(f) *Irreversible and irretrievable commitments of resources io the proposed action should it be implemented.* The EIS shall describe tne extent to which the proposed action requires commitment of construction materials, person-hours and funds to design and implement the project, as well as curtails the range of future uses of land and water resources. For example, induced growth in undeveloped areas may curtail alternative uses of that land. Also, irreversible environmental damage can result from equipment malfunctions or industrial accidents at the project site. Therefore, the need for any irretrievable and significant commitments of resources shall be explained fully.

(g) *Problems and objections raised by other Federal, State and local agencies and by interested persons in the review process.* Final EIS's (and draft EIS's if appropriate) shall summarize the comments and suggestions made by reviewing organizations and shall describe the disposition of issues raised, e.g., revisions to the proposed action to mitigate anticipated impacts or objections. In particular, the EIS shall address the major issues raised when the EPA position differs from most recommendations and explain the factors of overriding importance overruling the adoption of suggestions. Reviewer's statements should be set forth in a "comment" and discussed in a "response." In addition, the source of all comments should be clearly identified, and copies of the comments should be attached to the final EIS. Summaries of comments should be attached when a response has been exceptionally long or the same comments were received from many reviewers.

§ 6.306 Documentation.

All books, research reports, field study reports, correspondence and other documents which provided the data base for evaluating the proposed action and alternatives discussed in the EIS shall be used as references in the body of the EIS and shall be included in a bibliography attached to the EIS.

Subpart D—EPA Public Hearings on EIS's

§ 6.400 General.

While EPA is not required by statute to hold public hearings on EIS's, the responsible official should hold a public hearing on a draft EIS whenever a hearing may facilitate the resolution of conflicts or significant public controversy. This hearing may be in addition to public hearings held on facilities plans or section 209 plans. The responsible official may take special measures to involve interested persons through personal contact.

§ 6.402 Public hearing process.

(a) When public hearings are to be held, EPA shall inform the public of the hearing, for example, with a notice in the draft EIS. The notice should follow the summary sheet at the beginning of the EIS. The draft EIS shall be available for public review at least thirty (30) days before the public hearing. Public notice shall be given at least fifteen (15) working days before the public hearing and shall include:

(1) Publication of a public notice in a newspaper which covers the project area, identifying the project, announcing the date, time and place of the hearing and announcing the availability of detailed information on the proposed action for public inspection at one or more locations in the area in which the project will be located. "Detailed information" shall include a copy of the project application and the draft EIS.

(2) Notification of appropriate State and local agencies and appropriate State, regional and metropolitan clearinghouses.

(3) Notification of interested persons.

(b) A written record of the hearing shall be made. A stenographer may be used to record the hearing. As a minimum, the record shall contain a list of witnesses with the text of each presentation. A summary of the record, including the issues raised, conflicts resolved and unresolved, and any other significant portions of the record, shall be appended to the final EIS.

(c) When a public hearing has been held by another Federal, State, or local agency on an EPA action, additional hearings are not necessary. The responsible official shall decide if additional hearings are needed.

(d) When a program office is the originating office, the appropriate regional office will provide assistance to the originating office in holding any public hearing if assistance is requested.

Subpart E—Guidelines for Compliance With NEPA in the Title II Wastewater Treatment Works Construction Grants Program and the Areawide Waste Treatment Management Planning Program

394

§ 6.500 Purpose.

This subpart amplifies the general EPA policies and procedures described in Subparts A through D with detailed procedures for compliance with NEPA in the wastewater treatment works construction grants program and the areawide waste treatment management planning program.

§ 6.502 Definitions.

(a) "Step 1 grant." A grant for preparation of a facilities plan as described in 40 CFR 35.930-1.

(b) "Step 2 grant." A grant for preparation of construction drawings and specifications as described in 40 CFR 35.930-1.

(c) "Step 3 grant." A grant for fabrication and building of a publicly owned treatment works as described in 40 CFR 35.930-1.

§ 6.504 Applicability.

(a) *Administrative actions covered.* This subpart applies to the administrative actions listed below:

(1) Approval of all section 208 plans according to procedures in 40 CFR 35.1067-2;

(2) Approval of all facilities plans except those listed in paragraph (a)(5) of this section;

(3) Award of step 2 and step 3 grants, if an approved facilities plan was not required;

(4) Award of a step 2 or step 3 grant when either the project or its impact has changed significantly from that described in the approved facilities plan, except when the situation in paragraph (a)(5) of this section exists;

(5) Consultation during the NEPA review process. When there are overriding considerations of cost or impaired program effectiveness, the Regional Administrator may award a step 2 or a step 3 grant for a discrete segment of the project plans or construction before the NEPA review is completed if this project segment is noncontroversial. The remaining portion of the project shall be evaluated to determine if an EIS is required. In applying the criteria for this determination, the entire project shall be considered, including those parts permitted to proceed. In no case may these types of step 2 or step 3 grants be awarded unless both the Office of Federal Activities and CEQ have been consulted, a negative declaration has been issued on the segments permitted to proceed, and the grant award contains a specific agree-

ment prohibiting action on the segment of planning or construction for which the NEPA review is not complete. Examples of consultation during the NEPA review process are: award of a step 2 grant for preparation of plans and specifications for a large treatment plant, when the only unresolved NEPA issue is where to locate the sludge disposal site; or award of a step 3 grant for site clearance for a large treatment plant, when the unresolved NEPA issue is whether sludge from the plant should be incinerated at the site or disposed of elsewhere by other means.

(b) *Administrative actions excluded.* The actions listed below are not subject to the requirements of this part:

(1) Approval of State priority lists;

(2) Award of a step 1 grant;

(3) Award of a section 208 planning grant;

(4) Award of a step 2 or step 3 grant when no significant changes in the facilities plan have occurred;

(5) Approval of issuing an invitation for bid or awarding a construction contract;

(6) Actual physical commencement of building or fabrication;

(7) Award of a section 206 grant for reimbursement;

(8) Award of grant increases whenever § 6.504(a)(4) does not apply;

(9) Awards of training assistance under FWPCA, as amended, section 109(b).

(c) *Retroactive application.* The new criteria in § 6.510 of this subpart do not apply to step 2 or step 3 grants awarded before July 1, 1975. However, the Regional Administrator may apply the new criteria of this subpart when he considers it appropriate. Any negative declarations issued before the effective date of this regulation shall remain in effect.

§ 6.506   Completion of NEPA procedures before start of administrative actions.
See § 6.108 and § 6.504.

§ 6.510   Criteria for preparation of environmental impact statements.

In addition to considering the criteria in § 6.200, the Regional Administrator shall assure that an EIS will be prepared on a treatment works facilities plan, 208 plan or other appropriate water quality management plan when:

(a) The treatment works or plan will induce significant changes (either absolute changes or increases in the rate of change) in industrial, commercial, agricultural, or residential land use concentrations or distributions. Factors that should be considered in determining if these changes are significant include but are not limited to: the vacant land subject to increased development pressure as a result of the treatment works; the increases in population which may be induced; the faster rate of change of population; changes in population density; the potential for overloading sewage treatment works; the extent to which landowners may benefit from the areas subject to increased development; the nature of land use regulations in the affected area and their potential effects on development; and deleterious changes in the availability or demand for energy.

(b) Any major part of the treatment works will be located on productive wetlands or will have significant adverse effects on wetlands, including secondary effects.

(c) Any major part of the treatment works will be located on or significantly affect the habitat of wildlife on the Department of Interior's threatened and endangered species lists.

(d) Implementation of the treatment works or plan may directly cause or induce changes that significantly:

(1) Displace population;

(2) Deface an existing residential area; or

(3) Adversely affect significant amounts of prime agricultural land or agricultural operations on this land.

(e) The treatment works or plan will have significant adverse effects on parklands, other public lands or areas of recognized scenic, recreational, archaeological or historic value.

(f) The works or plan may directly or through induced development have a significant adverse effect upon local ambient air quality, local ambient noise levels, surface or groundwater quantity or quality, fish, wildlife, and their natural habitats.

(g) The treated effluent is being discharged into a body of water where the present classification is too lenient or is being challenged as too low to protect present or recent uses, and the effluent will not be of sufficient quality to meet the requirements of these uses.

§ 6.512 Procedures for implementing NEPA.

(a) *Environmental assessment.* An adequate environmental assessment must be an integral, though identifiable, part of any facilities or section 208 plan submitted to EPA. (See § 6.202 for a general description.) The information in the fa-

cilities plan, particularly the environmental assessment, will provide the substance of an EIS and shall be submitted by the applicant. The analyses that constitute an adequate environmental assessment shall include:

(1) *Description of the existing environment without the project.* This shall include for the delineated planning area a description of the present environmental conditions relevant to the analysis of alternatives or determinations of the environmental impacts of the proposed action. The description shall include, but not be limited to, discussions of whichever areas are applicable to a particular study: surface and groundwater quality; water supply and use; general hydrology; air quality; noise levels, energy production and consumption; land use trends; population projections, wetlands, floodplains, coastal zones and other environmentally sensitive areas; historic and archaeological sites; other related Federal or State projects in the area; and plant and animal communities which may be affected, especially those containing threatened or endangered species.

(2) *Description of the future environment without the project.* The future environmental conditions with the no project alternative shall be forecast, covering the same areas listed in § 6.512 (a) (1).

(3) *Documentation.* Sources of information used to describe the existing environment and to assess future environmental impacts should be documented. These sources should include regional, State and Federal agencies with responsibility or interest in the types of impacts listed in § 6.512(a)(1). In particular, the following agencies should be consulted:

(i) Local and regional land use planning agencies for assessments of land use trends and population projections, especially those affecting size, timing, and location of facilities, and planning activities funded under section 701 of the Housing and Community Development Act of 1974 (Pub. L. 93–383);

(ii) The HUD Regional Office if a project involves a flood risk area identified under the Flood Disaster Protection Act of 1973 (Pub. L. 93–234);

(iii) The State coastal zone management agency, if a coastal zone is affected;

(iv) The Secretary of the Interior or Secretary of Agriculture, if a wild and scenic river is affected;

(v) The Secretary of the Interior or Secretary of Commerce, if a threatened or endangered species is affected;

(vi) The Fish and Wildlife Service

(Department of Interior), the Department of Commerce, and the U.S. Army Corps of Engineers, if a wetland is affected.

(4) *Evaluation of alternatives.* This discussion shall include a comparative analysis of feasible options and a systematic development of wastewater treatment alternatives. The alternatives shall be screened with respect to capital and operating costs; significant primary and secondary environmental effects; physical, legal or institutional constraints; and whether or not they meet regulatory requirements. Special attention should be given to long term impacts, irreversible impacts and induced impacts such as development. The reasons for rejecting any alternatives shall be presented in addition to any significant environmental benefits precluded by rejection of an alternative. The analysis should consider, when relevant to the project:

(i) Flow and waste reduction measures, including infiltration/inflow reduction;

(ii) Alternative locations, capacities, and construction phasing of facilities;

(iii) Alternative waste management techniques, includign treatment and discharge, wastewater reuse and land application;

(iv) Alternative methods for disposal of sludge and other residual waste, including process options and final disposal options;

(v) Improving effluent quality through more efficient operation and maintenance;

(vi) For assessments associated with section 208 plans, the analysis of options shall include in addition:

(A) Land use and other regulatory controls, fiscal controls, non-point source controls, and institutional arrangements; and

(B) Land management practices.

(5) *Environmental impacts of the proposed action.* Primary and secondary impacts of the proposed action shall be described, giving special attention to unavoidable impacts, steps to mitigate adverse impacts, any irreversible or irretrievable commitments of resources to the project and the relationship between local short term uses of the environment and the maintenance and enhancement of long term productivity. See § 6.304 (c), (d), (e), and (f) for an explanation of these terms and examples. The significance of land use impacts shall be evaluated, based on current population of the planning area; design year population for the service area; percentage of the service area currently vacant; and plans for staging facilities. Special attention should be given to induced changes in population patterns and growth, particularly if a project involves some degree of regionalization. In addition to these items, the Regional Administrator may require that other analyses and data, which he determines are needed to comply with NEPA, be included with the facilities or section 208 plan. Such requirements should be discussed during preapplication conferences. The Regional Administrator also may require submission of supplementary information either before or after a step 2 grant or before a step 3 grant award if he determines it is needed for compliance with NEPA. Requests for supplementary information shall be made in writing.

(6) *Steps to minimize adverse effects.* This section shall describe structural and nonstructural measures, if any, in the facilities plan to mitigate or eliminate significant adverse effects on the human and natural environments. Structural provisions include changes in facility design, size, and location; nonstructural provisions include staging facilities as well as developing and enforcing land use regulations and environmentally protective regulations.

(b) *Public hearing.* The applicant shall hold at least one public hearing before a facilities plan is adopted, unless waived by the Regional Administrator before completion of the facilities plan according to § 35.917–5 of the Title II construction grants regulations. Hearings should be held on section 208 plans. A copy of the environmental assessment should be available for public review before the hearing and at the hearing, since these hearings provide an opportunity to accept public input on the environmental issues associated with the facilities plan or the 208 water quality management strategy. In addition, a Regional Administrator may elect to hold an EPA hearing if environmental issues remain unresolved. EPA hearings shall be held according to procedures in § 6.402.

(c) *Environmental review.* An environmental review of a facilities plan or section 208 plan shall be conducted according to the procedures in § 6.204 and applying the criteria of § 6.510. If deficiencies exist in the environmental assessment, they shall be identified in writing by the Regional Administrator and must be corrected before the plan can be approved.

(d) *Additional procedures.* (1) Historic and archaeological sites. If a facilities or section 208 plan may affect properties with historic, architectural, archaeological or cultural value which are listed in or eligible for listing in the National Register of Historic Places or may cause irreparable loss or destruction of significant scientific, prehistoric, historic or archaeological data, the applicant shall follow the procedures in § 6.214(a).

(2) If the facilities or section 208 plan may affect wetlands, floodplains, coastal zones, wild and scenic rivers, fish or wildlife, the Regional Administrator shall follow the appropriate procedures described in § 6.214(b).

(e) *Notice of intent.* The notice of intent on a facilities plan or section 208 plan shall be issued according to § 6.206.

(f) *Scope of EIS.* It is the Regional Administrator's responsibility to determine the scope of the EIS. He should determine if an EIS should be prepared on a facilities plan(s) or section 208 plan and which environmental areas should be discussed in greatest detail in the EIS. Once an EIS has been prepared for the designated section 208 area, another need not be prepared unless the significant impacts of individual facilities or other plan elements were not adequately treated in the EIS. The Regional Administrator should document his decision not to prepare an EIS on individual facilities.

(g) *Negative declaration.* A negative declaration on a facilities plan or section 208 plan shall be prepared according to § 6.212. Once a negative declaration and environmental appraisal have been prepared for the facilities plan for a certain area, grant awards may proceed without preparation of additional negative declarations, unless the project has changed significantly from that described in the facilities plan.

§ 6.514 Content of environmental impact statements.

EIS's for treatment works or plans shall be prepared according to § 6.304.

Subpart F—Guidelines for Compliance With NEPA in Research and Development Programs and Activities

§ 6.600 Purpose.

This subpart amplifies the general EPA policies and procedures described in Subparts A through D by providing procedures for compliance with NEPA on actions undertaken by the Office of Research and Development (ORD).

§ 6.602 Definitions.

(a) "Work plan." A document which defines and schedules all projects required to fulfill the objectives of the program plan.

(b) "Program plan." An overall planning document for a major research area which describes one or more research objectives, including outputs and target completion dates, as well as person-year and dollar resources.

(c) "Appropriate program official." The official at each decision level within ORD to whom the Assistant Administrator delegates responsibility for NEPA compliance.

(d) "Exemption certification." A certified statement delineating those actions specifically exempted from NEPA compliance by existing legislation.

§ 6.604 Applicability.

The requirements of this subpart are applicable to administrative actions undertaken to approve program plans, work plans, and projects, except those plans and projects excluded by existing legislation. However, no administrative actions are excluded from the additonal procedures in § 6.214 of this part concerning historic sites, wetlands, coastal zones, wild and scenic rivers, floodplains or fish and wildlife.

§ 6.608 Criteria for determining when to prepare EIS's.

(a) An EIS shall be prepared by ORD when any of the criteria in § 6.200 apply or when:

(1) The action will have significant adverse impacts on public parks, wetlands, floodplains, coastal zones, wildlife habitats, or areas of recognized scenic or recreational value.

(2) The action will significantly deface an existing residential area.

(3) The action may directly or through induced development have a significant adverse effect upon local ambient air quality, local ambient noise levels, surface or groundwater quality; and fish, wildlife or their natural habitats.

(4) The treated effluent is being discharged into a body of water where the present classification is being challenged as too low to protect present or recent uses, and the effluent will not be of

sufficient quality to meet the requirements of these uses.

(5) The project consists of field tests involving the introduction of significant quantities of: toxic or polluting agricultural chemicals, animal wastes, pesticides, radioactive materials, or other hazardous substances into the environment by ORD, its grantees or its contractors.

(6) The action may involve the introduction of species or subspecies not indigenous to an area.

(7) There is a high probability of an action ultimately being implemented on a large scale, and this implementation may result in significant environmental impacts.

(8) The project involves commitment to a new technology which is significant and may restrict future viable alternatives.

(b) An EIS will not usually be needed when:

(1) The project is conducted completely within a laboratory or other facility, and external environmental effects have been minimized by methods for disposal of laboratory wastes and safeguards to prevent hazardous materials entering the environment accidentally; or

(2) The project is a relatively small experiment or investigation that is part of a non-Federally funded activity of the private sector, and it makes no significant new or additional contribution to existing pollution.

§ 6.610 Procedures for compliance with NEPA.

EIS related activities for compliance with NEPA will be integrated into the decision levels of ORD's research planning system to assure managerial control. This control includes those administrative actions which do not come under the applicability of this subpart by assuring that they are made the subject of an exemption certification and filed with the Office of Public Affairs (OPA). ORD's internal procedures provide details for NEPA compliance.

(a) *Environmental assessment.* (1) Environmental assessments shall be submitted with all grant applications and all unsolicited contract proposals. The assessment shall contain the same information required for EIS's in § 6.304. Copies of § 6.304 (or more detailed guidance when available) and a notice of the requirement for assessment shall be included in all grant application kits and attached to letters concerning the submission of unsolicited proposals.

(2) In the case of competitive contracts, assessments need not be submitted by potential contractors since the NEPA procedures must be completed before a request for proposal (RFP) is issued. If there is a question concerning the need for an assessment, the potential contractor should contact the official responsible for the contract.

(b) *Environmental review.* (1) At the start of the planning year, an environmental review will be performed for each program plan with its supporting substructures (work plans and projects) before incorporating them into the ORD program planning system, unless they are excluded from review by existing legislation. This review is an evaluation of the potentially adverse environmental effects of the efforts required by the program plan. The criteria in § 6.608 shall be used in conducting this review. Each program plan with its supporting substructures which does not have significant adverse impacts may be dismissed from further current year environmental considerations with a single negative declaration. Any supporting substructures of a program plan which cannot be dismissed with the parent plan shall be reviewed at the appropriate subordinate levels of the planning system for NEPA compliance.

(i) All continuing program plans and supporting substructures, including those previously dismissed from consideration, will be reevaluted annually for NEPA compliance. An environmental review will coincide with the annual planning cycle and whenever a major redirection of a parent plan is undertaken. All NEPA-associated documents will be updated as appropriate.

(ii) All approved program plans and supporting substructures, less budgetary data, will be filed in the OPA with a notice of intent or negative declaration and environmental appraisal.

(iii) Later plans and/or projects, added to fulfill the mission objectives but not identified at the time the program plans were approved, will be subjected to the same NEPA requirements for environmental assessments and/or reviews.

(iv) Those projects subjected to environmental assessments as outlined in paragraph (a) of this section and not exempt under existing legislation also shall undergo an environmental review before work begins.

(c) *Notice of intent and EIS.*

(1) If the reviews conducted according to paragraph (b) of this section reveal a potentially significant adverse effect on the environment and the adverse impact cannot be eliminated by replanning, the appropriate program official shall, after making sure the project is to be funded, issue a notice of intent according to § 6.206, and through proper organizational channels, shall request the Regional Administrator to assist him in the preparation and distribution of the EIS.

(2) As soon as possible after release of the notice of intent, the appropriate program official shall prepare a draft EIS using the criteria in Subpart B, § 6.208 and Subpart C. Through proper organizational channels, he shall request the Regional Administrator to assist him in the preparation and distribution of the draft EIS.

(3) The appropriate program official shall prepare final EIS's according to criteria in Subpart B, § 6.210 and Subpart C.

(4) All draft and final EIS's shall be sent through the proper organizational channels to the Assistant Administrator for ORD for approval. The approved statements then will be distributed according to the procedures in Appendix C.

(d) *Negative declaration and environmental impact appraisal.* If an environmental review conducted according to paragraph (b) of this section reveals that proposed actions will not have significant adverse environmental impacts, the appropriate program official shall prepare a negative declaration and environmental impact appraisal according to Subpart B, § 6.212. Upon assurance that the program will be funded, the appropriate program official shall distribute the negative declaration as described in § 6.212 and make copies of the negative declaration and appraisal available in the OPA.

(e) *Project start.* As required by § 6.108, a contract or grant shall not be awarded for an extramural project, nor for continuation of what was previously an intramural project, until at least fifteen (15) working days after a negative declaration has been issued or thirty (30) days after forwarding the final EIS to the Council on Environmental Quality.

Subpart G—Guidelines for Compliance With NEPA in Solid Waste Management Activities

§ 6.700 Purpose.

This subpart amplifies the general policies and procedures described in Subparts A through D by providing additional procedures for compliance with NEPA on actions undertaken by the Office of Solid Waste Management Programs (OSWMP).

§ 6.702 Criteria for the preparation of environmental assessments and EIS's.

(a) *Assessment preparation criteria.* An environmental assessment need not be submitted with all grant applications and contract proposals. Studies and investigations do not require assessments. The following sections describe when an assessment is or is not required for other actions:

(1) *Grants.* (i) *Demonstration projects.* Environmental assessments must be submitted with all applications for demonstration grants that will involve construction, land use (temporary or permanent), transport, sea disposal, any discharges into the air or water, or any other activity having any direct or indirect effects on the environment external to the facility in which the work will be conducted. Preapplication proposals for these grants will not require environmental assessments.

(ii) *Training.* Grant applications for training of personnel will not require assessments.

(iii) *Plans.* Grant applications for the development of comprehensive State, interstate, or local solid waste management plans will not require environmental assessments. A detailed analysis of environmental problems and effects should be part of the planning process, however.

(2) *Contracts.* (i) *Sole-source contract proposals.* Before a sole-source contract can be awarded, an environmental assessment must be submitted with a bid proposal for a contract which will involve construction, land use (temporary or permanent), sea disposal, any discharges into the air or water, or any other activity that will directly or indirectly affect the environment external to the facility in which the work will be performed.

(ii) *Competitive contract proposals.* Assessments generally will not be required on competitive contract proposals.

(b) *EIS preparation criteria.* The responsible official shall conduct an environmental review on those OSWMP projects on which an assesment is required or which may have effects on the environment external to the facility in which the work will be performed. The criteria in § 6.200 shall be utilized in determining whether an EIS need be prepared.

§ 6.704 Procedures for compliance with NEPA.

(a) *Environmental assessment.* (1) Environmental assessments shall be submitted to EPA according to procedures in § 6.702. If there is a question concerning the need for an assessment, the potential contractor or grantee should consult with the appropriate project officer for the grant or contract.

(2) The assessment shall contain the same sections specified for EIS's in § 6.304. Copies of § 6.304 (or more detailed guidance when available) and a notice alerting potential grantees and contractors of the assessment requirements in § 6.702 shall be included in all grant application kits, attached to letters concerning the submission of unsolicited proposals, and included with all RFP's.

(b) *Environmental review.* An environmental review will be conducted on all projects which require assessments or which will affect the environment external to the facility in which the work will be performed. This review must be conducted before a grant or contract award is made on an extramural project or before an intramural project begins. The guidelines in § 6.200 will be used to determine if the project will have any significant environmental effects. This review will include an evaluation of the assessment by both the responsible official and the appropriate Regional Administrator. The Regional Administrator's comments will include his recommendations on the need for an EIS. No detailed review or documentation is required on projects for which assessments are not required and which will not affect the environment external to a facility.

(c) *Notice of intent and EIS.* If any of the criteria in § 6.200 apply, the responsible official will assure that a notice of intent and a draft EIS are prepared. The responsible official may request the appropriate Regional Administrator to assist him in the distribution of the NEPA-associated documents. Distribution procedures are listed in Appendix C.

(d) *Negative declaration and environmental impact appraisal.* If the environmental review indicated no significant environmental impacts, the responsible official will assure that a negative declaration and environmental appraisal are prepared. These documents need not be prepared for projects not requiring an environmental review.

(e) The EIS process for the Office of Solid Waste Management Programs is shown graphically in Exhibit 7.

Subpart H—Guidelines for Compliance With NEPA in Construction of Special Purpose Facilities and Facility Renovations

§ 6.800 Purpose.

This subpart amplifies general EPA policies and procedures described in Subparts A through D by providing detailed procedures for the preparation of EIS's on construction and renovation of special purpose facilities.

§ 6.802 Definitions.

(a) "Special purpose facility." A building or space, including land incidental to its use, which is wholly or predominantly utilized for the special purpose of an agency and not generally suitable for other uses, as determined by the General Services Administration.

(b) "Program of requirements." A comprehensive document (booklet) describing program activities to be accomplished in the new special purpose facility or improvement. It includes architectural, mechanical, structural, and space requirements.

(c) "Scope of work." A document similar in content to the program of requirements but substantially abbreviated. It is usually prepared for small-scale projects.

§ 6.804 Applicability.

(a) *Actions covered.* These guidelines apply to all new special purpose facility construction, activities related to this construction (e.g., site acquisition and clearing), and any improvements or modifications to facilities having potential environmental effects external to the facility, including new construction and improvements undertaken and funded by the Facilities Management Branch, Facilities and Support Services Division, Office of Administration; by a regional office; or by a National Environmental Research Center.

(b) *Actions excluded.* This subpart does not apply to those activities of the Facilities Management Branch, Facilities and Support Services Division, for which the branch does not have full fiscal responsibility for the entire project. This includes pilot plant construction, land acquisition, site clearing and access road construction where the Facilities

Management Branch's activity is only supporting a project financed by a program office. Responsibility for considering the environmental impacts of such projects rests with the office managing and funding the entire project. Other subparts of this regulation apply depending on the nature of the project.

§ 6.808 Criteria for the preparation of environmental assessments and EIS's.

(a) *Assessment preparation criteria.* The responsible official shall request an environmental assessment from a construction contractor or consulting architect/engineer employed by EPA if he is involved in the planning, construction or modification of special purpose facilities when his activities have potential environmental effects external to the facility. Such modifications include but are not limited to: facility additions, changes in central heating systems or wastewater treatment systems, and land clearing for access roads and parking lots.

(b) *EIS preparation criteria.* The responsible official shall conduct an environmental review of all actions involving construction of special purpose facilities and improvements to these facilities. The guidelines in § 6.200 shall be used to determine whether an EIS shall be prepared.

§ 6.810 Procedures for compliance with NEPA.

(a) *Environmental review and assessment.* (1) An environmental review shall be conducted when the program of requirements or scope of work has been completed for the construction, improvement, or modification of special purpose facilities. For special purpose facility construction, the Chief, Facilities Management Branch, shall request the assistance of the appropriate program office and Regional Administrator in the review. For modifications and improvements, the appropriate responsible official shall request assistance in making the review from other cognizant EPA offices.

(2) Any assessments requested shall contain the same sections listed for EIS's in § 6.304. Contractors and consultants shall be notified in contractual documents when an assessment must be prepared.

(b) *Notice of intent, EIS, and negative declaration.* The responsible official shall decide at the completion of the environmental review whether there may be any significant environmental impacts. If there could be significant environmental impacts, a notice of intent and an EIS shall be prepared according to the procedures in § 6.206. If there may not be any significant environmental impacts, a negative declaration and environmental impact appraisal shall be prepared according to the procedures in § 6.212.

(c) *Project start.* As required by § 6.108, a contract shall not be awarded or construction-related activities begun until at least fifteen (15) working days after release of a negative declaration, or until thirty (30) days after forwarding the final EIS to the Council on Environmental Quality.

Exhibit 1

NOTICE OF INTENT TRANSMITTAL MEMORANDUM SUGGESTED FORMAT

(Date)

ENVIRONMENTAL PROTECTION AGENCY

(Appropriate Office)

(Address, City, State, Zip Code)

To All Interested Government Agencies and Public Groups:

As required by guidelines for the preparation of environmental impact statements (EIS's), attached is a notice of intent to prepare an EIS for the proposed EPA action described below:

(Official Project Name and Number)

(City, State)

If your organization needs additional information or wishes to participate in the preparation of the draft EIS, please advise the (appropriate office, city, State).

Very truly yours,

(Appropriate EPA Official)

(List Federal, State, and local agencies to be solicited for comment.)

(List public action groups to be solicited for comment.)

NOTICE OF INTENT SUGGESTED FORMAT

NOTICE OF INTENT—ENVIRONMENTAL PROTECTION AGENCY

1. Project location:
   City -------------------------------------
   County -----------------------------------
   State,------------------------------------
2. Proposed EPA action:

   ----

3. Issues involved:

----

4. Estimated project costs:

Federal Share (total) _____ $_____

Contract $____ Grant $____ Other $_____

Applicant share (if any):

(Name) _____$_____

Other (specify)_____$_____

Total _____$_____

5. Period covered by project:

Start date:_____

(Original date, if project covers

more than one year)

Dates of different project phases:_____

* Approximate end date:_____

6. Estimated application filing date:_____

EXHIBIT 2

PUBLIC NOTICE AND NEWS RELEASE SUGGESTED
FORMAT

PUBLIC NOTICE

The Environmental Protection Agency
(originating office) (will prepare, will not
prepare, has prepared) a (draft, final) environmental impact statement on the following project:

(Official Project Name and Number)

(Purpose of Project)

(Project Location, City, County, State)

(Where EIS or negative declaration and environmental impact appraisal can be
obtained)

This notice is to implement EPA's policy
of encouraging public participation in the
decision-making process on proposed EPA
actions. Comments on this document may
be submitted to (full address of originating
office).

EXHIBIT 3

NEGATIVE DECLARATION SUGGESTED FORMAT

(Date)

ENVIRONMENTAL PROTECTION AGENCY

(Appropriate Office)

(Address, City, State, Zip
Code)

To All Interested Government Agencies and
Public Groups:

As required by guidelines for the preparation of environmental impact statements
(EIS's), an environmental review has been
performed on the proposed EPA action
below:

(Official Project Name and
Number)

(Potential Agency
Financial Share)

(Project Location: City,
County, State)

(Other Funds Included)

PROJECT DESCRIPTION, ORIGINATOR, AND
PURPOSE

(Include a map of the project area and a
brief narrative summarizing the growth the
project will serve, the percent of vacant land
the project will serve, major primary and
secondary impacts of the project, and the
purpose of the project.)

The review process did not indicate significant environmental impacts would result from the proposed action or significant
adverse impacts have been eliminated by
making changes in the project. Consequently, a preliminary decision not to prepare an EIS has been made.

This action is taken on the basis of a
careful review of the engineering report,
environmental impact assessment, and
other supporting data, which are on file in
the above office with the environmental impact appraisal and are available for public
scrutiny upon request. Copies of the environmental impact appraisal will be sent at cost
on your request.

Comments supporting or disagreeing with
this decision may be submitted for consideration by EPA. After evaluating the comments received, the Agency will make a final
decision; however, no administrative action
will be taken on the project for at least
fifteen (15) working days after release of
the negative declaration.

Sincerely,

(Appropriate EPA Official)

EXHIBIT 4

ENVIRONMENTAL IMPACT APPRAISAL
SUGGESTED FORMAT

A. Identify Project.

Name of Applicant:_____

Address: _____

Project Number:_____

B. Summarize Assessment.

1. Brief description of project:_____

----

2. Probable impact of the project on the
environment: _____

----

3. Any probable adverse environmental
effects which cannot be avoided:_____

----

4. Alternatives considered with evaluation
of each:_____

```
----------------------------------------
----------------------------------------
----------------------------------------
```

5. Relationship between local short-term
uses of man's environment and maintenance and enhancement of long-term productivity: `---------------------------------`

```
----------------------------------------
```

6. Steps to minimize harm to the environment: `-----------------------------------`

```
----------------------------------------
```

7. Any irreversible and irretrievable commitment of resources: `---------------------`

```
----------------------------------------
```

8. Public objections to project, if any, and
their resolution: `-----------------------`

```
----------------------------------------
```

9. Agencies consulted about the project: `--`

```
----------------------------------------
----------------------------------------
```

    State representative's name: `----------`
    Local representative's name: `----------`
    Other: `---------------------------------`
C. Reasons for concluding there will be no
significant impacts.

```
-------------------------
```
(Signature of
appropriate official)
(Date)

EXHIBIT 5

COVER SHEET FORMAT FOR ENVIRONMENTAL
IMPACT STATEMENTS
(Draft, Final)

ENVIRONMENTAL IMPACT STATEMENT

```
-------------------------------------------
-------------------------------------------
```

(Describe title of project plan and give
identifying number)

Prepared by: `---------------------------`
             (Responsible Agency Office)
Approved by: `---------------------------`
            (Responsible Agency Official)
```
-------------------------
```
(Date)

EXHIBIT 6

SUMMARY SHEET FORMAT FOR ENVIRONMENTAL
IMPACT STATEMENTS

(Check One)
    ( ) Draft
    ( ) Final

ENVIRONMENTAL PROTECTION AGENCY
```
-------------------------
```
(Responsible Agency Office)

1. Name of action. (Check one)
    ( ) Administrative action.
    ( ) Legislative action.
2. Brief description of action indicating what
States (and counties) are particularly
affected.
3. Summary of environmental impact and
adverse environmental effects.
4. List alternatives considered.
5. a. (for draft statements) List all Federal
State, and local agencies and other
comments have been requested.
   b. (for final statements) List all Federal

State, and local agencies and other
sources from which written comments have been received.
6. Dates draft statement and final statement made available to Council on Environmental Quality and public.

APPENDIX A

CHECKLIST FOR ENVIRONMENTAL REVIEWS

Areas to be considered, when appropriate,
during an environmental review include, but
are not limited to, the items on this checklist, based on Appendix II of the CEQ guidelines for the preparation of environmental
impact statements which appeared in the
FEDERAL REGISTER August 1, 1973. The classification of items is not mandatory.

I. *Natural environment.* Consider the impacts of a proposed action on air quality,
water supply and quality, soil conservation
and hydrology, fish, and wildlife populations,
fish and wildlife habitats, solid waste disposal, noise levels, radiation, and hazardous
substances use and disposal.

II. *Land use planning and management.*
Consider the impacts of a proposed action on
energy supply and natural resources development; protection of environmentally critical
areas, such as floodplains, wetlands, beaches
and dunes, unstable soils, steep slopes and
aquifer recharge areas, coastal area land use;
and redevelopment and construction in
built-up areas.

III. *Socioeconomic environment.* Consider
the impacts of a proposed action on population density changes, congestion mitigation,
neighborhood character and cohesion, low
income populations, outdoor recreation, in-
dustrial/commercial/residential development
and tax ratables, and historic, architectural
and archaeological preservation.

APPENDIX B

RESPONSIBILITIES

I. *General responsibilities.* (a) *Responsible
official.* (1) Requires contractors and grantees
to submit environmental assessments and related documents needed to comply with
NEPA, and assures environmental reviews are
conducted on proposed EPA projects at the
earliest possible point in EPA's decision-
making process.

(2) When required, assures that draft EIS's
are prepared and distributed at the earliest
possible + point in EPA's decision-making
process, their internal and external review is
coordinated, and final EIS's are prepared and
distributed.

(3) When an EIS is not prepared, assures
that negative declarations and environmental
appraisals are prepared and distributed for
those actions requiring them.

(4) Consults with appropriate officials
identified in § 6.214 of this part.

(5) Consults with the Office of Federal

EXHIBIT 7

FLOWCHART FOR OSWMP

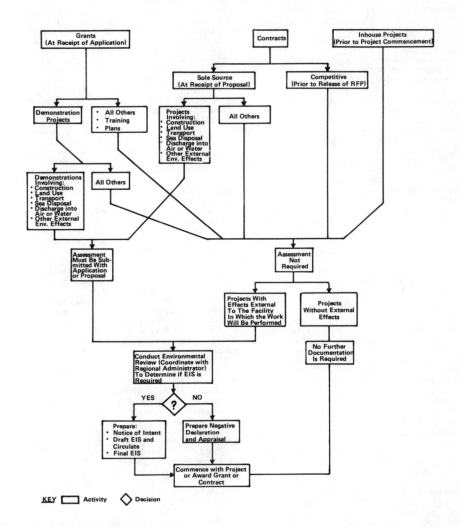

KEY ☐ Activity ◇ Decision

Activities on actions involving unresolved conflicts with other Federal agencies.

(b) *Office of Federal Activities.* (1) Provides EPA with policy guidance and assures that EPA offices establish and maintain adequate administrative procedures to comply with this part.

(2) Monitors the overall timeliness and quality of the EPA effort to comply with this part.

(3) Provides assistance to responsible officials as required.

(4) Coordinates the training of personnel involved in the review and preparation of EIS's and other NEPA-associated documents.

(5) Acts as EPA liaison with the Council on Environmental Quality and other Federal and State entities on matters of EPA policy and administrative mechanisms to facilitate external review of EIS's, to determine lead agency and to improve the uniformity of the NEPA procedures of Federal agencies.

(6) Advises the Administrator and Deputy Administrator on projects which involve more than one EPA office, are controversial, are nationally significant, or "pioneer" EPA policy.

when these projects have had or should have an EIS prepared on them.

(c) *Office of Public Inquiries.* Assists the Office of Federal Activities and responsible officials by answering the public's queries on the EIS process and on specific EIS's and by directing requests for copies of specific documents to the appropriate regional office or program.

(d) *Office of Public Affairs.* Analyzes the present procedures for public participation, and develops and recommends to the Office of Federal Activities a program to improve those procedures and increase public participation.

(e) *Regional Office Division of Public Affairs.* (1) Assists the responsible official or his designee on matters pertaining to negative declarations, notices of intent, press releases, and other public notification procedures.

(2) Assists the responsible official or his designee by answering the public's queries on the EIS process and on specific EIS's, and by filling requests for copies of specific documents.

(f) *Offices of the Assistant Administrators and Regional Administrators.* (1) Provides specific policy guidance to their respective offices and assures that those offices establish and maintain adequate administrative procedures to comply with this part.

(2) Monitors the overall timeliness and quality of their respective office's efforts to comply with this part.

(3) Acts as liaison between their offices and the Office of Federal Activities and between their offices and other Assistant Administrators or Regional Administrators on matters of agencywide policy and procedures.

(4) Advises the Administrator and Deputy Administrator through the Office of Federal Activities on projects or activities within their respective areas of responsibilities which involve more than one EPA office, are controversial, are nationally significant, or "pioneer" EPA policy, when these projects have had or should have an EIS prepared on them.

(g) *The Office of Legislation.* (1) Provides the necessary liaison with Congress.

(2) Coordinates the preparation of EIS's required on reports on legislation originating outside EPA. (See § 6.106(d)).

(h) *The Office of Planning and Evaluation.* Coordinates the preparation of EIS's required on EPA legislative proposals. (See § 6.106 (d)).

II. *Responsibilities for Title II Construction Grants Program (Subpart E).* (a) *Responsible official.* The responsible official for EPA actions covered by this subpart is the Regional Administrator. The responsibilities of the Regional Administrator in addition to those in Appendix B.I. are to:

(1) Assist the Office of Federal Activities in coordinating the training of personnel involved in the review and preparation of NEPA-associated documents.

(2) Require grant applicants and those who have submitted plans for approval to provide the information the regional office requires to comply with these guidelines.

(3) Consult with the Office of Federal Activities concerning works or plans which significantly affect more than one regional office, are controversial, are of national significance or "pioneer" EPA policy, when these works have had or should have had an EIS prepared on them.

(b) *Assistant Administrator.* The responsibilities of the Office of the Assistant Administrator, as described in Appendix B.I, shall be assumed by the Assistant Administrator for Water and Hazardous Materials for EPA actions covered by this subpart.

(c) Oil and Special Materials Control Division, Office of Water Program Operations, coordinates all activities and responsibilities of the Office of Water Program Operations concerned with preparation and review of EIS's. This includes providing technical assistance to the Regional Administrators on EIS's and assisting the Office of Federal Activities in coordinating the training of personnel involved in the review and preparation of NEPA-associated documents.

(d) *Public Affairs Division, Regional Offices.* The responsibilities of the regions' Public Affairs Divisions, in addition to those in Appendix B.I, are to:

(1) Assist the Regional Administrator in the preparation and dissemination of NEPA-associated documents.

(2) Collaborate with the Headquarters Office of Public Affairs to analyze procedures in the regions for public participation and to develop and recommend to the Office of Federal Activities a program to improve those procedures.

III. *Responsibilities for Research and Development Programs (Subpart F).* The Assistant Administrator for Research and Development, in addition to those responsibilities outlined in Appendix B.I(a), will also assume the responsibilities described in Appendix B.I(f).

IV. *Responsibilities for Solid Waste Management Programs (Subpart G).* (a) *Responsible Official.* The responsible official for EPA actions covered by this subpart is the Deputy Assistant Administrator for Solid Waste Management Programs. The responsibilities of this official, in addition to those in Appendix B.I(a), are to:

(1) Assist the Office of Federal Activities in coordinating the training of personnel involved in the review and preparation of all NEPA-associated documents.

(2) Advise the Assistant Administrator for Air and Waste Management concerning projects which significantly affect more than one regional office, are controversial, are nationally significant, or "pioneer" EPA policy.

V. *Responsibilities for Special Purpose Facilities and Facility Renovation Programs (Subpart H).*

(a) *Responsible official.* The responsible official for new construction and modification of special purpose facilities is as follows:

(1) The Chief, Facilities Management Branch, Data and Support Systems Division, shall be the responsible official on all new construction of special purpose facilities and on all improvement and modification projects for which the Facilities Management Branch has received a funding allowance.

(2) The Regional Administrator shall be the responsible official on all improvement and modification projects for which the regional office has received the funding allowance.

(3) The Center Directors shall be the responsible officials on all improvement and modification projects for which the National Environmental Research Centers have received the funding allowance.

(b) The responsibilities of the responsible officials, in addition to those in Appendix B.I, are to:

(1) Ensure that environmental assessments are submitted when requested, that environmental reviews are conducted on all projects, and EIS's are prepared and circulated when there will be significant impacts.

(2) Assist the Office of Federal Activities in coordinating the training of personnel involved in the review and preparation of NEPA-associated documents.

Appendix C

DISTRIBUTION AND AVAILABILITY OF DOCUMENTS

I. *Negative Declaration.* (a) The responsible official shall distribute two copies of each negative declaration to:

(1) The appropriate Federal, State and local agencies and to the appropriate State and areawide clearinghouses.

(2) The Office of Legislation, the Office of Public Affairs and the Office of Federal Activities.

(3) The headquarters EIS coordinator for the program office originating the document. When the originating office is a regional office and the action is related to water quality management, one copy should be forwarded to the Oil and Special Materials Control Division, Office of Water Program Operations.

(b) The responsible official shall distribute one copy of each negative declaration to:

(1) Local newspapers and other local mass media.

(2) *Interested persons on request.* If it is not practical to send copies to all interested persons, make the document available through local libraries or post offices, and notify individuals that this action has been taken.

(c) The responsible official · shall have a copy of the negative declaration and any documents supporting the negative declaration available for public review at the originating office.

II. *Environmental Impact Appraisal.* (a) The responsible official shall have the environmental impact appraisal available when the negative declaration is distributed and shall forward one copy to the headquarters EIS coordinator for the program office originating the document and to any other Federal or State agency which requests a copy.

(b) The responsible official shall have a copy of the environmental impact appraisal available for public review at the originating office and shall provide copies at cost to persons who request them.

III. *Notice of Intent.* (a) The responsible official shall forward one copy of the notice of intent to:

(1) The appropriate Federal, State and local agencies and to the appropriate State, regional and metropolitan clearing houses.

(2) Potentially interested persons.

(3) The Offices of Federal Activities, Public Affairs and Legislation.

(4). The headquarters Grants Administration Division, Grants Information Branch.

(5) The headquarters EIS coordinator for the program office originating the notice. When the originating office is a regional office and the action is related to water quality management, one copy should be forwarded to the Oil and Special Materials Control Division, Office of Water Program Operations.

IV. *Draft EIS's.* (a) The responsible official shall send two copies of the draft EIS to:

(1) The Office of Federal Activities.

(2) The headquarters EIS coordinator for the program office originating the document. When the originating office is a regional office and the project is related to water quality management, send two copies to the Oil and Special Materials Control Division, Office of Water Program Operations.

(b) If none of the above offices requests any changes within ten (10) working days after notification, the responsible official shall:

(1) Send five copies of the draft EIS to CEQ.

(2) Send two copies of the draft EIS to the Office of Public Affairs and to the Office of Legislation.

(3) Send two copies of the draft EIS to the appropriate offices of reviewing Federal agencies that have special expertise or jurisdiction by law with respect to any impacts involved. CEQ's guidelines (40 CFR 1500.9 and Appendices II and III) list those agencies to which draft EIS's will be sent for official review and comment.

(4) Send two copies of the draft EIS to the appropriate Federal, State, regional and metropolitan clearinghouses.

(5) Send one copy of the draft EIS to public libraries in the project area and interested persons. Post offices, city halls or courthouses may be used as distribution points if public library facilities are not available.

(c) The responsible official shall make a copy of the draft EIS available for public review at the originating office and at the Office of Public Affairs.

V. *Final EIS.* (a) The responsible official shall distribute the final EIS to the following offices, agencies and interested persons:

(1) Five copies to CEQ.

(2) Two copies to the Office of Public Affairs, Legislation and Federal Activities.

(3) Two copies to the headquarters EIS

coordinator for the program office originating the document.

(4) One copy to Federal, State and local agencies and interested persons who made substantive comments on the draft EIS or requested a copy of the final EIS.

(5) One copy to a grant applicant.

(b) The responsible official shall make a copy of the final EIS available for public review at the originating office and at the Office of Public Affairs.

VI. *Legislative EIS*. Copies of the legislative EIS shall be distributed by the responsible official according to the procedures in section IV(b) of this appendix. In addition, the responsible official shall send two copies of the EIS to the Office of Federal Activities and the EIS coordinator of the originating office.

[FR Doc.75-9553 Filed 4-11-75;8:45 am]

ENVIRONMENTAL PROTECTION AGENCY

[40 CFR Part 6]

[FRL 349-7]

NEW SOURCE NPDES PERMITS

Preparation of Environmental Impact Statements

The National Environmental Policy Act of 1969 (NEPA), 42 U.S.C. 4321 et seq., implemented by Executive Order 11514 of March 5, 1970, and the Council on Environmental Quality's (CEQ's) Guidelines of August 1, 1973, requires that all agencies of the Federal Government prepare detailed environmental statements on proposals for legislation and other major Federal actions significantly affecting the quality of the human environment. The objective of the Act is to build into the agency decision-making process an appropriate and careful consideration of all environmental aspects of proposed actions, explain potential environmental effects of proposed actions and their alternatives for public understanding, avoid or minimize adverse effects of proposed actions and restore or enhance environmental quality as much as possible.

Section 511(c)(1) of the Federal Water Pollution Control Act as amended (FWPCA) (Pub. L. 92-500) requires that NEPA apply to the issuance of a permit under section 402 of FWPCA for the discharge of any pollutant by a new source as defined in section 306 of FWPCA. The discharge of a pollutant, as defined in section 502(12) of FWPCA, means an addition of any pollutant to navigable waters, the contiguous zone, or the ocean from any point source.

This proposed regulation provides procedures for applying NEPA to the issuance of new source National Pollutant Discharge Elimination System (NPDES) permits as authorized by § 301 and § 402 of the Federal Water Pollution Control Act as amended. This regulation shall apply only to the issuance of a new source NPDES permit by the U.S. Environmental Protection Agency and not to the issuance of a new source NPDES permit from any State which has an approved NPDES program in accordance with section 402(b) of FWPCA. The regulation, when used in conjunction with the references to 40 CFR Part 125 (the National Pollutant Discharge Elimination System (NPDES)), provides the EPA procedures for processing new source NPDES permit applications. A final regulation will be published after receipt and consideration of the comments. Upon the date of promulgation of this regulation in final form, the FEDERAL REGISTER notice of September 30, 1974, "Requirements for Environmental Assessments" shall no longer be effective. This notice requested that potential new source applicants request a pre-application conference with the appropriate Regional Administrator twenty-four (24) months prior to discharge.

The new source NPDES regulation is published separately from the regulation applying NEPA to EPA's nonregulatory programs which was promulgated in final form in the FEDERAL REGISTER (40 FR 16814) on April 14, 1975. EPA also issued a separate notice in the October 21, 1974, FEDERAL REGISTER (39 FR 37419) which gave Agency procedures for voluntarily preparing EIS's on certain other EPA regulatory activities.

The Environmental Protection Agency invites all interested persons who desire to submit written comments or suggestions concerning the preparation of final regulations to do so in triplicate to the Office of Federal Activities, (A-104), Environmental Protection Agency, Washington, D.C. 20460. Such submissions should be received not later than November 24, 1975, to allow time for appropriate consideration. Copies of the submissions will be available for inspection and copying at the U.S. Environmental Protection Agency, Public Information Reference Unit, Room 2922 (EPA Library), 401 M Street, S.W., Washington, D.C. 20460.

In consideration of the foregoing, it is proposed to amend Chapter I of Title 40 of the Code of Federal Regulations by adding a new Subpart I to Part 6 as set forth below.

Dated: October 1, 1975.

JOHN QUARLES,
Acting Administrator.

Subpart I—Preparation of Environmental Impact Statements on New Source NPDES Permits

Sec.
6.900   Purpose and policy.
6.902   Definitions.
6.904   Administrative activity subject to this part.
6.906   New source determination procedures.
6.908   Procedures for environmental review.
6.910   Guidelines for determining whether to prepare an EIS.
6.912   Draft environmental impact statement.
6.914   Public hearing.
6.916   Final environmental impact statement.
6.918   Decision on the Federal action.
6.920   Additional procedures.
6.922   Availability of documents.

6.924 Content of an environmental impact statement.

AUTHORITY: Sec. 102, 103, 83 Stat. 854 (The National Environmental Policy Act of 1969); Sec. 301, 306, 402, 86 Stat. 816 et seq., (The Federal Water Pollution Control Act as amended).

§ 6.900 Purpose and policy.

(a) The National Environmental Policy Act of 1969 (NEPA), 42 U.S.C. 4321 et seq., implemented by Executive Order 11514 and the Council on Environmental Quality's (CEQ's) Guidelines (40 CFR 1500) requires that all agencies of the Federal Government prepare detailed environmental statements on proposals for legislation and other major Federal actions significantly affecting the quality of the human environment. The objective of NEPA is to include in the agency decision-making process appropriate and careful consideration of all environmental effects of proposed actions, explain potential environmental effects of proposed actions and their alternatives for public understanding, avoid or minimize adverse effects of proposed actions and restore or enhance environmental quality as much as possible.

(b) This part provides procedures for compliance with NEPA in the issuance of new source National Pollutant Discharge Elimination System (NPDES) discharge permits as authorized by section 301 and section 402 of the Federal Water Pollution Control Act as amended, (FWPCA) (33 U.S.C. 1151 et seq.).

(c) All references in this part to Part 125 shall mean Part 125 of Title 40 of the Code of Federal Regulations (CFR).

(d) EPA hereby reserves all odd numbers beginning with § 6.901 et seq. for future modifications and additions.

§ 6.902 Definitions.

(a) The abbreviated term "EPA" means the United States Environmental Protection Agency.

(b) The term "Source," as defined in section 306(a)(3) of FWPCA, means "any building, structure, facility or installation from which there is or may be the discharge of pollutants."

(c) The term "New Source," as defined in section 306(a)(2) of FWPCA, means "any source, the construction of which is commenced after the publication of proposed regulations prescribing a standard of performance under this section which will be applicable to such source, if such standard is thereafter promulgated in accordance with this section." (See Appendix A for guidance.)

(d) The term "Construction," as defined in section 306(a)(5) of FWPCA, means "any placement, assembly, or installation of facilities or equipment (including contractual obligations to purchase such facilities or equipment) at the premises where such equipment will be used, including preparation work at such premises."

(e) The term "Administrative Action" means the issuance by EPA of an NPDES permit to discharge as a new source.

(f) "Responsible Official" means the Regional Administrator of EPA or his designee.

(g) The term "Environmental Assessment" means the report prepared by the applicant for an NPDES permit to discharge as a new source which identifies and analyzes the environmental impacts of the applicant's proposed source and feasible alternatives as provided in § 6.908 of this part.

(h) The term "Environmental Review" means the formal evaluation undertaken by EPA to determine whether a proposed administrative action will be a major Federal action significantly affecting the quality of the human environment.

(i) The term "Environmental Impacts" shall refer to both the adverse and the beneficial impacts associated with a new source.

(j) The term "Notice of Intent" means the written announcement to Federal, State, and local agencies, and to interested persons, that a draft environmental impact statement will be prepared. The notice shall briefly describe the EPA action, its location, and the issues involved. (Exhibit 1.) The purpose of a notice of intent is to involve other government agencies and interested persons as early as possible in the planning and evaluation of actions which may have significant environmental impacts. This notice should encourage public in-

410

put in the preparation of a draft EIS and assure that environmental values will be identified and weighed from the outset, rather than accommodated by adjustments at the end of the decisionmaking process.

(k) The term "Draft Environmental Impact Statement" means the document, prepared by EPA, which attempts to identify and analyze the environmental impacts of a proposed EPA action and feasible alternatives, and is circulated for public comment prior to preparation of the final environmental impact statement.

(l) The term "Final Environmental Impact Statement" means the document prepared by EPA which identifies and analyzes in detail the environmental impacts of a proposed EPA action and incorporates comments made on the draft EIS.

(m) The term "Negative Declaration" means the written announcement, prepared subsequent to the environmental review, which states that EPA has decided not to prepare a draft environmental impact statement. The negative declaration shall describe the proposed project, its location, any potential primary and secondary impacts of the project, and the procedures whereby interested persons may comment on the decision not to prepare an EIS. (Exhibit 2)

(n) The term "Environmental Impact Appraisal" means a document, based on the environmental review, which supports a negative declaration. (Exhibit 3)

(o) The term "New Source and Environmental Questionnaire" means a document which EPA furnishes to a potential new source applicant to obtain information on the status and potential impact of the proposed source.

(p) The term "Interested Persons" means any individuals, Federal or State agencies, conservation groups, organizations, corporations, or other nongovernmental units, including any applicant for a new source NPDES permit, issued by the U.S. Environmental Protection Agency, who may be interested in, affected by, or technically competent to comment on the environmental impact of the proposed action.

(q) The term "Potential New Source Applicant" means the prospective owner or operator of an anticipated point source, as defined in section 502(14) of the FWPCA, who arguably falls within a proposed standard of performance category.

§ 6.904   Administrative activity subject to this part.

(a) This part shall apply solely to the issuance of a new source NPDES permit by the EPA with the following exceptions:

(1) These detailed procedures shall not apply to the issuance of a new source NPDES permit to a Federal facility, as defined in Executive Order 11752 of December 18 1973. The official of any Federal agency making application for an EPA new source NPDES permit shall be responsible for determining whether the Agency's proposed activity necessitating the permit will constitute a major Federal action significantly affecting the quality of the human environment in accordance with its own regulations. Documentation of the Federal agency's determination shall be communicated to EPA prior to EPA's public notice of the issuance of a permit under 40 CFR 125.32.

(2) These detailed procedures shall not apply where another Federal agency has agreed to be "lead agency" or has been designated by the Council on Environmental Quality (CEQ) to be "lead agency" in accordance with the CEQ Guidelines, 40 CFR 1500.7(b). These procedures shall be supplemented by the provisions of an interagency agreement which has been established between EPA and any other Federal agency, or agencies, to designate "lead" and "nonlead" agency responsibilities in the preparation of an environmental impact statement. Prior to the establishment of a lead agency agreement, EPA will assume responsibility for consulting with those Federal agencies that are also responsible for performing a NEPA review on their own Federal actions affecting an applicant who has been determined by EPA to be a new source in order to determine which agency shall be "lead agency."

§ 6.906   New source determination procedures.

(a) Any person who may require an NPDES permit under the FWPCA shall so notify the EPA responsible official having jurisdiction over the area in which the discharge is proposed to be located.

(b) The responsible official, upon receipt of such notice or of his own accord, shall provide any potential new source applicant with the new source and environmental questionnaire (NS/EQ).

(c) The potential new source applicant shall return the completed NS/EQ at least 9 months prior to commencement of construction of the facility, as defined in § 306 of the FWPCA. (It is to the applicant's advantage to return the questionnaire as early as possible, so that if the facility is determined to be a new source, and therefore subject to an environmental review, construction will not be unnecessarily delayed pending completion of the environmental review.)

(d) Upon receipt of the NS/EQ, the responsible official shall make an initial determination of whether the facility is a "new source" (see Appendix A for guidance) unless there is insufficient information to make this determination.

(e) If additional information is needed to make the initial new source determination, the responsible official shall obtain such additional information. The applicant shall provide additional information as requested by the responsible official. The applicant may request confidential treatment of such information in accordance with procedures in 40 CFR 125.37.

(f) If the facility is initially determined to be an existing source, the responsible official shall:

(1) Notify the applicant of this initial determination and of his right to have the initial determination reconsidered at an adjudicatory hearing held pursuant to 40 CFR 125.36.

(2) Provide the applicant with an application for a permit to discharge as an existing source.

(3) Notify the public of such decision no later than the public notice of the issuance of a permit pursuant to 40 CFR 125.32.

(g) If the facility is initially determined to be a new source, the responsible official shall:

(1) Notify the applicant of this initial determination and of his right to have the initial determination reconsidered at an adjudicatory hearing held pursuant to 40 CFR 125.36.

(2) Provide the applicant with an application for a permit to discharge as a new source.

(3) Notify the public of such decision no later than the public notice of the issuance of a permit pursuant to 40 CFR 125.32.

(4) Notify the applicant that he must submit an adequate environmental assessment unless the responsible official determines that the new source and environmental questionnaire is an adequate environmental assessment. A suggested format for the contents of the environmental assessment is found in § 6.924(c) of this Part.

(h) If the applicant or any interested person, within 20 days of the date of mailing the notice of initial determination, requests an adjudicatory hearing, the responsible official shall act upon the request for the adjudicatory hearing in accordance with procedures prescribed in 40 CFR 125.36.

(i) If no hearing is requested in accordance with (h) above, the initial new source determination of the responsible official shall become the final new source determination of EPA.

§ 6.908 Procedures for environmental review.

(a) If EPA's final new source determination under § 6.906 is that the facility is a new source, the responsible official shall conduct the environmental review to determine whether the issuance of the permit is likely to have significant impact on the quality of the human environment, whether any feasible alternatives can be adopted or changes can be made in project design to eliminate or minimize significant adverse impacts, and whether an EIS or a negative declaration is required.

(b) The responsible official shall base his decision on the need for preparing an EIS on the guidelines in § 6.910 of this Part.

(c) The responsible official may require that the applicant submit environmental assessment information in addition to the NS/EQ containing the additional information that the responsible official deems necessary to conduct the environmental review. The responsible official shall determine the proper scope of the environmental review and the applicant's environmental assessment and shall specify to the applicant what information is required. In determining the scope of the environmental assessment, the responsible official shall consider the size of the new source, the potential environmental impacts of the new source, and the extent to which the applicant or his designee is capable of providing the required information. The responsible official shall not require the applicant to gather raw data or to perform analyses either of which duplicate existing data or the results of existing analyses available to EPA. The responsible official shall keep requests for data to a minimum consistent with his responsibilities under NEPA.

(d) If the environmental review reveals that the preparation of an environmental impact statement is required, the responsible official may require reports, data and other information for the EIS to be compiled by the applicant

412

or a third party under contract with the applicant and furnished directly to the responsible official. In all cases, the responsible official shall specify the type of information to be developed and shall maintain control of the information throughout the gathering and presentation of this information. The responsible official shall keep requests for data to a minimum consistent with his responsibilities under NEPA. When the third party approach is taken, the responsible official shall approve the selection of this third party contractor after consulting with interested Federal, State, and local agencies, public interest groups, and members of the general public as he deems appropriate to assure objectivity in this selection.

(e) Upon completion of the environmental review, the responsible official shall make known his determination regarding the need for a draft EIS. If a draft EIS is to be prepared and circulated, the responsible official shall issue a notice of intent (Exhibit 1); if the determination is made not to prepare a draft EIS, the responsible official shall issue a negative declaration (Exhibit 2).

(1) Such notice of intent shall be issued prior to the public notice of the issuance of a permit under 40 CFR 125.-32. Such negative declaration shall be issued prior to or simultaneously with the public notice of the issuance of a permit under 40 CFR 125.32.

(2) Such notice of intent or negative declaration shall be distributed in accordance with procedures described in 40 CFR 125.32(a). Potentially appropriate agencies referred to in 40 CFR 125.32 (a) are found in the Council on Environmental Quality's Guidelines, 40 CFR 1500, Appendices II and III. Additional distribution procedures are provided in Appendix B.

(3) Any negative declaration shall state that interested persons wishing to comment on the decision may submit comments for consideration by the responsible official.

(4) For any negative declaration, the responsible official shall prepare an environmental impact appraisal which states EPA's reasons for concluding that there will be no significant impact resulting from the issuance of the applicable new source NPDES permit or that significant adverse impacts have been mitigated by making changes in the proposed new source. (Exhibit 3). This document shall briefly describe the proposed action and feasible alternatives, environmental impacts of the proposed new source, steps to minimize harm to the environment, the relationship between short term uses of man's environment and the maintenance and enhancement of long term beneficial uses, the irreversible and irretrievable commitments of resources for the new source, comments and consultations on the new source and reasons for concluding there will be no significant adverse impacts. The environmental impact appraisal shall be available for public inspection at the time of the issuance of the negative declaration and shall remain with the internal records of the permit.

§ 6.910 Guidelines for determining whether to prepare an EIS.

The following guidelines shall be used when performing the environmental review:

(a) *General guidelines.* (1) When determining the significance of a proposed new source's impact, the responsible official shall consider both its short term and long term effects as well as its primary and secondary effects as defined in § 6.924(c). However, EIS's should be prepared first on those proposed actions with the most adverse effects which are scheduled for earliest implementation and on other proposed actions according to priorities assigned by the responsible official.

(2) If EPA is proposing to issue a number of minor, environmentally insignificant new source NPDES permits, during a limited time span and in the same general geographic area the responsible official may determine that the cumulative impact of the issuance of all these permits may have a significant environmental effect.

(3) In determining the significance of a proposed new source NPDES permit, the unique characteristics of the new source area should be carefully considered. For example, proximity to historic sites, parklands, wetlands or wild and scenic rivers may make the impact significant.

(b) *Specific criteria.* An EIS will be prepared when: (1) The new source will induce or accelerate significant changes in industrial, commercial, agricultural, or residential land use concentrations or distributions which have the potential for significant environmental effects. Factors that should be considered in determining if these changes are environmentally significant include but are not limited to: the nature and extent of the vacant land subject to increased develop-

413

ment pressure as a result of the new source; the increases in population or population density which may be induced and the ramifications of such changes; the nature of land use regulations in the affected area and their potential effects on development and the environment; and the changes in the availability or demand for energy and the resulting environmental consequences.

(2) The new source may directly or through induced development have a significant adverse effect upon local ambient air quality, local ambient noise levels, surface or groundwater quality or quantity, fish, wildlife, and their natural habitats.

(3) Any major part of the new source will be located on wetlands or will have significant adverse effects on wetlands, including secondary effects.

(4) Any major part of the new source will be located on or significantly affect the habitat of threatened or endangered species on the Department of Interior's lists of threatened and endangered species.

(5) The environmental impact of the issuance of new source NPDES permit is likely to be highly controversial.

(6) The environmental impact of the issuance of a new source NPDES permit will have significant direct and adverse effect on a property listed in or eligible for listing in the National Register of Historic Places or will cause irreparable loss or destruction of significant scientific, prehistoric, historic or archaeological data.

§ 6.912 Draft environmental impact statement.

(a) The responsible official shall assure that a draft environmental impact statement is prepared as soon as practicable after the release of the notice of intent. The draft EIS shall be published not later than the publication of public notice of the issuance of a permit pursuant to 40 CFR 125.32.

(b) The content of the draft EIS shall be as specified according to § 6.924 of this Part.

(c) The specific procedures that should be taken with respect to distribution and availability of the draft EIS's are listed in Appendix B.

(d) Parties who wish to comment have at least forty-five (45) days to reply after the date of publication in the FEDERAL REGISTER of the listing of the draft EIS by CEQ.

§ 6.914 Public hearing.

(a) If there is a significant degree of public interest, the responsible official may convene a public hearing after publication and circulation of the draft EIS. He shall issue public notice of such hearing in accordance with 40 CFR 125.32(d). The public hearing shall be conducted in accordance with 40 CFR 125.34.

(b) In addition to the procedures provided in § 6.914(a), the following shall also apply:

(1) If the responsible official determines, prior to publication and distribution of the draft EIS, that a public hearing shall be held, he shall place such notice of such hearing in the draft EIS following the summary sheet.

(2) A written record of the hearing shall be made. As a minimum, the record shall contain a list of witnesses together with the text of each presentation. A summary of the record including the issues raised, conflicts resolved and any other significant portions of the record shall be appended to the final EIS.

§ 6.916 Final environmental impact statement.

(a) The responsible official shall prepare a final environmental impact statement, which shall contain responses to substantive comments received on the draft EIS, a summary of the record of any public hearing, and any other relevant information.

(b) The final EIS shall be published not later than the responsible official's determination containing the proposed permit pursuant to 40 CFR 125.35.

(c) The final EIS shall include the responsible official's recommendation on whether the permit is to be issued or denied.

(1) If the recommendation is to deny the permit, the final EIS shall contain the reason(s) for such a recommendation and the measures that EPA recommends the applicant take in order to receive a permit.

(2) If the recommendation is to issue the permit, the final EIS shall, when appropriate, also recommend the actions the permittee shall take to prevent or minimize any adverse environmental impacts identified in the analysis.

(d) The specific procedures that should be followed with respect to the distribution and availability of the final EIS are provided in Appendix B.

(e) In addition to the requirements defined in 40 CFR 125.35, no administrative action shall be taken by EPA until thirty (30) days after the publication of the final EIS and not until a minimum of ninety (90) days after the publication of the draft EIS.

§ 6.918   Decision on the Federal action.

The responsible official may approve or deny the new source NPDES permit following a complete evaluation of any significant beneficial and adverse environmental impacts on the human environment consistent with EPA's legal authority, including, but not limited to the Federal Water Pollution Control Act (33 U.S.C. 1151 et seq.), the National Environmental Policy Act of 1969 (42 U.S.C. 4321 et seq.), the Clean Air Act of 1970 (42 U.S.C. 1857 et seq.), Solid Waste Disposal Act (42 U.S.C. 3254 et seq.), the Federal Insecticide, Fungicide, and Rodenticide Act (7 U.S.C. 136 et seq.), the 1954 Atomic Energy Act as amended (42 U.S.C. 201 et seq.), and the Safe Drinking Water Act of 1974 (42 U.S.C. 300f).

§ 6.920   Additional procedures.

(a) Historic and archaeological sites. EPA is subject to the requirements of § 106 of the National Historic Preservation Act of 1966, 16 U.S.C. 470 et seq., Executive Order 11593 and the Archaeological and Historic Preservation Act of 1974, 16 U.S.C. 469 et seq., and the regulations promulgated thereunder. These statutes and regulations establish environmental review procedures to follow independently of the requirements of NEPA.

(1) If the new source may affect properties with historic, architectural, archaeological or cultural value which are listed in or eligible for listing in the National Register of Historic Places (published in the FEDERAL REGISTER each February with supplements on the first Tuesday of each month), the responsible official shall comply with the procedures of the Advisory Council on Historic Preservation (36 CFR 800) including determining the need for a memorandum of agreement among EPA, the Advisory Council, and the State Historic Preservation Officer.

(2) Whenever a memorandum of agreement has been executed in accordance with 36 CFR 800, it shall be included in the EIS if one is prepared on that new source NPDES permit. Copies of the draft and final EIS's should be sent to the appropriate State Historic Preservation Officer and the Executive Director of the Advisory Council on Historic Preservation for their comment according to the Advisory Council's procedures (36 CFR 800).

(3) In order to adequately complete his environmental review and his responsibilities under 36 CFR 800, the responsible official may request that the applicant for a new source NPDES permit consult with the State Historic Preservation Officer to determine if the new source will have a significant adverse effect on properties with historic, architectural, archaeological or cultural value which are listed in or eligible for listing in the National Register of Historic Places. If the new source will not have an adverse effect, the applicant may be requested to submit a determination of no-effect in a memorandum to the responsible official in accordance with 36 CFR 800.4(c). If the new source will have an adverse effect, the applicant may be requested by the responsible official to work with the State Historic Preservation Officer to develop alternatives to avoid or mitigate the adverse effect(s). The responsible official may request further assistance of the new source NPDES applicant in order to comply with EPA's requirements under 36 CFR 800 prior to the responsible official's determination containing the proposed permit pursuant to 40 CFR 125.35.

(4) If the new source may cause irreparable loss or destruction of significant scientific, prehistoric, historic or archaeological data, the responsible official shall consult with the Secretary of Interior in compliance with the Archaeological and Historic Preservation Act of 1974, 6 U.S.C. 469.

(b) Wetlands, coastal zones, floodplains, fish and wildlife, threatened and endangered species, and wild and scenic rivers. The following procedures shall be applied to the EPA administrative activities covered by this part that may affect these environmentally sensitive areas.

(1) If the new source may affect wetlands, the responsible official shall consult with the appropriate offices of the Department of the Interior, the Department of Commerce, U.S. Army Corps of Engineers, and the states involved, during the environmental review to determine the probable impact of the new source on the fish and wildlife resources and land use of these areas.

(2) If the new source may affect coastal zones or coastal waters as defined in Title III of the Coastal Zone Management Act of 1972, 16 U.S.C. 1451 et seq., the responsible official shall consult with

415

the appropriate State offices and with the appropriate office of the Department of Commerce during the environmental review to determine the probable impact of the new source on coastal zone or coastal water resources.

(3) If the proposed new source will encourage new industrial, commercial, and residential development in currently undeveloped floodplains which are of significant value for agricultural production, recreation, or wildlife habitat, the responsible official shall act pursuant to Executive Order 11296.

(4) If the new source may affect portions of rivers designated wild and scenic or being considered for this designation under the Wild and Scenic Rivers Act, 16 U.S.C. 28, the responsible official shall consult with appropriate State offices and with the Secretary of the Interior, or where national forest lands are involved, with the Secretary of Agriculture, during the environmental review to determine the probable impact of the new source on eligible rivers or portions thereof.

(5) Whenever the new source will result in the control or structural modification of any stream or other body of water for any purpose, including navigation and drainage, the responsible official shall consult with the United States Fish and Wildlife Service (Department of the Interior), the National Marine Fisheries Service of the National Oceanic and Atmospheric Administration (Department of Commerce), the U.S. Army Corps of Engineers, and the head of the agency administering the wildlife resources of the particular state in which the action will take place, to determine any steps which may be taken to conserve wildlife resources.

(6) If the new source may affect threatened or endangered species, defined under section 4 of the Endangered Species Act of 1973, 16 U.S.C. 35, the responsible official shall consult with the Secretary of the Interior or the Secretary of Commerce according to the procedures of section 7 of that Act.

(7) Requests for consultation and the results of such consultation shall be documented in writing. The agencies should be given thirty (30) days to comment as measured from the date of the written request. If an EIS is to be prepared on a new source and wetlands, coastal zones, floodplains, fish and wildlife, threatened or endangered species or wild and scenic rivers may be affected, the required consultation may be deferred until the preparation of the draft EIS. In all cases where consultation has occurred, the

agencies consulted shall receive copies of either the notice of intent and EIS or the negative declaration and environmental appraisal prepared on the proposed action.

§ 6.922 Availability of documents.

(a) EPA will print copies of draft and final EIS's for agency and public distribution. A nominal fee may be charged for copies requested by the public.

(b) When EPA no longer has copies of an EIS to distribute, copies shall be made available for public inspection at regional and headquarters Offices of Public Affairs. Interested persons also should be advised of the availability (at cost) of the EIS from the Environmental Law Institute, 1346 Connecticut Avenue, N.W., Washington, D.C. 20036.

(c) Lists of EIS's prepared or under preparation and lists of negative declarations prepared will be available at both the regional and headquarters Offices of Public Affairs.

§ 6.924 Content of an environmental impact statement.

(a) *Cover sheet.* The cover sheet shall indicate the type of EIS (draft or final), the nature of the proposed EPA action, the name of the permit applicant, the responsible EPA office, the date, and the signature of the responsible official. The format is shown in Exhibit 4.

(b) *Summary sheet.* The summary sheet shall conform to the format prescribed in Appendix I of the August 1, 1973 Council on Environmental Quality's Guidelines (40 CFR 1500). The format is shown in Exhibit 5.

(c) *Body of statement.* The body of the EIS shall identify, develop, and analyze the pertinent issues included in the seven sections below. Each section need not be a separate chapter in the statement. The EIS shall serve as a means for the responsible official and the public to assess the environmental impacts of the proposed issuance of a new source NPDES permit, rather than as a justification for decisions already made. Environmental impact statements should be prepared using a systematic, interdisciplinary approach. Statements should incorporate all relevant analytical disciplines and should provide meaningful and factual data, information, and analyses. The presentation should be simple and concise, yet include all facts necessary to permit independent evaluation and appraisal of the beneficial and adverse environmental effects on the human environment of alternative actions. The amount of detail provided

416

should be commensurate both with the extent and expected impact of the actions, and with the amount of information required at the particular level of decisionmaking. To the extent possible, statements shall not be drafted in a style which requires extensive scientific or technical expertise to comprehend and evaluate the environmental impact of the proposed EPA action.

(1) Background and description of the proposed new source. The EIS shall describe the proposed source, its product or purpose, its location, its construction and operation time schedule. To prevent piecemeal decision making, the new source should be described in as broad a context as necessary. The relationship of the proposed new source project to other projects and proposals directly affected by or stemming from the construction and the operation of the new source shall be discussed, including not only other EPA activities, but also those of other Governmental and private organizations. Development and population trends in the project area and the assumptions on which they are based shall also be included. Maps, photos, and artist sketches should be incorporated if available when they help depict the environmental setting. If not enclosed, supporting documents should be referenced.

(2) Alternatives available to the proposed new source. The feasible alternatives available to the proposed new source shall be described, developed and objectively weighed against the proposed new source. The analysis should be sufficiently detailed to reveal the EPA's comparative evaluation of the environmental impacts on the human environment, costs, and risks of each feasible alternative. The analysis of alternatives shall include the alternative of not constructing or operating the new source or postponing construction or operation. Feasible design, process, and site alternatives must be described. This analysis should be written in such a manner that the general public independently can judge the relative desirability of the various alternatives.

(3) Environmental impacts of the proposed new source. This shall be a description of the primary and secondary environmental impacts, both beneficial and adverse, anticipated from the new source. The scope of the description shall include both short and long-term impacts. Emphasis should be given to discussing those factors most directly impacted by the proposed activity.

(i) Primary impacts are those that can be attributed directly to the construction or operation of the new source.

(ii) Secondary impacts are indirect or induced impacts. Construction of a facility such as a large industrial facility may stimulate or induce secondary effects in the form of associated investments and changed patterns of social and economic activities. Particular attention should be paid to potential changes in population patterns or growth. When such changes are significant, their effect on the resource base, including land use, water quality and quantity and air quality should be determined. A discussion of how these impacts conform or conflict with the objectives and specific terms of approved or proposed Federal, State, and local land use plans, policies, and controls for the area should be included.

(4) Adverse impacts which cannot be avoided should the new source permit be issued. The EIS shall describe the kinds and magnitudes of adverse impacts which cannot be reduced in severity, give the remedial and protective measures which shall be taken, describe the adverse impacts which can be reduced to an acceptable level, and the mitigative measures which should be taken. These adverse impacts may include water or air pollution, undesirable land use patterns, damage to ecological systems, urban congestion, threats to health or other consequences adverse to the environmental goals set out in section 101(b) of the National Environmental Policy Act.

(5) Relationship between local short term uses of the environment and the maintenance and enhancement of long term beneficial uses. This shall be a description of the extent to which the proposed activity involves trade offs between short term environmental gains at the expense of long term losses, or vice-versa, and the extent to which the proposed action forecloses future options. Special attention should be given to effects which narrow the range of beneficial uses of the environment or pose long term risks to health or safety.

(6) Irreversible and irretrievable commitment of resources which would result if a new source permit were issued. This shall be a description of the extent to which the proposed activity curtails the diversity and range of beneficial uses of the environment. Secondary impacts, such as induced growth in undeveloped areas, may make alternative uses of that land impossible. Also, irreversible damage can result from environmental acci-

dents associated with the new source and this possibility should be evaluated.

(7) A discussion of problems and objections raised by other Federal, State, and local agencies and by interested persons in this review process. Final EIS's (and draft EIS's if appropriate) shall summarize the comments and suggestions made by reviewing organizations and shall describe the disposition of issues raised, e.g., changes to the proposed new source to mitigate anticipated impacts or objections. In particular, the EIS shall address any major issues in which the EPA position differs from reviewers' recommendations and objections, giving reasons why specific comments and suggestions could not be adopted. Reviewers' statements should be set forth in a list of "comments" and accompanied by EPA's "responses." In addition, the source of all comments should be clearly identified and copies of the comments (or summaries where a response has been exceptionally long) should be attached to the final EIS.

(d) *Documentation.* Any books, research reports, field study reports, correspondence and other documents which provided the data base for evaluating the impact of the proposed new source and alternatives discussed in the EIS shall be cited in the body of the EIS and included in a bibliography attached to the EIS.

EXHIBIT 1

NOTICE OF INTENT TRANSMITTAL MEMO-
RANDUM—SUGGESTED FORMAT

(Date)

ENVIRONMENTAL PROTECTION AGENCY,

(Appropriate Office)

(Address, City, State, Zip Code)

To All Interested Government Agencies and Public Groups.

Gentlemen: As required by the EPA regulations, "Preparation of Environmental Impact Statements (EIS's) for New Source NPDES Permits" (40 CFR 6.900), attached is a Notice of Intent to prepare an EIS for the proposed EPA action described below:

(Nature of EPA Action and NPDES Permit Application Number)

(Name of Applicant and Nature of Project)

(City, County, State)

If your organization needs additional information or wishes to participate in the preparation of the draft EIS, please advise the (appropriate office, city, state).

Very truly yours,

(Appropriate EPA Official)

(List Federal, State, and local agencies to be solicited for comment.)

(List public action groups to be solicited for comment.)

NOTICE OF INTENT—SUGGESTED FORMAT

NOTICE OF INTENT—ENVIRONMENTAL
PROTECTION AGENCY

1. Proposed EPA Action:

----

2. Type of Facility:

----

3. Location of Facility:
City --------------------------------
County ------------------------------
State ------------------------------

4. Issues Involved:

----

5. Proposed Starting Date of Discharge:

EXHIBIT 2

NEGATIVE DECLARATION—SUGGESTED FORMAT

NEGATIVE DECLARATION

(Date)

ENVIRONMENTAL PROTECTION AGENCY,

(Appropriate Office)

(Address, City, State, Zip Code)

To All Interested Government Agencies and Public Groups.

Gentlemen: As required by the EPA regulations, "Preparation of Environmental Impact Statements (EIS's) for New Source NPDES Permits" (40 CFR 6.900), an environmental review has been performed on the proposed EPA action below:

(Name of Applicant and Type of Facility)

(Facility Location: City, County, State)

(Nature of EPA Action)

(NPDES Permit Application Number)

Project Description, Originator and Purpose
(Include a map of the project area and a brief narrative describing the primary and secondary impacts of the project, purpose of the project, and other data in support of the negative declaration.)

The review process did not indicate significant environmental impacts would result from the proposed action, or that significant adverse impacts have been mitigated by making changes in the project. Consequently, a preliminary decision not to prepare an EIS has been made.

This action is taken on the basis of a careful review of the environmental assessment, and other supporting data, which are on file in the above office and will be available for public review upon request.

Comments on this decision may be submitted for consideration by EPA. After evaluating the comments received, the Agency will make a final decision on the need for an EIS.

Sincerely,

(Appropriate EPA Official)

EXHIBIT 3

ENVIRONMENTAL IMPACT APPRAISAL—
SUGGESTED FORMAT

A. Identity Project:
Name of Applicant_____
Type of Facility_____
. Address _____
B, Summarize Assessment:
1. Brief description of the facility:_____
----
2. Probable impact of the issuance of an NPDES New Source permit on the environment: _____

----
3. Any probable adverse environmental effects which cannot be avoided:_____
----
4. Alternatives considered with evaluation of each:_____

----
5. Relationship between local short-term uses of the environment and maintenance and enhancement of long-term beneficial uses: _____

----
6. Any irreversible and irretrievable commitment of resources:_____
----
7. Public objections to the facility, if any, and their resolution:_____
----
8. Agencies consulted about the facility:__

State representative's name_____
Local representative's name_____
Other _____
C. Reasons for concluding there will be no significant impacts.
(Discuss topics 2, 3, 5, 6, and 7 above, and how the alternative (topic 4) selected is the most appropriate.

(Signature of appropriate official)

                            (Date)
EXHIBIT 4

COVER SHEET FORMAT FOR ENVIRONMENTAL
IMPACT STATEMENTS (DRAFT, FINAL)

Environmental Impact Statement

(Provide Name of Facility and Type of EPA Action)

(Provide Identifying NPDES Permit Application Number)

Prepared by _____
          (Responsible Agency Office)
Approved by _____
          (Responsible Agency Official)

                            (Date)
EXHIBIT 5

SUMMARY SHEET FORMAT FOR ENVIRONMENTAL
IMPACT STATEMENTS

(Check one)
( )   Draft.
( )   Final.

Environmental Protection Agency

(Responsible Agency Office)
1. Name of action. (Check one)
( )   Administrative action.
( )   Legislative action.
2. Brief description of action indicating what States (and counties) are particularly affected.
3. Summary of environmental impact and adverse environmental effects.
4. List alternatives considered.
5. a. (for draft statements) List all Federal, State, and local agencies from which comments have been requested.
b. (for final statements) List all Federal, State, and local agencies and other sources from which written comments have been received.
6. Dates draft statement and final statement made available to Council on Environmental Quality and public.

EXHIBIT 6

PUBLIC NOTICE AND NEWS RELEASE—
SUGGESTED FORMAT

PUBLIC
NOTICE

The Environmental Protection Agency (originating office) (will prepare, will not prepare, has prepared) a (draft, final) environmental impact statement on the following project:

(Name of Applicant and Type of Facility)

(Nature of EPA Action)

(Facility Location, City, County, State)

(Where EIS or Negative Declaration can be obtained)

This notice is to implement EPA's policy of encouraging public participation in the decision-making process on proposed EPA actions. Comments on this document may be submitted to (full address of originating office).

419

GUIDANCE ON DETERMINING A NEW SOURCE

(1) A source should be considered a new source provided that at the time of proposal of the applicable new source standard of performance, there has not been any:

(i) Significant site preparation work, such as major clearing or excavation; or

(ii) Placement, assembly, or installation of unique facilities or equipment at the premises where such facilities or equipment will be used; or

(iii) Contractual obligation to purchase such unique facilities or equipment. Facilities and equipment shall include only the major items listed below, provided that the value of such items represents a substantial commitment to construct the facility:

(a) structures; or
(b) structural materials; or
(c) machinery; or
(d) process equipment; or
(e) construction equipment.

(iv) Contractual obligation with a firm to design, engineer and erect a completed facility (i.e., a "turnkey" plant).

(2) Modifications to existing sources will be controlled through the permit modification procedures. A new source is a totally new source (i.e., all of which has yet to be constructed), or a major alteration to an existing source. A major alteration will be considered a new source if the alteration is of such magnitude to, in effect, create a new facility. In making such a determination, the responsible official shall find that the permit modification procedures are not appropriate and shall consider, among other relevant factors, whether as a result of the alteration, the source can reasonably achieve the standard of performance. (Only those portions of a facility determined to be a new source shall be required to achieve the Standard of Performance promulgated under Section 306 of the FWPCA.)

Appendix B

DOCUMENT DISTRIBUTION AND AVAILABILITY PROCEDURES

I. Distribution of Documents—Suggested Guidance

(a) The responsible official should distribute notices of intent and negative declarations according to procedures listed in 40 CFR 125.32(a) and as follows:

(1) The Office of Federal Activities (one copy).

(2) The Office of Public Affairs (one copy).

(3) The Office of Legislation (one copy).

(4) The Office of Enforcement (one copy).

(5) A brief news release may be submitted to a local newspaper, which has adequate circulation to cover the area that will be affected by the proposed facility, informing the public that an impact statement will be or will not be prepared on a particular project and that the agency is requesting public comment (see Exhibit 3).

(b) Draft environmental impact statements. The specific procedures that should be taken with respect to draft environmental impact statements are as follows:

(1) Before transmitting the draft statement to the Council on Environmental Quality, the responsible official should:

(i) Notify by phone the Office of Federal Activities (OFA) that the draft impact statement has been prepared.

(ii) Send two (2) copies of the draft statement to the Office of Federal Activities (OFA) for their review and comment. OFA may seek assistance from other Agency components to provide their review and comment on all or individual environmental impact statements.

(2) If neither OFA nor one of the offices requested by OFA for comment requests any changes within a ten (10) working day period after notification, the responsible official should:

(i) Send five (5) copies of the draft environmental impact statement to the Council on Environmental Quality.

(ii) Inform the Office of Public Affairs of the transmittal to the Council on Environmental Quality and the plans for local press release.

(iii) Notify the Office of Legislation of the transmittal.

(3) The responsible official should provide copies of the draft statement to:

(i) The appropriate offices of reviewing Federal agencies that have special expertise or jurisdiction by law with respect to any environmental impacts. The Council on Environmental Quality's Guidelines (40 CFR 1500.9 and Appendices II and III thereof) list those potential agencies to which draft EIS's may be sent for official review and comment. Two (2) copies of the impact statement should be provided each agency unless they have made a specific request for more copies. The agencies are expected to reply directly to the originating EPA office. Commenting agencies shall have at least forty-five (45) calendar days to reply (the reply period shall commence from the date of publication in the FEDERAL REGISTER of lists of statements received by the Council on Environmental Quality); thereafter, it should be presumed that, unless a time extension has been requested, the agency has no comment to make. EPA may grant extensions where practical of fifteen (15) or more calendar days.

(ii) The Office of Legislation if they request copies (two copies).

(iii) The Office of Public Affairs (two copies).

(iv) The Office of Enforcement (two copies).

(v) The Office of Federal Activities (two copies).

(4) The appropriate State and local agencies and to the appropriate State and metropolitan clearinghouses. The time limits for review and extensions should be the same as those available to Federal agencies.

(5) Interested persons and public libraries. The time limits for review and extensions

should be the same as those available to Federal agencies.

(c) The responsible official should submit to the local newspapers and other appropriate media a news release (see Exhibit 6 of this Part) that the draft statement is available for comment and where copies may be obtained.

(d) Final environmental impact statements. Distribution and other specific actions will be as specified for draft statements. In the case of Federal and State agencies and interested persons, only those who made substantive comments on the draft statement or request a copy of the final statement should be sent a copy. The applicant should be sent a copy. Where the number of comments on the draft statement is such that distribution of the final statement to all commenting entities appears impracticable, the responsible official preparing the statement should consult with the OFA, who will discuss with the Council on Environmental Quality alternative arrangements for distribution of the statement.

II. Availability of Documents

Draft and final EIS's, negative declarations and environmental impact appraisals should be made available for public review at the following locations:

(1) The originating office; (2) Public libraries within the project area. Post offices, city halls or courthouses may be used as distribution points if public library facilities are not available; (3) The Office of Public Affairs for draft and final EIS's only.

[FR Doc.75–26945 Filed 10–8–75;8:45 am]

§ 51.20, 10 CFR PART 51,
APPLICANT'S ENVIRONMENTAL REPORT–
CONSTRUCTION PERMIT STAGE

(a) Environmental considerations. Each applicant [1] for a permit to construct a production or utilization facility covered by § 51.5(a) shall submit with its application a separate document, entitled "Applicant's Environmental Report–Construction Permit Stage," which contains a description of the proposed action, a statement of its purposes, and a description of the environment affected, and which discusses the following considerations:

(1) The probable impact of the proposed action on the environment;

(2) Any probable adverse environmental effects which cannot be avoided should the proposal be implemented;

(3) Alternatives to the proposed section;

(4) The relationship between local short-term uses of man's environment and the maintenance and enhancement of long-term productivity; and

(5) Any irreversible and irretrievable commitments of resources which would be involved in the proposed action should it be implemented. The discussion of alternatives to the proposed action required by paragraph (a) (3) shall be sufficiently complete to aid the Commission in developing and exploring, pursuant to section 102(2) (D) of NEPA, "appropriate alternatives . . . in any proposal which involves unresolved conflicts concerning alternative uses of available resources."

(b) Cost-benefit analysis. The Environmental Report required by paragraph (a) shall include a cost-benefit analysis which considers and balances the environmental effects of the facility and the alternatives available for reducing or avoiding adverse environmental effects, as well as the environmental, economic, technical and other benefits of the facility. The cost-benefit analysis shall, to the fullest extent practicable, quantify the various factors considered. To the extent that such factors cannot be quantified, they shall be discussed in qualitative terms. The Environmental Report should contain sufficient data to aid the Commission in its development of an independent cost-benefit analysis.

(c) Status of compliance. The Environmental Report required by paragraph (a) shall include a discussion of the status of compliance of the facility with applicable environmental quality standards and requirements (including, but not limited to, applicable zoning and land-use regulations and thermal and other water pollution limitations or requirements promulgated or imposed pursuant to the Federal Water Pollution Control Act) which have been imposed by Federal, State, regional, and local agencies having responsibility for environmental protection. The discussion of alternatives in the Report shall include a discussion whether the alternatives will comply with such applicable environmental quality standards and requirements. The environmental impact of the

---

[1] Where the "applicant," as used in this part, is a Federal agency, different arrangements for implementing NEPA may be made, pursuant to the Guidelines established by the Council on Environmental Quality.

Table S–3. —Summary of environmental considerations for uranium fuel cycle
[Normalized to model LWR annual fuel requirement]

Natural resource use	Total	Maximum effect per annual fuel requirement of model 1,000 MWe LWR
Land (acres):		
Temporarily committed	63	
Undisturbed area	45	
Disturbed area	18	Equivalent to 90 MWe coal-fired powerplant.
Permanently committed	4.6	
Overburden moved(millions of MT)	2.7	Equivalent to 90 MWe coal-fired powerplant.
Water (millions of gallons):		
Discharged to air	156	$\approx$ 2 percent model 1,000 MWe LWR with cooling tower.
Discharged to water bodies	11,040	
Discharged to ground	123	
Total	11,319	$<$4 percent of model 1,000 MWe LWR with once-through cooling.
Fossil fuel:		
Electrical energy (thousands of MW- hour)	317	$<$5 percent of model 1,000 MWe LWR output.
Equivalent coal (thousands of MT)	115	Equivalent to the consumption of a 45 MWe coal-fired powerplant.
Natural gas (millions of sci)	92	$<$0.2 percent of model 1,000 MWe energy output.
Effluents–chemical (MT):		
Gases (including entrainment): [1]		
SO_{4_2}	4,400	
NO_3^2	1,177	Equivalent to emissions from 45 MWe coal-fired plant for a year.
Hydrocarbons	13.5	
CO	28.7	
Particulates	1,156	
Other gases:		
F^-	.72	Principally from UF_6 production enrichment and reprocessing. Concentration within range of state standards–below level that has effects on human health.
Liquids:		
SO_4^-	10.3	From enrichment, fuel fabrication, and reprocessing steps.
NO_3^-	26.7	Components that constitute a potential for adverse environmental effect are present in dilute concentrations
Fluoride	12.9	and receive additional dilution by receiving bodies of
Ca^{++}	5.4	water to levels below permissible standards. The con-
Cl^-	8.6	stitutents that require dilution and the flow of dilution
Na^+	16.9	water are:
NH_3	11.5	NH_3–600 cfs.
Fe	.4	NO_3–20 cfs.
		Fluoride–70 cfs.
Tailings solutions (thousands of MT)	240	From mills only–no significant effluents to environment.
Solids	91,000	Principally from mills–no significant effluents to environment.
Effluents-radiological (curies):		
Gases (including entrainment):		
Rn-222	75	Principally from mills–maximum annual dose rate $<$ 4
Ra-226	.02	percent of average natural background within 5 mi of
Th-230	.02	mill. Results in 0.06 man-rem per annual fuel require-
Uranium	.032	ment.
Tritium (thousand)	16.7	Principally from fuel reprocessing plants–Whole body
Kr-85 (thousands)	850	dose is 6 man-rem per annual fuel requirements for
I-129	.0024	population within 50 mi radius. This is $<$0.007 percent
I-131	.024	of average natural background dose to this population.
Fission products and transuranics	1.01	Release from Federal Waste Repository of 0.005 Ci/yn has been included in fission products and transuranics total.
Liquids:		
Uranium and daughters	2.1	Principally from milling–included in tailings liquor and returned to ground–no effluents; therefore, no effect on environment.
Ra-226	.0034	From UF production-concentration 5 percent of 10 CFR
Th-230	.0015	20 for total processing of 27.5 model LWR annual fuel requirements.
Th-234	.01	From fuel fabrication plants–concentration 10 percent of 10 CFR 20 for total processing 26 annual fuel requirements for model LWR.
Ru-106 [3]	.13	From reprocessing plants–maximum concentration 4 per-
Tritium (thousands)	2.5	cent of 10 CFR 20 for total reprocessing of 26 annual fuel requirements for model LWR.
Solids (buried):		
Other than high level	601	All except 1 Ci comes from mills–included in tailings returned to ground–no significant effluent to the environment. 1 Ci from conversion and fuel fabrication is buried.
*Effluents-thermal (billions of British thermal units)	3,360	7 percent of model 1,000 MWe LWR.
Transportation (man-rem): Exposure of workers and general public.	.834	

[1] Estimated effluents based upon combustion of equivalent coal for power generation.
[2] 1.2 percent from natural gas use and process.
[3] Ca-187 (0.075 Cl/AFR) and Sr-90 (0.004 Cl/AFR) are also emitted.

* Amended 40 FR 31593.

424

facility and alternatives shall be fully discussed with respect to matters covered by such standards and requirements irrespective of whether a certification or license from the appropriate authority has been obtained (including, but not limited to, any certification obtained pursuant to section 401 of the Federal Water Pollution Control Act[2]). Such discussion shall be reflected in the cost-benefit analysis prescribed in paragraph (b). While satisfaction of Commission standards and criteria pertaining to radiological effects will be necessary to meet the licensing requirements of the Atomic Energy Act, the cost-benefit analysis prescribed in paragraph (b) shall, for the purposes of NEPA, consider the radiological effects, together with the other effects, of the facility and alternatives.

(d)   The information submitted pursuant to paragraphs (a)–(c) of this section should not be confined to data supporting the proposed action but should include adverse data as well.

(e)   In the Environmental Report required by paragraph (a) for light-water-cooled nuclear power reactors, the contribution of the environmental effects of uranium mining and milling, the production of uranium hexafluoride, isotopic enrichment, fuel fabrication, reprocessing of irradiated fuel, transportation of radioactive materials and management of low level wastes and high level wastes related to uranium fuel cycle activities to the environmental costs of licensing the nuclear power reactor, shall be as set forth in Table S–3. No further discussion of such environmental effects shall be required. This paragraph does not apply to any applicant's environmental report submitted prior to June 6, 1974.

(f)   Number of copies. Each applicant for a permit to construct a production or utilization facility covered by § 51.5(a) shall submit the number of copies, as specified in § 51.40, of the Environmental Report required by § 51.5(a).

(g)(1) The Environmental Report required by paragraph (a) for light-water-cooled nuclear power reactors shall contain either (i) a statement that the transportation of cold fuel to the reactor and irradiated fuel from the reactor to a fuel reprocessing plant and the transportation of solid radioactive wastes from the reactor to waste burial grounds is within the scope of this paragraph, and as the contribution of the environmental effects of such transportation to the environmental costs of licensing the nuclear power reactor, the values set forth in the following Summary Table S–4; or (ii) if such transportation does not fall within the scope of this paragraph, a full description and detailed analysis of the environmental effects of such transportation and, as the contribution of such effects to the environmental costs of licensing the nuclear power reactor, the values determined by such analyses for the environmental impact under normal conditions of transport and the environmental risk from accidents in transport.

(2)   This paragraph applies to the transportation of fuel and wastes to and from a nuclear power reactor only if:

(i)   The reactor is a light-water-cooled nuclear power reactor with a core thermal power level not exceeding 3,800 megawatts;

(ii)   The reactor fuel is in the form of sintered uranium dioxide pellets encapsulated in zircaloy rods with a uranium-235 enrichment not exceeding 4% by weight;

(iii)   The average level of irradiation of the irradiated fuel from the reactor does not exceed 33,000 megawatt days per metric ton and no irradiated fuel assembly is shipped until at least 90 days have elapsed after the fuel assembly was discharged from the reactor;

(iv)   Waste (other than irradiated fuel) shipped from the reactor is in the form of packaged, solid wastes; and

(v)   Unirradiated fuel is shipped to the reactor by truck; irradiated fuel is shipped from the reactor by truck, rail, or barge; and waste other than irradiated fuel is shipped from the reactor by truck or rail.

---

[2]No permit or license will, of course, be issued with respect to an activity for which a certification required by section 401 of the Federal Water Pollution Control Act has not been obtained.

(3)   This paragraph does not apply to any applicant's environmental report submitted prior to February 5, 1975.

Each applicant for a license to operate a production of utilization facility covered by § 51.5(a) shall submit with its application the number of copies, as specified in § 51.40, of a separate document (Amended 41 1 R 15832) to be entitled "Applicant's Environmental Report—Operating License Stage," which discusses the same matters described in § 51.20 but only to the extent that they differ from those discussed or reflect new information in addition to that discussed in the final environmental impact statement prepared by the Commission in connection with the construction permit. The "Applicant's Environmental Report—Operating License Stage" may incorporate by reference any information contained in the Applicant's Environmental Report or final environmental impact statement previously prepared in connection with the construction permit. With respect to the operation of nuclear reactors, the applicant, unless otherwise required by the Commission, shall submit the "Applicant's Environmental Report—Operating License Stage" only in connection with the first licensing action that would authorize full power operation of the facility.

Summary Table S—4. — Environmental impact of transportation of fuel and waste to and from 1 light-water-cooled nuclear power reactor[1]

[Normal conditions of transport]

	Environmental impact
Heat (per irradiated fuel cask in transit)	250,000 Btu/hr.
Weight (governed by Federal or State restrictions)	73,000 lbs. per truck; 100 tons per cask per rail car.
Traffic density:	
Truck	Less than 1 per day.
Rail	Less than 3 per month.

Exposed population	Estimated number of persons exposed	Range of doses to exposed individuals[2] (per reactor year)	Cumulative dose to exposed population (per reactor year)[3]
Transportation workers	200	0.01 to 300 mrem	4 man-rem.
General public:			
Onlookers	1,100	0.003 to 1.3 millirem . . . . }	3 man-rem.
Along Route	600,000	0.0001 to 0.06 millirem . . . }	

Accidents in Transport

	Environmental risk
Radiological effects	Small[4]
Common (nonradiological) causes	1 fatal injury in 100 reactor years; 1 nonfatal injury in 10 reactor years; $475 property damage per reactor year.

[1] Data supporting this table are given in the Commission's "Environmental Survey of Transportation of Radioactive Materials to and from Nuclear Power Plants," WASH-1238, December 1972, and Supp. 1, NUREG-75/038, April 1975. Both documents are available for inspection and copying at the Commission's Public Document Room, 1717 H St. N.W., Washington, D.C., and may be obtained from National Technical Information Service, Springfield, Va. 22161. WASH-1238 is available from NTIS at a cost of $5.45 (microfiche, $2.25) and NUREG-75,038 is available at a cost of $3.25 (microfiche, $2.25).

[2] The Federal Radiation Council has recommended that the radiation doses from all sources of radiation other than natural background and medical exposures should be limited to 5,000 millirem per year for individuals as a result of occupational exposure and should be limited to 500 millirem per year for individuals in the general population. The dose to individuals due to average natural background radiation is about 130 millirem per year.

[3] Man-rem is an expression for the summation of whole body doses to individuals in a group. Thus, if each member of a population group of 1,000 people were to receive a dose of 0.001 rem (1 millirem), or if 2 people were to receive a dose of 0.5 rem (500 millirem) each, the total man-rem dose in each case would be 1 man-rem.

[4] Although the environmental risk of radiological effects stemming from transportation accidents is currently incapable of being numerically quantified, the risk remains small regardless of whether it is being applied to a single reactor or a multireactor site.

REFERENCES

1. Rasmussen, K. H., R. L. Hobel and M. Taheri. "Sources and Natural Removal Processes for Some Atmospheric Pollutants," NITS, U.S. Department of Commerce (June 1974).
2. *Environmental Health Letter* (April 15, 1976), p. 6.
3. "Antipollution Rules Haven't Hurt Industry," *Chem. Eng. News* 54(20):5 (May 10, 1976).
4. *Air/Water Pollution Report* (May 31, 1976), p. 7.
5. "International Cooperation with the Soviets," *Environ. Sci. Technol.* 10(5) (May 1976).
6. U.S. Environmental Protection Agency "In Productive Harmony," Washington, D.C. (1973).
7. "EPA, Preparation of EIS's—Final Regulations," *Federal Register* 40, No. 72, Part III, Washington, D.C. (April 14, 1975).
8. "CEQ, Preparation of EIS's—Guidelines," *Federal Register*, 38, No. 147, Part II, Washington, D.C. (August 1, 1973).
9. "Save Crystal Beach Assoc. vs. Callaway," *Environmental Reporter* No. 45, Bureau of National Affairs (March 5, 1976).
10. "City of Davis vs. Coleman," *Environmental Reporter*, No. 29 Bureau of National Affairs, (November 14, 1975).
11. Final Environmental Statement. "1976 Outer Continental Shelf Oil and Gas Lease Sale Offshore the Mid-Atlantic States," U.S. Department of the Interior, Bureau of Land Management, New York, Vol. 1 (1976).
12. Final Environmental Statement. "1976 Outer Continental Shelf Oil and Gas Lease Sale Offshore the Mid-Atlantic States," U.S. Department of the Interior, Bureau of Land Management, New York, Vol. 5 (1976).
13. U.S. Environmental Protection Agency. "Environmental Impact Assessment Guidelines for Selected New Source Industries," Office of Federal Activities, Washington, D.C. (October 1975).
14. Draft Environmental Statement. "El Paso Coal Gasification Project, New Mexico," U.S. Department of the Interior, Bureau of Reclamation, Upper Colorado Region (July 1974).
15. El Paso Natural Gas Company. "Revised Report on Environmental Factors Burnham Coal Gasification Project" (January 1974).
16. ASME. "Prediction of the Dispersion of Airborne Effluents," New York (1973).

17. U.S. Department of Commerce.   Summary of Hourly Observations, Galveston, Texas, Decennial Census of United States Climate 1951-1960, Weather Bureau, Washington, D.C. (1971).
18. Draft Environmental Statement for the Proposed Lake Hackensack, Bergen County, New Jersey Board of Chosen Freeholders, New Jersey, (February 1976).
19. Perry, W. J. *et al.* "Stratigrophy of the Atlantic Continental Margin North of Cape Hatteras—A Brief Survey,"   USGS Open File Report, U.S. Department of the Interior, Washington, D.C. (1974).
20. Frank, O. L. and N. G. McClymonds. "Summary of the Hydrologic Situation on Long Island, New York," USGS Paper 627-F, U.S. Department of Interior, Washington, D.C. (1972).
21. Dee, N. *et al.* "Water Quality Management Planning Environmental Assessments for Effective Water Quality Management Planning," PB-228593, NTIS, U.S. Department of Commerce, Springfield, Virginia (April 1972).
22. Billings, W. D. "The Environmental Complex in Relation to Plant Growth and Distribution," *Quart. Rev. Biol.* 27:251-265 (1952).
23. Billings, W. D. *Plants, Man, and the Ecosystem.* (Belmont, California: Wadsworth Publishing Co. Inc., 1970).
24. Kerbec, M. J., Ed. "Your Government and the Environment," Output Systems Corp., Arlington, Virginia (1971).
25. Merriam, C. H. "Results of a Biological Survey of the San Francisco Mountain Region and the Desert of the Little Colorado, Arizona," *North American Fauna* 3:1-136 (1890).
26. Niering, W. A. "Terrestrial Ecology of Kapingamarangi Atoll, Caroline Islands," *Ecological Monographs* 33:131-160 (1963).
27. Dee, N. *et al.* "Environmental Evaluation System for Water Resource Planning," Final Report to the Bureau of Reclamation, U.S. Department of the Interior, Contract #14-06D-7182, Battelle-Columbus, Columbus, Ohio (January 1972).
28. Smith, M. E., Ed. "Prediction of the Dispersion of Airborne Effluents," American Society of Mechanical Engineers, New York (1973).
29. Cheremisinoff, P. N. and R. A. Young. *Pollution Engineering Practice Handbook.* (Ann Arbor, Michigan: Ann Arbor Science Publishers, Inc., 1975).
30. Pasquill, F. *Atmospheric Diffusion* (London: D. Van Nostrand, 1962).
31. Turner, D. B. "Workbook of Atmospheric Dispersion Estimates," U.S. Department of Health, Education and Welfare, Cincinnati, Ohio (1969).
32. Bussi, A. D. and J. R. Zimmerman. "User's Guide for the Climatological Dispersion Model, National Environmental Research Center, Research Triangle Park, North Carolina (December 1973).
33. U. S. Water Resources Council "The Nation's Water Resources," Washington, D. C. (1968).
34. U.S. Environmental Protection Agency. "Handbook for Monitoring Industrial Wastewater," Technology Transfer, Washington, D.C. (August 1973).

35. Sawyer, C. N. and P. L. McCarty. *Chemistry for Sanitary Engineers* (New York: McGraw-Hill Book Company, 1967).

36. Office of Science and Technology, Executive Office of the President, "Solid Waste Management," Washington, D.C. (May 1969).

37. U.S. Environmental Protection Agency. "Hazardous Waste Management Facilities in the U.S.," EPA 530-SW-146, Cincinnati, Ohio.

38. Cheremisinoff, P. N. and A. C. Morresi. *Energy from Solid Wastes* (New York: Marcel Dekker, 1976).

39. U.S. Environmental Protection Agency. "Residential Collection Systems," Vol. 1, EPA 530-SW-976.1, Cincinnati, Ohio (March 1975).

40. U.S. Environmental Protection Agency. "Liners for Disposal Sites," EPA 530-SW-137, Cincinnati, Ohio (March 1975).

41. Cheremisinoff, P. N., P. P. Cheremisinoff, A. C. Morresi and R. A. Young. *Woodwastes Utilization & Disposal.* (Westport, Connecticut: Technomic Publishing Co., 1976).

42. Arella, D. G. "Recovering Resources from Solid Waste Using Wet-Processing," U.S. Environmental Protection Agency, Washington, D.C. (1974).

43. "The Noise Around Us," U.S. Department of Commerce, Washington D.C. (September 1970).

44. American Petroleum Institute. "Guidelines on Noise," Medical Research Report #EA7301, Washington, D.C. (1973).

45. "Noise Standards," *Federal Register*, 41, No. 80 (April 23, 1976); General Services Administration, Washington, D.C., 16933-16972.

46. Beyaert, B. "Analysis of Oil Spill Accidents for EISs," 1975 Conference on Prevention and Control of Oil Pollution, sponsored by API, EPA and U.S. Coast Guard, San Francisco, California, (March 1975), pp. 39-45.

47. Boesch, D. F., C. H. Hershner, J. H. Milgram. *Oil Spills and the Marine Environment.* (Cambridge, Massachusetts: Ballinger Publishing Co., 1974).

48. Porricelli, J. D., V. F. Keith, R. L. Storch. "Tankers and the Ecology," *Trans. Soc. Naval Architects and Marine Engineers* 79:169-221 (1971).

49. Draft Environmental Impact Statement on Deepwater Ports, U.S. Department of the Interior, Washington, D.C. (June 1973).

50. Baptist, C. *Tanker Handbook for Deck Officers* (Glasgow, England: Brown, Son & Ferguson, Ltd., 1975).

51. Card, J. C., P. V. Ponce, N. D. Snider. "Tankship Accidents and Resulting Oil Outflows, 1969-1973," 1975 Conference on Prevention and Control of Oil Pollution, sponsored by API, EPA and U.S. Coast Guard, San Francisco, California (March 1975), pp. 205-221.

52. Wilcox, J. D. "A Hydrodynamically Effective Horizontal Oil Book," 1975 Conference on Prevention and Control of Oil Pollution, sponsored by API, EPA and U.S. Coast Guard, San Francisco, California (March 1975), pp. 363-364.

53. Norton, R. W. and D. W. Lerch. "An Oil Recovery System for San Francisco Bay Area," 1975 Conference on Prevention and Control of Oil Pollution, sponsored by API, EPA and U.S. Coast Guard, San Francisco, California (March 1975), pp. 317-322.

54. Neal, R. W., R. A. Bianchi and E. E. Johanson. "The Design and Demonstration of a Remotely-Controlled High Seas Oil Recovery System," 1975 Conference on Prevention and Control of Oil Pollution, sponsored by API, EPA and U.S. Coast Guard, San Francisco, California (March 1975), pp. 395-400.

55. U.S. Nuclear Regulatory Commission. "Preparation of Environmental Reports for Nuclear Power Stations," Regulatory Guide 4.2, Revision 2, NUREG-0099, Washington, D.C. (July 1976).

56. U.S. Nuclear Regulatory Commission. Final Environmental Statement, Sterling Power Project Nuclear Unit 1, Docket No. STN 50-485, Office of Nuclear Reactor Regulation (June 1976).

57. Air Force Systems Command. "Nuclear Radiation Guide," Technical Documentary Report No. MRL-TDR-62-61, Wright-Patterson Air Force Base, Ohio (November 1962).

58. Yamomoto, Y., N. Mitsuishi and S. Kadoya. "Design and Operation of Evaporators for Radioactive Wastes," Technical Reports Series No. 87, International Atomic Energy Agency, Vienna (1968), pp. 1-22.

59. "Operation and Control of Ion-Exchange Processes for Treatment of Radioactive Wastes," Technical Reports Series No. 78, International Atomic Energy Agency, Vienna (1967), pp. 38-50.

60. Brooks, N. S. and E. R. Chow. "Cartridge Filtration in Pressurized Water Reactors," paper presented at AIChE Filtration Society Annual Meeting, Houston, Texas (March 17-21, 1975).

61. Lancey, T. and T. Vandenberg. "Control of Corrosion Products in Power Plant Waters," Materials Performance 13(2):39-42 (February 1974).

62. Kniazewycz, B. G. "The Treatment of Liquid Radwaste by Reverse Osmosis—Conception to Operation," paper presented at ASME-IEEE Joint Power Generation Conference, Portland, Oregon (September 28-October 1, 1975).

63. Kibbey, A. H. and H. W. Godbee. "Solid Radioactive Waste Practices at Nuclear Power Plants," Nuclear Safety 16(5):581-592 (September-October 1975).

64. Fitzgerald, C. L. et al. "The Feasibility of Incorporating Radioactive Wastes in Asphalt or Polyethylene," Nuclear Appl. Technol. (September 1970), pp. 821-829.

65. Stewart, J. E. and R. Herter. "Solid Radwaste Experience in Europe Using Asphalt," paper presented at ASME-IEEE Joint Power Generation Conference, Portland, Oregon (September 28-October 1, 1975).

66. Werner & Pfleiderer Corp. "Asphalt Extruder-Evaporator Solidification System," paper presented at UCLA Seminar on Radioactive Waste Management for Nuclear Power Reactors, (October 20-24, 1975).

67. Quaka, T. E. "Dresden Station Liquid Radwaste System Modifications," paper presented at ASME-IEEE Joint Power Generation Conference, Portland, Oregon (September 28-October 1, 1975).

68. Fey, F. L. "Monticello Nuclear Plant Radwaste System Operating Experience," paper presented at ASME-IEEE Joint Power Generation Conference, Portland, Oregon (September 28-October 1, 1975).

69. Burns, R. A. "Liquid and Solid Radwaste Experience at the Nine Mile Point BWR," paper presented at ASME-IEEE Joint Power Generation Conference, Portland, Oregon (September 28-October 1, 1975).

70. Letter, Rochester Gas & Electric Co., to Atomic Energy Commission, Waste Evaporator Operating History at R. E. Ginna Nuclear Plant (June 3, 1971).

71. Letter, Carolina Power & Light Co., to Atomic Energy Commission, H. B. Robinson Plant Unit 2—System Design Report LB-74-2 for Installation of Radwaste Evaporator, Revision 3 (October 17, 1974).

72. Letter, Commonwealth Edison Co., to Atomic Energy Commission, Zion Nuclear Plant Unit 1—Special Report No. 7, Modification to the Radwaste System (February 19, 1974).

73. Oyen, L. C. "Solving Radwaste Problems," *Power* 120:75-77 (March 1976).

74. U.S. Atomic Energy Commission. "Final Environmental Statement Concerning Proposed Rulemaking Action: Numerical Guides for Design Objectives and Limiting Conditions for Operation to Meet the Criteria (As Low As Practicable) for Radioactive Material in Light-Water-Cooled Nuclear Power Reactor Effluents," WASH-1258 (July 1973).

75. Casto, W. R., Ed. "Summary of Radioactivity Released in Effluents from Nuclear Power Plants During 1973," *Nuclear Safety* 16(6): 734-738 (November-December 1975).

76. Kozorowski, B. and J. Kucharski. *Industrial Waste Disposal.* (New York: Pergamon Press, 1972).

77. U.S. Department of the Interior, "The Cost of Clean Water, Vol. III, Industrial Waste Profiles No. 5—Petroleum Refining," Federal Water Pollution Control Administration, Washington, D.C. (1967).

78. Mencher, S. K. "Minimizing Waste in the Petrochemical Industry," *Chemical Engineering Progress* 63:80-88 (October 1967).

79. Gloyna, E. F. *et al.* "The Characteristics and Pollutional Problems Associated with Petrochemical Wastes," Federal Water Pollution Control Administration, Washington, D.C. (1970).

80. U.S. Department of Health, Education and Welfare. "Atmospheric Emissions from Petroleum Refineries," Public Health Service Publication No. 763, Washington, D.C. (1960).

81. Armistead, G., Jr. *Safety in Petroleum Refining and Related Industries* (New York: John G. Simmons & Co., Inc., 1950).

82. National Technical Information Services, "Compilation of Air Pollutant Emission Factors," Publication No. PB-223-966 AP-42.

83. American Petroleum Institute, "Hydrocarbon Emissions from Refineries," Publication 0.928, Washington, D.C. (July 1973).

84. U.S. Department of Interior. "The Cost of Clean Water and Its Economic Impact, Vol. IV, Projected Wastewater Treatment Costs in the Organic Chemicals Industry," Federal Water Pollution Control Administration, Washington, D.C. (1969).

85. Orford, H. E. "Breaking in a Plant for Treating Gum Wastes," *Sewage Industrial Wastes* 23 (1951).

86.   U.S. Environmental Protection Agency.   "Development Document for Effluent Limitations Guidelines and New Source Performance Standards for the Rubber Processing Industry,"   Washington, D.C. (1973).

87.   U.S. Environmental Protection Agency.   "Rubber Reuse and Solid Waste Management,"   Part I, Solid Waste Series SW-22C, Washington, D.C. (1972).

88.   Reiter, W. M.   "Handling Liquid Effluents," in *Pollution Control—73* (New York: McGraw-Hill, Inc., 1973).

89.   Buckley, C. F.   "Design for Pollution Control," *Industrial Process Design for Pollution Control,* Vol. 4 (New York: AIChE, 1971), pp. 10-18.

90.   Celenza, C. J.   "The Effect of Effluent Control Standards in Designing Chemical Process Plants," *Industrial Process Design for Pollution Control,* Vol. 4 (New York: AIChE, 1971), pp. 71-83.